REALISM AND THE AIM OF SCIENCE

REALISM
AND THE AIM OF SCIENCE

KARL R. POPPER

From the POSTSCRIPT TO THE LOGIC OF
SCIENTIFIC DISCOVERY
Edited by W. W. Bartley, III

ROWMAN AND LITTLEFIELD
Totowa, New Jersey

Copyright © Karl Raimund Popper, 1956, 1983

First published in the United States 1983 by Rowman and
Littlefield, 81 Adams Drive, Totowa, New Jersey 07512

Library of Congress Cataloging in Publication Data

Popper, Karl Raimund, Sir, 1902-
   Realism and the aim of science.

   (The Postscript to The logic of scientific dis-
covery/as edited by W. W. Bartley, III)
   Bibliography: p.
   Includes index.
   1. Science—Philosophy. I. Bartley, William Warren,
1934-   . II. Title. III. Series: Popper, Karl
Raimund, Sir, 1902-   . Postscript to The logic
of scientific discovery.
Q175.P8643 1982   501   82-501
ISBN 0-8476-7015-5   AACR2

Printed in the United States of America

TO MY EDITOR
For his rescue of the *Postscript*

CONTENTS

Note on Numbering of Sections. The sections in each of the three volumes of the *Postscript*, and in the two parts of the first volume, are numbered consecutively, beginning with section 1. The original section numbers, indicating the order of the sections within the *Postscript* as a whole, are given in starred brackets in the Tables of Contents. Ed.

PART II. THE PROPENSITY INTERPRETATION OF PROBABILITY

EDITOR'S FOREWORD

THIS book, *Realism and the Aim of Science,* is the first volume of Sir Karl Popper's long-awaited *Postscript* to *The Logic of Scientific Discovery.* Although it was written some twenty-five years ago, it has never before been published. It contains a new and highly expanded development of Popper's views on induction, demarcation, and corroboration, and also presents his propensity theory of probability. This book also contains a detailed consideration of and reply to numerous criticisms and objections that have been made to Popper's views over the years since *The Logic of Scientific Discovery* was first published.

Together with other parts of the *Postscript* (all of which are now being published), this volume was written mainly during the years 1951-56, at the time when *Logik der Forschung,* Popper's first published book (1934), was being translated into English as *The Logic of Scientific Discovery.*

The different volumes of the *Postscript* were originally part of a series of Appendices to *The Logic of Scientific Discovery,* in which Popper proposed to correct, expand, and develop the ideas of his first book. Some of these Appendices were in fact included in *The Logic of Scientific Discovery* when it was published in 1959. But one group of Appendices took on a life of its own, and gradually grew into a single, closely-integrated work—far exceeding the original *Logik der Forschung* in length. It was decided to publish this new work—called the *Postscript: After Twenty Years*—as a sequel or companion volume to *The Logic of Scientific Discovery.* And it was accordingly set in type, in galley proofs, in 1956-57.

Within a few months of the anticipated publication, however, the project came grinding to a halt. In *Unended Quest,* his intellectual autobiography, Sir Karl has reported of these galley proofs: 'Proof reading turned into a nightmare. . . . I then had to have operations on both eyes. After this I could not start proof reading again for some time, and as a result the *Postscript* is still unpublished.'

I remember this time vividly: I went to Vienna to visit Popper in the hospital there shortly after his operation for several detachments of both retinas; and we worked on the *Postscript* as he was recuperating. For a long time he could barely see, and we were very much afraid that he would become blind.

When he was able to see again, a great deal of work was done on the *Postscript*: several sections were added, and thousands of corrections were made to the galleys. But the pressure of other work had now become too great, and virtually nothing was added to the text after 1962. During the next, highly productive decade, after publishing *Conjectures and Refutations* (1963), Popper completed and published three new books: *Objective Knowledge: An Evolutionary Approach* (1972), *Unended Quest* (1974 and 1976), and (with Sir John Eccles) *The Self and Its Brain* (1977), as well as many papers. These were the years, and the works, in which his now famous theory of objective mind (and of Worlds 1, 2, and 3) was developed, and in which his approach was extended into the biological sciences.

Meanwhile, the *Postscript,* which represented the culmination of Sir Karl's work in the philosophy of physics, went unpublished. But not unread: most of Popper's closest students and colleagues have studied this work, and several have had copies of the galley proofs over the years. It is a source of great satisfaction to those like myself, who have known this book and been deeply influenced by it, to see it finally completed and shared with the general public.

The text that has now been edited for publication is essentially that which existed in 1962. Except in a few places, as marked, no major alterations have been made. It was felt that this was the appropriate approach to a work that had now acquired, through its influence on Popper's students and colleagues, an historical character—some twenty-five years having passed since its composition, and forty-five years since the writing of the original *Logik der Forschung.* Obviously, many points would have been put differently today. But a complete revision by the author would have delayed publication indefinitely.

The editing has included bringing together the different versions of some parts of the text, as they had accumulated over the years; copy-editing the book; and adding bibliographical and other notes for the reader's assistance. A few new additions made by Popper himself are clearly marked: they are presented in brackets and marked with a star:*. My own brief editorial and bibliographical notes are also in brackets, followed by the abbreviation 'Ed.'. Here I have in general followed the practice established by Troels Eggers Hansen, the editor of Popper's *Die beiden Grundprobleme der Erkenntnistheorie* (written in 1930-32 and published in 1979). Popper has been able to check the editorial work at a series of meetings which we have held at various places over the past four years—in Heidelberg, Guelph, Toronto, Washington D. C., Schloss Kronberg, and at his home in Buckinghamshire. He has also added new prefaces to all of the volumes, and new afterwords to the second and third volumes.

One major alteration in presentation has been made, at my own suggestion. To publish this large work under one cover would have been possible, but would have meant a heavy and unwieldy book beyond the means of many students of philosophy. Parts of the *Postscript*—including *Realism and the Aim of Science*—will be of wide interest, of concern not only to philosophers and students of philosophy but also to a wider public.

These parts are also, on the whole, independent of one another. This led me to suggest that the work be published in three separate volumes, in matching format, the whole constituting the *Postscript*. After some hesitation, Sir Karl agreed with this proposal, and also with the titles which I suggested for the three volumes.

Thus the *Postscript* is being published as follows:

Realism and the Aim of Science   (Volume I)
The Open Universe: An Argument for Indeterminism   (Volume II)
Quantum Theory and the Schism in Physics   (Volume III).

Although these three volumes of the *Postscript* can be read separately, the reader should be aware that they build a connected argument. Each volume of the *Postscript* attacks one or another of the subjectivist or idealist approaches to knowledge; each constructs one or more components of an objective, realist approach to knowledge.

Thus in the present volume, Popper pursues 'Inductivism', which he sees as the chief source of subjectivism and idealism, through four stages: logical, methodological, epistemological, and metaphysical. He develops his theory of falsifiability, and charts its effects in demarcating scientific, non-scientific, and pseudo-scientific views from one another. And he presents his theory of corroboration as a way to express rational preference for one theory over another without resorting either to the subjective 'certainties' or to the objective 'justification' of conventional philosophies. In this first volume Popper also discusses his relationship to those historical figures in philosophy, such as Berkeley, Hume, Kant, Mach, and Russell, who have contributed importantly to the subjectivist tradition; and he gives detailed replies to contemporary philosophical and scientific critics. Popper also attacks the subjective interpretation of the probability calculus, an interpretation that is rooted in the belief that probability measures a subjective state of insufficient knowledge. In *The Logic of Scientific Discovery*, Popper had championed an objective interpretation of the probability calculus, using for this purpose the frequency interpretation. Now he also criticizes the frequency interpretation; and in its place he presents in detail his own propensity interpretation—an interpretation which has, during the past twenty years, found many champions. These ideas and arguments are applied and developed in the remaining volumes.

In *The Open Universe: An Argument for Indeterminism*, Popper develops his own indeterministic perspective, and presents a critique of both 'scientific' and metaphysical forms of determinism, arguing that classical physics does not, contrary to common opinion, presuppose or imply determinism any more than quantum physics does. Yet he finds that metaphysical determinism continues to underlie the work of many contemporary quantum theorists, opponents of determinism included. Popper traces the continuing role played within physics by subjective interpretations of probability to these metaphysical deterministic presuppositions.

There is a deep connection between the arguments of the first and second volumes, in their mutual concern with the freedom, creativity and rationality of man.

The first volume, in its consideration of justification and rationality, rebuts a subjectivist and sceptical claim about the limits of criticism—and therewith the limits of rationality. If such a limit existed, then serious argument would be futile; and the appearance of it would be illusory.

The second volume, in its treatment of determinism, champions the claim that our rationality is limited in respect to the prediction of the future growth of human knowledge. If such a limit did *not* exist, then serious argument would be futile; and the appearance of it would be illusory.

Popper thus argues that human reason is unlimited with regard to criticism yet limited with regard to its powers of prediction; and shows that both the lack of limitation and the limitation are, in their respective places, necessary for human rationality to exist at all.

In Volume III, *Quantum Theory and the Schism in Physics*, Popper reviews and rebuts an array of arguments and 'paradoxes' that are widely used to defend an idealist outlook. Conjecturing that the problems of interpretation of quantum mechanics can be traced to problems of the interpretation of the calculus of probability, Popper develops his own propensity interpretation of probability further. And then he gives a sweeping critique of some of the leading interpretations of quantum theory, attempting to resolve their well-known paradoxes and to exorcise 'the Observer' from quantum physics. His concluding 'Metaphysical Epilogue' weaves together the themes of the entire *Postscript* in an historical and programmatic study of the role of metaphysical research programmes or interpretations in the history of physics.

The Editor wishes to express his gratitude to the American Council of Learned Societies and to the American Philosophical Society for their generous support of his editorial work on these volumes. He also wishes to thank his secretary, Nancy Artis Sadoyama, for her devoted and unfailing assistance.

ACKNOWLEDGEMENTS

I w i s h to take this opportunity to thank my colleague John W. N. Watkins for the great encouragement which his unflagging interest has been to me. He has read this volume in manuscript and in proof, and has made the most helpful suggestions for improvements. It was at his suggestion that I decided to publish this *Postscript* as a separate work rather than, as originally planned, as a series of Appendices to *The Logic of Scientific Discovery*. But even more important for the completion of the work than these suggestions was his interest in its ideas.

I also wish to thank the co-translators of the *The Logic of Scientific Discovery*, Dr. Julius Freed and Lan Freed, who read most of the volume in galley proof, and made a great number of suggestions for improving its style. [They both died many years before its publication. Ed.]

Joseph Agassi was, during the period that this book was written, first my research student and later my research assistant. I discussed almost every section with him in detail, very often with the result that, on his advice, I expanded a statement or two into a whole new section—or, in one case, into a whole new part. [It became Part 2 of this volume: *Realism and the Aim of Science*.] His co-operation was of the greatest value.

I also wish to thank the London School of Economics and Political Science, which made it possible for me to benefit from Dr Agassi's assistance, and the Center for Advanced Study in the Behavioral Sciences (Ford Foundation) in Stanford, California, for giving me the opportunity of working uninterruptedly, from October 1956 to July 1957, on the galley proofs of this book, and for making it possible for Dr Agassi to help me during this period.

PENN, BUCKINGHAMSHIRE, *1959*.

Professor W. W. Bartley, III, was my student, and then my colleague at the London School of Economics, from 1958–63, and worked closely with me during 1960–62 on this book. In 1978 he kindly consented to act as Editor of the *Postscript*. I am grateful to him for his assistance and for undertaking this arduous task. I am indebted to him more than I can say.

It is also a pleasure to thank several other persons who have in the intervening years worked with me on this *Postscript*, in particular, Alan E. Musgrave, David Miller, Arne F. Petersen, Tom Settle, Jeremy Shearmur, and Tyrrell Burgess. Of these, David Miller and Arne Petersen should be specially mentioned because of the immense amount of work which they both did at various periods before 1970. Miller made valuable additional corrections and suggestions in 1980 and 1981.

The London School has continued through all these years to help me by appointing a research assistant. For the thirteen years since my retirement in 1969 it has done so with the help of a grant from the Nuffield Foundation, to whom I wish to express my thanks. Chiefly responsible for this arrangement were my friend and successor, Professor John Watkins; the late Sir Walter Adams, Director of the School; and the present Director, Professor Ralf Dahrendorf, to whose warm friendship and great interest in my work I am deeply indebted.

Had the *Postscript* been published in the 1950s, I should have dedicated it to Bertrand Russell: Professor Bartley has told me that a letter to this effect exists in the Russell Archives at McMaster University.

I may mention, finally, that this *Postscript* (together with the translation of *The Logic of Scientific Discovery*) seemed to me almost ready in 1954. It was then that I chose its original title, '*Postscript: After Twenty Years*', with an allusion to the publication of *Logik der Forschung* in 1934.

PENN, BUCKINGHAMSHIRE, *1981*.

INTRODUCTION, 1982

In this Introduction to the first volume of the *Postscript*, I should like to discuss very briefly several issues that have been raised, during the three decades since this book was written, against the views which I present herein.

I

The first has to do with the technical terms 'falsifiable' ('empirically refutable') and 'falsifiability' ('empirical refutability'). I first introduced these in *Erkenntnis* 3, 1933, and in *Logik der Forschung*, 1934, in connection with my solution of the problem of demarcation (discussed at length in Part I, Chapter 2, of the present volume). The problem of demarcation is to find a criterion that permits us to distinguish between statements that belong to the empirical sciences (theories, hypotheses) and other statements, particularly pseudo-scientific, prescientific, and metaphysical statements; but also mathematical and logical statements. The problem of demarcation is to be distinguished from the far more important problem of truth: theories which have been shown to be false—as for example the radiation formulae of Rayleigh-Jeans and of Wien, or Bohr's atom model of 1913—can nevertheless retain the character of empirical, scientific hypotheses.

Although, following Tarski, I do not believe that a criterion of truth is possible, I have proposed a criterion of demarcation—the criterion of falsifiability. My proposal was that a statement (a theory, a conjecture) has the status of belonging to the empirical sciences if and only if it is falsifiable.

But when is a statement falsifiable? It is of great importance to current discussion to notice that falsifiability in the sense of my demarcation criterion is a purely logical affair. It has to do only with the logical structure of statements and of classes of statements. And it has *nothing* to do with the question whether or not certain possible experimental results would be accepted as falsifications.

A statement or theory is, according to my criterion, falsifiable if and only if there exists at least one potential falsifier—at least one possible basic statement that conflicts with it logically. It is important not to demand that the basic statement in question be *true*. The class of basic statements is designated so that a basic statement describes a logically possible event of which it is logically possible that it might be observed.

To make these matters less abstract, I shall give four examples here: two of falsifiable statements, and two of unfalsifiable statements.

(1) 'All swans are white'. This theory is falsifiable since, for example, it contradicts the following basic statement (which is, incidentally, false): 'On the 16th of May, 1934, a black swan stood between 10 and 11 o'clock in the morning in front of the statue of Empress Elizabeth in the Volksgarten in Vienna.'

(2) Einstein's principle of proportionality of inert and (passively) heavy mass. This equivalence principle conflicts with many potential falsifiers: events whose observation is logically possible. Yet despite all attempts (the experiments by Eötvös, more recently refined by Dicke) to realize such a falsification experimentally, the experiments have so far corroborated the principle of equivalence.

(3) 'All human actions are egotistic, motivated by self-interest.' This theory is widely held: it has variants in behaviourism, psychoanalysis, individual psychology, utilitarianism, vulgar-marxism, religion, and sociology of knowledge. Clearly this theory, with all its variants, is not falsifiable: no example of an altruistic action can refute the view that there was an egotistic motive hidden behind it.

(4) Purely existential statements are not falsifiable—as in Rudolf Carnap's famous example: 'There is a colour ('Trumpet-red') which incites terror in those who look at it.' Another example is: 'There is a ceremony whose exact performance forces the devil to appear.' Such statements are not falsifiable. (They are, in principle, verifiable: it is logically possible to find a ceremony whose performance leads to

the appearance of a human-like form with horns and hooves. And if a repetition of the ceremony fails to achieve the same result, that would be no falsification, for perhaps an unnoticed yet essential aspect of the correct ceremony was omitted.)

As these examples show, falsifiability in the sense of the demarcation criterion does not mean that a falsification can in practice be carried out, or that, if it is carried out, it will be unproblematic. Falsifiability in the sense of the demarcation criterion signifies nothing more than a logical relation between the theory in question and the class of basic statements, or the class of the events described by them: the potential falsifiers. Falsifiability is thus relative to these two classes: if one of these classes is given, then falsifiability is a matter of pure logic—the logical character of the theory in question.

That the class of potential falsifiers (or of basic statements) must be given can best be shown by our first example—'All swans are white'.

As I have already said, this statement is falsifiable. Suppose, however, that there is someone who, when a non-white swan is shown to him, takes the position that it cannot be a swan, since it is 'essential' for a swan to be white.

Such a position amounts to holding non-white swans as logically impossible structures (and thus also as unobservable). It excludes them from the class of potential falsifiers.

Relative to this *altered* class of potential falsifiers the statement 'All swans are white' is of course unfalsifiable. In order to avoid such a move, we can demand that anyone who advocates the empirical-scientific character of a theory must be able to specify under what conditions he would be prepared to regard it as falsified; i.e., he should be able to describe at least some potential falsifiers.

We now come to a second sense of 'falsifiable' and 'falsifiability' which has to be distinguished very clearly from my purely logical criterion of demarcation in order to avoid gross confusion.

One can raise the question whether an actual falsification is ever so compelling that one *must* regard the theory in question as falsified (and thus as false). Is there not always a way out for one who wishes to save the theory in question?

I have always maintained, even in the first edition of *Logik der Forschung* (1934), and also in my earlier yet only recently published

xxi

book *Die beiden Grundprobleme der Erkenntnistheorie* (1979, written 1930–33), that it is never possible to prove conclusively that an empirical scientific theory is false. In *this sense*, such theories are *not falsifiable*. 'Every theoretical system can in various ways be protected from an empirical falsification.' (*Grundprobleme*, p. 353). 'It is always possible to find some way of evading falsification, for example by introducing *ad hoc* an auxiliary hypothesis . . .' (*Logic of Scientific Discovery* (*L.Sc.D.*), p. 42, in the same section in which falsifiability is introduced). 'No conclusive disproof of a theory can ever be produced . . . ' (*L.Sc.D.*, p. 50).

Hence, to repeat, we must distinguish two meanings of the expressions 'falsifiable' and 'falsifiability':

(1) 'Falsifiable' as a logical-technical term, in the sense of the demarcation criterion of falsifiability. This purely logical concept— falsifiable in principle, one might say—rests on a logical relation between the theory in question and the class of basic statements (or the potential falsifiers described by them).

(2) 'Falsifiable' in the sense that the theory in question can *definitively* or *conclusively* or *demonstrably* be falsified ('demonstrably falsifiable'). I have always stressed that even a theory which is obviously falsifiable in the first sense is never falsifiable in this second sense. (For this reason I have used the expression 'falsifiable' as a rule only in the first, technical sense. In the second sense I have as a rule spoken not of 'falsifiability' but rather of 'falsification' and of its problems.)

It is clear that the suffixes 'able' and 'ability' are used somewhat differently in these two senses. Although the first sense refers to the logical possibility of a falsification in principle, the second sense refers to a *conclusive practical experimental proof* of falsity. But anything like conclusive proof to settle an empirical question does not exist.

An entire literature rests on the failure to observe this distinction. It is often said that my criterion of demarcation is inapplicable because empirical scientific theories cannot be definitively falsified. Less importantly, it is often said (see section IV below) that the discovery of the unfalsifiability of scientific theories in the second sense is an achievement that contradicts my theory, despite the fact that I myself have pointed this out over and over again. (Instead of distinguishing the two meanings—'falsifiability$_1$', the possibility that certain theories can in principle be falsified, because they have some potential falsifiers, and 'falsifiability$_2$', the *always* problematic possibility that a theory can be shown to be false, since final empirical proofs do not exist—the ironic distinctions of 'Popper$_0$', 'Popper$_1$', and 'Popper$_2$', and so on, have been made (i.e., of various stages of 'Popper' that flagrantly contradict one another and cannot be brought into harmony).[1] And the difficulties, in many cases the impossibility, of a conclusive practical falsification are put forward as a difficulty or even impossibility of the proposed criterion of demarcation.

This would all be of little importance but for the fact that it has led some people to abandon rationalism in the theory of science, and to tumble into irrationalism. For if science does not advance rationally and critically, how can we hope that rational decisions will be made anywhere else? A flippant attack on a misunderstood logical-technical term has thus led some people to far-reaching and disastrous philosophical and even political conclusions.

It should be stressed that the uncertainty of every empirical falsification (which I have myself repeatedly pointed out) should not be taken too seriously (as I have also pointed out). There are a number of important falsifications which are as 'definitive' as general human fallibility permits. Moreover, every falsification may, in its turn, be tested again. An example of a falsification—the falsification of Thomson's model of the atom, which led Ernest Rutherford to propose the nuclear model—should be mentioned here, to illustrate the force which a falsification may have. In Thomson's model of the atom, the positive charge was distributed over the entire space which the atom occupied. Rutherford had accepted this model. But then came the famous experiments of his students Geiger and Marsden. They found that alpha particles which were shot on to a very thin piece of gold leaf were sometimes reflected from the golf leaf, instead of being only deflected. The reflected particles were rare—approximately one among twenty thousand—but they occurred with statistical regularity. Rutherford was astonished. He wrote about this a quarter of a century later: 'It was quite the most incredible event that has ever happened to me in my life. It was

[1][See the works of Imre Lakatos, especially 'Criticism and the Methodology of Scientific Research Programmes', *Proc. Arist. Soc.* **69**, pp. 149–86, and *The Methodology of Scientific Research Programmes*, 1978, pp. 93–101. Ed.]

almost as incredible as if you fired a fifteen-inch shell at a piece of tissue paper and it came back and hit you.'[2]

Rutherford's formulation is excellent. It is not impossible—certainly not *logically* impossible—that a shot from a giant cannon onto a piece of tissue paper is reflected by it—even with a regular statistical probability of 1 in 20,000. This is not logically impossible; and hence Thomson's theory (according to which the atoms form a wall like tissue paper) is not definitively refuted. But Rutherford and some other physicists, among them Niels Bohr, were satisfied that another theory was needed. Therefore they proposed that Thomson's theory be regarded as falsified and be replaced by the nuclear model of Rutherford; and a little later (since this had its own problems) by the marvellous atom model of Bohr which, after about twelve years, was in its turn superseded by quantum mechanics.

Often it takes a long time before a falsification is accepted. It is usually not accepted until the falsified theory is replaced by a proposal for a new and better theory. As Max Planck remarked, one must often wait until a new generation of scientists has grown up; that, however, is not always necessary. It was so neither with Rutherford's new model of the atom (1912), nor with the recognition of J. J. Thomson (1897) of subatomic particles such as the electron, which meant that the theory of indivisible atoms was falsified. (Atoms had been regarded as indivisible by definition since about 460 B.C.) Nor was it so with the falsification by Carl Anderson (1932) of the powerful theory that there were only two elementary particles—the proton and the electron—and the rejection by Hideki Yukawa of the electro-magnetic theory of matter.

These are only four of many examples of scientific revolutions which were introduced through successful falsifications.

The misunderstood logical-technical meaning of falsifiability in the first sense, in the sense of the criterion of demarcation, has led to two historical legends. The first, unimportant legend is that I overlooked the non-conclusiveness of the falsifiability of theories—the fact that theories are never conclusively falsifiable in the second sense. Whereas in fact, I had repeatedly stressed this since 1932. The

[2]Lord Rutherford: 'The Development of the Theory of Atomic Structure', in J. Needham and W. Pagel, eds.: *Background of Modern Science*, 1938, p. 68.

second legend (and it is a far more important legend) is that falsification plays no role in the history of science. In fact, it plays a leading
role, despite its non-definitive character. The examples already
given provide some evidence of this, but I shall provide some
further examples in the next section.

II

It has been alleged by some people—even by some of my former
students—that my theory of science is refuted by the facts of the
history of science. This is a mistake: it is a mistake about the facts of
the history of science, and it is also a mistake concerning the claims
of my methodology.

As I tried to make clear in 1934 (*L.Sc.D.*, p. 37; and sections 10
and 11), I do not regard methodology as an empirical discipline, to
be tested, perhaps, by the facts of the history of science. It is, rather,
a philosophical—a metaphysical—discipline, perhaps partly even a
normative proposal. It is largely based on metaphysical realism, and
on the logic of the situation: the situation of a scientist probing into
the unknown reality behind the appearances, and anxious to learn
from mistakes.

Nevertheless, I have always thought that my theory—of refutation, followed by the emergence of a new problem, followed, in its
turn, by a new and perhaps revolutionary theory—was of the
greatest interest for the historian of science, since it led to a revision
of the way historians should look at history; especially as most
historians in those days (1934) believed in an inductivist theory of
science.[3] (They have now largely given this up—even my critics.)

That my theory, to the extent to which it is accurate, should be of
interest to scientists and historians is hardly surprising; for many of
them—I believe most of them—share my realist view of the world
and also understand the aims of science as I do: to achieve better and
better explanations.

Some examples may be useful.

A list is offered here, of interesting cases in which refutations led
to revolutionary theoretical reconstructions. This list goes back

---

[3][See Joseph Agassi: *Towards an Historiography of Science*, 1963. Ed.]

largely to the 1930s, and to my New Zealand days, when I gave a series of lectures to the Christchurch branch of the Royal Society of New Zealand, illustrating my theory with examples from the history of physics. I have written about some of these cases in various places; and I do not think that this list contains all the cases of falsification to which I have referred in my various writings. In so doing, I have always relied mainly on my memory: I do not pretend to be a historian of science myself. And I have, because of the pressure of other urgent work, never had time to survey the history of physics systematically in search of further examples: I have no doubt that there are hundreds. But I think that the list submitted here—a list of a few striking cases which can only be understood as examples of refutation—is sufficiently impressive. (I should even be inclined to suggest that, historically, a science becomes a science when it has accepted an empirical refutation; but I do not really propose this as a serious hypothesis; and the case of Copernicus *may* be a counter example: a great scientific theory that was not instigated by an empirical refutation.)

A List of Examples Chosen Almost at Random

(1) Parmenides-Leucippus: Leucippus takes the *existence of motion* as a partial refutation of Parmenides's theory that the world is full and motionless. This leads to the theory of 'atoms and the void'. It is the foundation of atomic theory.[4]

(2) Galileo refutes Aristotle's theory of motion: this leads to the foundation of the theory of acceleration, and later of Newtonian forces. Also, Galileo takes the moons of Jupiter and the phases of Venus as a refutation of Ptolemy,[5] and thus as empirical support of the rival theory of Copernicus.

---

[4]*Cp.* my *Conjectures and Refutations*, Chapter 2, sections vi and vii. [See also Volume III of the *Postscript*, 'Metaphysical Epilogue'. Ed.]

[5]In a very interesting letter to me, Allan Franklin has raised some doubts about this example, suggesting that although Galileo argued against the Aristotelian law of motion as if it dealt only with constant velocities, Aristotle himself, as well as later commentators, recognized that falling bodies accelerate. Franklin finds the kinematics of uniformly accelerated motion worked out in the fourteenth century at Merton College, Oxford, and also in Paris (by Oresme). He also refers to Buridan's 'impetus theory' and Domingo de Soto's argument that falling bodies exemplified uniform acceleration. I am grateful to Mr. Franklin for these comments. See also my *Conjectures and Refutations*, Chapter 3, section 1.

(3) Toricelli (and predecessors): the refutation of 'nature abhors a vacuum'. This prepares for a mechanistic world view.

(4) Kepler's refutation of the hypothesis of circular motion upheld till then (even by Tycho and Galileo), leads to Kepler's laws and so to Newton's theory.[6]

(5) Lavoisier's refutation of the phlogiston theory leads to modern chemistry.

(6) The falsification of Newton's theory of light (Young's two-slit experiment). This leads to the Young-Fresnel theory of light. The velocity of light in moving water is another refutation. It prepares for special relativity.

(7) Oersted's experiment is interpreted by Faraday as a refutation of the universal theory of Newtonian central forces and thus leads to the Faraday-Maxwell *field theory*.

(8) Atomic theory: the atomicity of the atom is refuted by the

---

[6]Historians have often claimed that there was a *prejudice* in favour of circular motion which Kepler and others had to overcome. But circular motion was not simply a prejudice; there was in effect a conservation law for circularity—not only for the rotations of the planets, but also for the wheel: the conservation of angular momentum. Of course the conservation laws were not clear in those days. But Galileo's conservation law, his form of the law of inertia, allowed for circular motion; and all of this lay at the back of the metaphysical ideas and principles of explanation accepted at that time. In this context circular motion was explicable, but elliptical motion was felt to be quite irrational. Thus the *rationality* of the pre-Keplerian metaphysical attitude (or research programme) should be taken into account: there was the well-established continuation of the wheel to rotate; if we have a freely suspended wheel, the angular momentum is as well supported by observation as are inertial forces. It took Kepler a great effort to get over this view—but not because it was a prejudice; rather, precisely because it formed an important part of the rational background. Feeling as he did at first that elliptical motion was irrational, Kepler needed a new type of explanation for it. It may be here that the sun came in: in Kepler's account there are forces emanating from the sun, whereas in Galileo's theory, the circular motion of the planets is not really dependent on the sun. Of course Kepler's theory is different from our own: he spoke, in the main, not of the sun's attraction, but of the pushing of rays from the sun. Only with Newton did it begin to become clear that the attraction of the sun influences the planets just as the attraction of the earth influences the moon. Galileo, however, continued to oppose Kepler's theory because of its astrological overtones—that is, its irrational overtones: the 'influence' of the planets on other planets. Kepler was indeed an astrologist; and astrology maintains that the heavenly bodies exert forces on one another. Thus one can understand both Galileo and Kepler. Galileo's metaphysical framework forced circular motion on him, and prevented him from accepting influences from the sun and moon. See also *Objective Knowledge*, Chapter 4, section 9.

Thomson electron. This leads to the electromagnetic theory of matter, and, in time, to the rise of electronics. See Einstein's and Weyl's attempts at a monistic ('unified') theory of gravitation and electromagnetics.[7]

(9) Michelson's experiment (1881–1887–1902, etc.) leads to Lorentz's *Versuch einer Theorie der elektrischen und optischen Erscheinungen in bewegten Körpern* (1895: see §89). Lorentz's book was crucially important to Einstein, who alluded to it twice in §9 of his relativity paper of 1905. (Einstein himself did not regard the Michelson experiment as very important.) Einstein's special relativity theory is (a) a development of the formalism founded by Lorentz and (b) a different—that is, relativistic—interpretation of that formalism. There is no crucial experiment so far to decide between Lorentz's and Einstein's interpretations; but if we have to adopt action at a distance (non-locality: see *Quantum Theory and the Schism in Physics*, Vol. III of the *Postscript*, Preface 1982), then we would have to return to Lorentz.

Incidentally, it took years before physicists began to come to some agreement about the importance of Michelson's experiments: I do not contend that falsifications are usually accepted at once (see the preceding section)—not even that they are immediately recognised as *potential* falsifications.

(10) The 'chance-discoveries' of Roentgen and of Becquerel refuted certain (unconsciously held) expectations; especially Becquerel's expectations. They had, of course, revolutionary consequences.

(11) Wilhelm Wien's (partially) successful theory of black body radiation conflicted with the (partially) also very successful theories of Sir James Jeans and Lord Rayleigh. (See preceding section.) The refutation by Lummer and Pringsheim of the radiation formula of Rayleigh and Jeans, together with Wien's work, leads to Planck's quantum theory (see *L.Sc.D.*, p. 108). In this, Planck refutes his own theory, the absolutistic interpretation of the entropy law, as opposed to a probabilistic interpretation similar to Boltzmann's.

[7] *Cp.* my 'The Rationality of Scientific Revolutions', in R. Harré, ed.: *Problems of Scientific Revolution*, 1975, pp. 72–101; see section XII.

(12) Philipp Lenard's experiments concerning the photoelectric effect conflicted, as Lenard himself insisted, with what was to be expected from Maxwell's theory. They led to Einstein's theory of light-quanta or photons (which were of course also in conflict with Maxwell), and thus, much later, to particle-wave dualism.

(13) The refutation of the Mach-Ostwald anti-atomistic and phenomenalistic theory of matter: Einstein's great paper on Brownian motion of 1905 suggested that Brownian motion may be interpreted as a refutation of this theory. Thus this paper did much to establish the reality of molecules and atoms.

(14) Rutherford's refutation of the vortex model of the atom.[8] This leads directly to Bohr's 1913 theory of the hydrogen atom, and thus, in the end, to quantum mechanics.

(15) Rutherford's refutation (in 1919) of the theory that chemical elements cannot be changed artificially (though they may disintegrate spontaneously).

(16) The theory of Bohr, Kramers and Slater (see *L.Sc.D.*, pp. 250, 243): this theory was refuted by Compton and Simon. The refutation leads almost at once to the Heisenberg-Born-Jordan quantum mechanics.

(17) Schrödinger's interpretation of his (and de Broglie's) theory is refuted by the statistical interpretation of matter waves (experiments of Davisson and Germer, and of George Thomson, for instance). This leads to Born's statistical interpretation.

(18) Anderson's discovery of the positron (1932) refutes a lot: the theory of two elementary particles—protons and electrons—is refuted; conservation of particles is refuted; and Dirac's own original interpretation of his predicted positive particles (he thought they were protons) is refuted. Some theoretical work of about 1930-31 is thereby corroborated. (For some details see Norwood Russell Hanson: *The Concept of the Positron*, 1963; an excellent book.)

(19) The electrical theory of matter[9] elaborated by Einstein and

---

[8] *Cp.* my 'The Rationality of Scientific Revolutions', p. 90.

[9] *Cp.* The Introduction to Volume III of the *Postscript*. Cp. also my 'The Rationality of Scientific Revolutions', p. 90, first new paragraph.

Weyl, and held implicitly—and at any rate, pursued—by Einstein to the end of his life (since he interpreted the unified field theory as a theory of *two* fields, gravitation and electromagnetics), is refuted by the neutron and by Yukawa's theory of *nuclear forces*: the Yukawa Meson. This gives rise to the theory of the nucleus.

(20) The refutation of parity conservation. (See Allan Franklin, *Stud. Hist. Philos. Sci.* **10**, 1979, p. 201.)

III

Of course, it is understood that these refutations merely created new problem situations which in their turn stimulated imaginative and critical thought. The new theories which developed were thus not direct results of the refutations: they were the achievements of creative thought, of thinking men.

Another obvious remark is that, in a number of these cases, it took time before the refutation was accepted as such: there often were rearguard actions, sometimes even prolonged ones, before the refutation was accepted as a matter of course *as a refutation* by all competent persons, rather than interpreted in some other way. But this was by no means always so: for example not in the cases (12) (Lenard's results were fairly quickly accepted), (13) to (17) and even (18), though the refuted theory in this last case had a long afterlife.

There are of course exceptions to this analysis in terms of refutation followed by reconstruction. The greatest exception seems to be Copernicus, whose aim was to give an alternative explanation of the empirical facts explained by Ptolemy.[10] In order to make sure that

---

[10]What happened in the case of Copernicus was the reinterpretation of the facts in terms of the old theory of Aristarchus, which had largely been forgotten, without any intention of having a crucial experiment. Copernicus wanted to say that the same facts can be reinterpreted in the light of Aristarchus's theory. Only later did others notice that the facts may possibly be *better* interpreted in the light of Copernicus's theory, and that there are other advantages too. Another way of putting this—although using different terminology from that which was employed by Copernicus—is to say that he proposed to substitute a different metaphysical background; and that the Ptolemaic theory was a metaphysical theory that suffered from grave difficulties: for example, the shell mechanism is violated by the comets which penetrate it. It was only when it was recognized that the two alternative theories have different empirical consequences that the matter became scientific, or

this is really an exception one would have to study the case in more detail, and especially the *acceptance* of the theory by scientists, which may have been delayed until Galileo's new empirical findings mentioned under (2), which can be claimed to be refutations of Ptolemy.

My theory of science was not intended to be an historical theory, or to be a theory supported by historical or other empirical facts, as I said before. *Yet I doubt whether there exists any theory of science which can throw so much light on the history of science as the theory of refutation followed by revolutionary and yet conservative reconstruction.*[11]

IV

This may be the place to mention, and to refute, the legend that Thomas S. Kuhn, in his capacity as a historian of science, is the one who has shown that my views on science (sometimes, but not by me, called 'falsification*ism*') can be refuted by the facts; that is to say, by the history of science.

I do not think that Kuhn has even attempted to show this. In any case, he has done no such thing. Moreover, on the question of the significance of falsification for the history of science, Kuhn's and my views coincide almost completely.

This does not mean that there are not great differences between Kuhn's and my views on science. I uphold the ancient theory of truth (almost explicit in Xenophanes and Democritus and Plato, and quite explicit in Aristotle) according to which truth is the agreement with the facts of what is being asserted. Kuhn's views on this fundamental question seem to me affected by relativism; more specifically, by some form of subjectivism and of elitism, as proposed for example by Polanyi. Kuhn seems to me also affected by

---

at least became a living research programme open to becoming scientific. Such tests in fact became available with Galileo: with the phases of Venus, the moons of Jupiter, and the differences in size of Venus, Mars, and Mercury. (The apparent sizes of the planets should be constant if the earth is at the center.) (See *Objective Knowledge*, 1972, p. 173.)

[11]See for example 'The Rationality of Scientific Revolutions', in Rom Harré, ed., *Problems of Scientific Revolution, op. cit.*, section VIII, pp. 82f.

Polanyi's fideism: the theory that a scientist *must* have faith in the theory he proposes (while I think that scientists—like Einstein in 1916 or Bohr in 1913—often realize that they are proposing conjectures that will, sooner or later, be superseded). There are many other such points of difference, of which perhaps the most important is my emphasis on objective rational criticism: I regard as characteristic of ancient and modern science the critical approach towards theories, from the point of view of whether they are true or false. Another important point seems to me that Kuhn does not seem to see the great importance of the many purely *scientific* revolutions that are *not* connected with *ideological* revolutions. In fact, he almost seems to identify these two.[12]

But concerning either falsifiability or the impossibility of conclusive proofs of falsification, and the part these play in the history of science and of scientific revolutions, there does not seem any significant difference whatever between Kuhn and me.

Kuhn, however, appears to see great differences between us here, although he himself also stresses many similarities between his views and mine. To explain these similarities he mentions that he attended my William James Lectures at Harvard in 1950. He was also, at the same time, I might add, one of the most active and most critical members of my seminars (eight two-hour lectures and eight two-hour seminars, I think). But it seems clear that he does not fully remember what happened during these sessions; and by the time he wrote his first book, *The Copernican Revolution* (1957; paperback edition, 1959) he evidently retained only a very schematic memory of my views, and considered me to be a 'naive falsificationist'. Yet in this book, Kuhn *practically* accepted my real views on the revolutionary character of the evolution of science. He deviates from my views only in upholding what I described above as 'fideism'; for he asserts 'that a scientist *must believe* [my italics] in his system before he will trust it as a guide to fruitful investigations of the *unknown*.'[13] But Kuhn follows me fairly closely when he continues: 'But the

---

[12]See my 'The Rationality of Scientific Revolutions', *op. cit.*, pp. 72–101, esp. 87–93.

[13]T. S. Kuhn, *The Copernican Revolution*, pp. 75f. of the paperback edition, 1959. I have italicized the words 'must believe', because fideism is the only point in this passage where Kuhn deviates from me: I should have said the scientist *may* believe; alternatively, he may accept 'his system' only tentatively (as we know from Einstein, for example; or from Niels Bohr—at any rate before 1926).

scientist pays a price for his commitment. . . . A single observation incompatible with his theory [may demonstrate][14] that he has been employing the wrong theory all along. His conceptual scheme must then be abandoned and replaced.'[15]

This is, obviously, 'falsificationism'; in fact, something like a 'methodological stereotype of falsification', to cite Kuhn's allusion to me in his later book, *The Structure of Scientific Revolutions* (1962, p. 77). But in his earlier book on Copernicus, Kuhn continues: 'That, in outline, is the logical structure of a scientific revolution. A conceptual scheme . . . finally leads to results that are incompatible with observation . . It is a useful outline, because the incompatibility of theory and observation is the ultimate source of every revolution in the sciences.'

This 'useful outline' of the logic of a scientific revolution is not only falsificationist; it is a far more simplistic stereotype of falsificationism than anything I myself ever said in my writings, my lectures, or my seminars; in fact, I have always been in full agreement with the following more critical remark that Kuhn adds: 'But historically, the process of revolution is never [I should say: 'hardly ever'], and could not possibly [Rutherford, see above?] be, so simple as the logical outline indicates. As we have already begun to discover observation is never *absolutely* incompatible with a [theory].'

Of course I had already been stressing this point in 1934 (I had always pointed out that 'observation is theory impregnated', just as I also pointed out that it is impossible to produce an unquestionable 'disproof' of an empirically scientific theory. See section I above.) I was therefore puzzled when I read, in Kuhn's second book, *The Structure of Scientific Revolutions* (p. 77): 'No process yet disclosed by the historical study of scientific development *at all resembles* the methodological stereotype of falsification by direct comparison with nature' [italics mine].

What did Kuhn mean by this? That the historical process does not at all resemble a process of falsification, or that it does not resemble that 'stereotype' which he characterizes as the 'direct comparison with nature', which he elsewhere calls 'naive falsificationism', and which I, for one, had always rejected?

---

[14]*Loc. cit.* Kuhn writes 'demonstrates'.
[15]I dislike the term 'his conceptual scheme'; I should have said: 'his theory'.

Now it turns out that it is my own theory that Kuhn has in mind when he speaks of a 'stereotype' of falsification. For elsewhere (in P. A. Schilpp, ed., *The Philosophy of Karl Popper*, p. 808), he writes: 'Sir Karl is not, of course a naive falsificationist. He knows all that has just been said and has emphasized it from the beginning of his career. . . . *Though he is not a naive falsificationist, Sir Karl may, I suggest, legitimately be treated as one.'* [The italics are mine.]

This passage is really astonishing. It is exactly like saying: 'Although Popper is not a murderer, he may, I suggest, legitimately be treated as one.'

There are no real arguments in Kuhn's paper leading up to this astonishing verdict; except that he believes that my own arguments against what he calls 'naive falsificationism' threaten 'the integrity of [my] basic position'. But what he wrongly believes to be my 'basic position' is only the *legend or paradigm* according to which Popper is a naive falsificationist. This argument (if one can call it that) is circular.

I believe that Kuhn really believes what he writes. Yet how can he? I have tried out several explanatory theories. Only one of them seems at all plausible to me. It is that Kuhn, early in his career, formed a theory of my views which became his paradigm of Popper: Popper was the man who replaced verificationism by ('naive') falsificationism. Kuhn formed this paradigm (according to his own indications) before he ever read any of my writings. When at last he read *The Logic of Scientific Discovery*, he read it in the light of this paradigm. Many passages in this book (one on the page immediately after my introduction of the idea of falsification) showed that I did not conform to his paradigm. But, as we have learnt from Kuhn, paradigms are not given up so easily.[16]

The issue now is this. Am I really the man who had naive

---

[16]Kuhn's contribution to P. A. Schilpp, ed., *The Philosophy of Karl Popper* (Vol. II, pp. 798–819), is a very pleasantly written criticism of the legendary naive falsificationist K.R.P. He is so convinced that he knows my opinions and their weaknesses that, with my books in his hands, he tells me of 'locutions' such as 'falsification' or 'refutation' which are 'antonyms of "proof" '. But in the *Index of Subjects* of my *L.Sc.D.* he could have found: 'Disproof, no conclusive disproof [of a theory] can be produced, 42, 50, 81–87.' This and many other remarks in his contribution show what happens to a reader of a book if he has a 'paradigm' of what must be found in it and what not. Altogether I find that a lot of historians of science are very bad (that is, prejudiced) readers.

falsificationism as the linchpin of his thought? Is the Kuhnian paradigm true? May I 'legitimately be treated as' a 'naive falsificationist', even though Kuhn admits, *after* looking at *The Logic of Scientific Discovery*, that, as early as 1934, I was *not* so?

It so happens that the real linchpin of my thought about human knowledge is fallibilism and the critical approach; and that I see, and saw even before 1934 (see my *Die beiden Grundprobleme der Erkenntnistheorie*), that human knowledge is a very special case of animal knowledge. My central idea in the field of animal knowledge (including human knowledge) is that it is based on inherited knowledge. It is of the character of unconscious expectations. It always develops as the result of modification of previous knowledge. The modification is (or is like) a mutation: it comes from inside, it is of the nature of a trial balloon, it is intuitive or boldly imaginative. It is thus of a conjectural character: the expectation may be disappointed, the balloon or bubble may be pricked: all the information received from outside is eliminative, selective.

The special thing about *human* knowledge is that it may be formulated in language, in propositions. This makes it possible for knowledge to become conscious and to be objectively criticizable by arguments and by tests. In this way we arrive at science. Tests are attempted refutations. All knowledge remains fallible, conjectural. There is no justification, including, of course, no final justification of a refutation. Nevertheless we learn by refutations, i.e., by the elimination of errors, by feedback.[17] In this account there is no room at all for 'naive falsification'.

V

Another objection to my theory of knowledge is better founded, even though its impact on my theory is negligible. It is the admitted failure of a definition (of verisimilitude, or approximation to truth) which I proposed in 1963.[18]

Let me first explain with two examples, (1) and (2), the only kind of use of the idea of verisimilitude that is likely to be made in my theory of knowledge (or in that of anybody else).

---

[17]I do not think that Norbert Wiener, in his *Cybernetics*, referred to Darwinism or to the elimination of error (trial and error).

[18]See *Conjectures and Refutations*, Chapter 10.

(1) The statement that the earth is at rest and that the starry heavens rotate round it is further from the truth than the statement that the earth rotates round its own axis; that it is the sun that is at rest; and that the earth and the other planets move in circular orbits round the sun (as Copernicus and Galileo proposed). The statement due to Kepler that the planets do not move in circles but in (not very elongated) ellipses with the sun in their common focus (and with the sun at rest, or spinning round its axis) is a further approximation to the truth. The statement (due to Newton) that there exists a space at rest but that, apart from rotation, its position cannot be found by observation of the stars or of mechanical effects, is a further step towards the truth.

(2) Gregor Mendel's ideas about heredity were nearer to the truth, it seems, than Charles Darwin's views. The later breeding experiments with fruit flies led to further improvements of the verisimilitude of the theory of heredity. The idea of a gene pool of a population (a species) was a further step. But the greatest steps by far were those that culminated in the discovery of the genetic code.

These examples, (1) and (2), show, I believe, that a formal definition of verisimilitude is not needed for talking sensibly about it. (See also below, pp. 261–278.)

Why, then, did I try to give a formal definition?

I have often argued against the need for definitions. They are never really needed, and rarely of any use, except in the following sort of situation: we may by introducing a definition show that not only are fewer basic assumptions needed for a good theory but that our theory can explain more than without the definition. In other words, a new definition is of interest only if it strengthens a theory. I thought that I could do this with my theory of the aims of science: the theory that science aims at truth *and* the solving of problems of explanation, that is, at theories of greater explanatory power, greater content, and greater testability. The hope further to strengthen this theory of the aims of science by the definition of verisimilitude in terms of truth and of content was, unfortunately, vain. But the widely held view that scrapping this definition weakens my theory is completely baseless. I may add that I accepted the criticism of my definition within minutes of its presentation, wondering why I had not seen the mistake before; but nobody has ever shown that my theory of knowledge, which I developed at least as early as 1933 and which has been growing lustily ever since and which is much used by working scientists, is shaken in the least by this unfortunate mistaken definition, or why the idea of verisimilitude (which is not an essential part of my theory) should not be used further within my theory as an undefined concept.

The assertion that my *authority* is damaged by this incident is obviously true, but I have never claimed or wished to have any authority. The assertion that *my theory* is damaged has been advanced without even attempting to give a reason, and seems to me just incompetent.

VI

It is also sometimes objected to my theory that it cannot answer Nelson Goodman's paradox.[19]

That this is not so will be seen from the following considerations which show, by a simple calculation, that the evidence statement e, 'all emeralds observed before the 1st of January of the year 2000 are green' does not make the hypothesis h_1, 'all emeralds are green, at least until February 2000' more probable than the hypothesis 'all emeralds are blue, for ever and ever, with the exception of those that were observed before the year 2000, which are green'. This is not a paradox, to be formulated and dissolved by linguistic investigations, but it is a demonstrable theorem of the calculus of probability. The theorem can be formulated as follows:

The calculus of probability is incompatible with the conjecture that probability is ampliative (and therefore inductive).

The idea that probability is ampliative is widely held. It is the idea that evidence e—say, that all swans in Austria are white—will somehow increase the probability of a statement that goes beyond e, such as h_2, 'all (or most) swans in regions bordering Austria are white'. In other words, the idea is that the evidence makes things

---

[19][See W. W. Bartley, III, 'Eine Lösung des Goodman-Paradoxons', in G. Radnitzky und Gunnar Andersson: *Voraussetzungen und Grenzen der Wissenschaft* (Tübingen, 1982), pp. 347–358; and in 'Rationality, Criticism and Logic,' *Philosophia* 11, February 1982, esp. pp. 169–173, and the references given therein. Ed.]

beyond what it actually asserts at least a little more probable. (This view was strongly defended by Carnap, for example.)

The view that probability is ampliative was suggested especially by the following theorem (h = hypothesis; e = empirical evidence; b = background knowledge):

Let $p(h,b) \neq 0$. Further, let e be favourable evidence (that is, e follows from h in the presence of b, so that $p(e,b) \neq 1$ and $p(e,hb) = 1$). Then $p(h,eb) > p(h,b)$. That is to say, the favourable evidence e makes h more probable, even though h says more than e. And this holds for every new $e_1, e_2, \ldots$, which satisfy similar conditions.

It therefore seems that increasing favourable evidence goes on supporting h; and so it seems that the support is ampliative.

But this is an illusion, as can be shown as follows:

Let h_1 and h_2 be any two hypotheses supported by e in the presence of b, so that

$$p(e,b) \neq 1 \text{ and } p(e,h_1b) = p(e,h_2b) = 1.$$

Let $R_{1,2}$ *(prior)* $= p(h_1b)/p(h_2b)$ be the ratio of the probabilities of h_1 and h_2 prior to the evidence e, and let

$$R_{1,2} \text{ (posterior)} = p(h_1,eb)/p(h_2,eb)$$

be the ratio of the two probabilities posterior to the evidence e.

Then we have, for any h_1, h_2 and e that satisfy the above conditions:

$$R_{1,2} \text{ (prior)} = R_{1,2} \text{ (posterior)}.$$

This follows almost immediately from

$$p(a,bc) = p(ab,c)/p(b,c),$$

that is, from Bayes's theorem.

What does

$$R_{1,2} \text{ (posterior)} = R_{1,2} \text{ (prior)}$$

signify? It says that the evidence does not change the ratio of the prior probabilities, whether we have calculated them or freely assumed them, provided the two hypotheses can both explain the evidence e. But this means that if we let

h_1 = all swans in *some* region greater than Austria are white;

$h_2 =$ all swans in the world are non-white except those in Austria which are white;

$e =$ all swans in Austria are white,

then, assuming any prior probability for h_1 and h_2 you like: their ratio $R_{1,2}$ *(prior)* remains unaffected by the evidence. Thus there is no spill-over, no ampliative support: there is no ampliative probability, neither for swans nor for emeralds. And this is not absurd, but tautological (and it is unaffected by translation).

REALISM AND THE AIM OF SCIENCE

So much is certain: that nothing is better adapted to form a mind which is capable of a great development, than living and participating in great scientific revolutions. I would therefore counsel all those whom the period they live in has not naturally presented with this advantage, to procure it artificially for themselves, by reading the writings of those periods in which the sciences have suffered great changes. To peruse the writings of the most opposite systems, and to extract their hidden truth, to answer questions raised by these opposite systems, to transfer the chief theories of the one system into the other, is an exercise which cannot be sufficiently recommended to the student. He would certainly be rewarded for this labour, by becoming as independent as possible of the narrow opinions of his age.

HANS CHRISTIAN OERSTED[1]

Mathematicians may flatter themselves that they possess new ideas which mere human language is as yet unable to express. Let them make the effort to express these ideas in appropriate words without the aid of symbols, and if they succeed, they will not only lay us laymen under a lasting obligation, but, we venture to say, they will find themselves very much enlightened during the process, and will even be doubtful whether the ideas as expressed in symbols had ever quite found their way out of the equations into their minds.

JAMES CLERK MAXWELL[2]

[1]Hans Christian Oersted: 'Observations on the History of Chemistry: A Lecture, 1805–1807', in *The Soul in Nature, with Supplementary Contributions* (London: Henry G. Bohn; 1852), p. 322.

[2]*Scientific Papers of James Clark Maxwell*, ed. W. D. Niven, Vol. II (Cambridge, 1890), p. 328; reprinted from *Nature 7*, March 27, 1873, p. 400: Review of Thomson's & Taite's *Elements of Natural Philosophy*.

PREFACE, 1956

ON THE NON-EXISTENCE OF SCIENTIFIC METHOD

> But in fact, we know nothing from having seen it; for the truth is hidden in the deep.
>
> DEMOCRITUS [1]

As a rule, I begin my lectures on Scientific Method by telling my students that scientific method does not exist. I add that I ought to know, having been, for a time at least, the one and only professor of this non-existent subject within the British Commonwealth.

It is in several senses that my subject does not exist, and I shall mention a few of them.

First, my subject does not exist because subject matters in general do not exist. There are no subject matters; no branches of learning—or, rather, of inquiry: there are only problems, and the urge to solve them. A science such as botany or chemistry (or say, physical chemistry, or electrochemistry) is, I contend, merely an administrative unit. University administrators have a difficult job anyway, and it is a great convenience to them to work on the assumption that there are some named subjects, with chairs attached to them to be filled by the experts in these subjects. It has been said that the subjects are also a convenience to the student. I do not agree: even serious students are misled by the myth of the subject. And I should be reluctant to call anything that misleads a person a convenience to that person.

[1] See Hermann Diels: *Die Fragmente der Vorsokratiker,* ed. Walther Kranz, 6. Auflage, 1951, Vol. II, p. 166; 68 B 117.

5

So much about the non-existence of subjects in general. But Scientific Method holds a somewhat peculiar position in being even less existent than some other non-existent subjects.

What I mean is this. The founders of the subject, Plato, Aristotle, Bacon and Descartes, as well as most of their successors, for example John Stuart Mill, believed that there existed a method of finding scientific truth. In a later and slightly more sceptical period there were methodologists who believed that there existed a method, if not of *finding* a true theory, then at least of ascertaining whether or not some *given* hypothesis was true; or (even more sceptical) whether some given hypothesis was at least 'probable' to some ascertainable degree.

I assert that no scientific method exists in any of these three senses. To put it in a more direct way:

(1) There is no method of discovering a scientific theory.
(2) There is no method of ascertaining the truth of a scientific hypothesis, i.e., no method of verification.
(3) There is no method of ascertaining whether a hypothesis is 'probable', or probably true.

Having thus explained to my students that there is no such thing as scientific method, I hasten to begin my discourse, and we get very busy. For one year is hardly enough to scratch the surface of even a non-existent subject.

What do I teach my students? And how can I teach them?

I am a rationalist. By a rationalist I mean a man who wishes to understand the world, and to learn by arguing with others. (Note that I do not say a rationalist holds the mistaken theory that men are wholly or mainly rational.) By 'arguing with others' I mean, more especially, criticizing them; inviting their criticism; and trying to learn from it. The art of argument is a peculiar form of the art of fighting—with words instead of swords, and inspired by the interest of getting nearer to the truth about the world.

I do not believe in the current theory that in order to make an argument fruitful, the arguers must have a great deal in common.[2]

<hr>

[2][See 'The Myth of the Framework', in *The Abdication of Philosophy: Philosophy and the Public Good. Essays in Honor of Paul Arthur Schilpp*, ed. Eugene Freeman, 1976, pp. 23–48; and 'Addendum: Facts, Standards, and Truth: A Further Criticism of Relativism', in *The Open Society and Its Enemies*, fourth edition, 1962, pp. 369–396. Ed.]

On the contrary, I believe that the more different their back-grounds, the more fruitful the argument. There is not even a need for a common language to begin with: had there been no tower of Babel, we should have had to build one. Diversity makes critical argument fruitful. The only things which the partners in an argument must share are the wish to know, and the readiness to learn from the other fellow, by severely criticizing his views—in the strongest possible version that can be given to his views—and hearing what he has to say in reply.

I believe that *the so-called method of science consists in this kind of criticism*. Scientific theories are distinguished from myths merely in being criticizable, and in being open to modifications in the light of criticism. They can be neither verified nor probabilified.

My critical—or, if you prefer, my heretical—attitude influences, of course, my attitude towards my fellow philosophers.

You will all know the story of the soldier who found that his whole battalion (except himself, of course) was out of step. I constantly find myself in this entertaining position. And I am very lucky, for, as a rule, a few of the other members of the battalion are quite ready to fall into step. This adds to the confusion; and since I am not an admirer of philosophical discipline, I am content as long as enough members of the battalion are sufficiently out of step with one another.

Some of the things which put me out of step and which I like to criticize are:

(1) *Fashions*: I do not believe in fashions, trends, tendencies, or schools, either in science or in philosophy. In fact, I think that the history of mankind could well be described as a history of outbreaks of fashionable philosophical and religious maladies. These fashions can have only one serious function—that of evoking criticism. Nonetheless I do believe in the rationalist tradition of a common-wealth of learning, and in the urgent need to preserve this tradition.

(2) *The aping of physical science*: I dislike the attempt, made in fields outside the physical sciences, to ape the physical sciences by practising their alleged 'methods'—measurement and 'induction from observation'. The doctrine that there is as much science in a subject as there is mathematics in it, or as much as there is measurement or 'precision' in it, rests upon a complete misunderstanding. On the contrary, the following maxim holds for all sciences: Never aim at more precision than is required by the problem in hand.

7

Thus I have no faith in precision: I believe that simplicity and clarity are values in themselves, but not that precision or exactness is a value in itself. Clarity and precision are different and sometimes even incompatible aims. I do not believe in what is often called an 'exact terminology': I do not believe in definitions,[3] and I do not believe that definitions add to exactness; and I especially dislike pretentious terminology and the pseudo-exactness concerned with it. What can be said can and should always be said more and more simply and clearly.

(3) *The authority of the specialist:* I disbelieve in specialization and in experts. By paying too much respect to the specialist, we are destroying the commonwealth of learning, the rationalist tradition, and science itself.

To conclude, I think that there is only one way to science—or to philosophy, for that matter: to meet a problem, to see its beauty and fall in love with it; to get married to it, and to live with it happily, till death do ye part—unless you should meet another and even more fascinating problem, or unless, indeed, you should obtain a solution. But even if you do obtain a solution, you may then discover, to your delight, the existence of a whole family of enchanting though perhaps difficult problem children for whose welfare you may work, with a purpose, to the end of your days.[4]

[3][See *The Open Society and Its Enemies*, 1945, Chapter 11; and *Unended Quest*, section 7. Ed.]

[4][This 'Preface, 1956' was read at a meeting of the Fellows of the Center for Advanced Study in the Behavioral Sciences at Stanford, California, in November 1956. Ed.]

PART I

THE CRITICAL APPROACH

CHAPTER I

INDUCTION

But as for certain truth, no man has known it,
Nor will he know it; neither of the gods,
Nor yet of all the things of which I speak.
And even if by chance he were to utter
The perfect truth, he would himself not know it:
For all is but a woven web of guesses.

XENOPHANES

IN THIS introductory chapter, the problem of induction is treated more fully and in a wider setting than in my *Logic of Scientific Discovery* (*L.Sc.D.*, for short) to which the present work is a sequel: here I shall discuss all its more interesting ramifications of which I am aware.

In section 2, I try to give an outline of my theory of knowledge, to re-formulate the problem of induction, and to re-state its solution. The discussion of the views—largely metaphysical views—which tend to prevent this solution from being accepted is carried on to section 16, entitled 'Difficulties of Metaphysical Realism. By a Metaphysical Realist'.

Chapter Two (sections 17 to 26) is concerned with the *problem of demarcation*—the demarcation between science and metaphysics. (I do not attempt to demarcate between sense and nonsense.) There I try to show that the problem of demarcation, and its solution by a *testability criterion of demarcation*, have a significance which reaches far beyond the borders of philosophy.

Chapter Three (sections 27 to 32) is concerned with the *problem of corroboration*, and of introducing the technical term '*degree of corroboration*'. I try to show that this problem is of some interest, in view of the problem situation existing in the philosophy of science;

for its solution corrects the widespread but *mistaken* belief that scientific induction can help us to assess the probability of a hypothesis.

1. *A Puzzled Philosopher Abroad.*

Not long after I first came to London, in the autumn of 1935, about a year after the publication of my *Logik der Forschung* (*L.d.F.*, for short), I was taken to a meeting of the Aristotelian Society. Bertrand Russell, whom I had long admired as the greatest philosopher since Kant, read a paper on 'The Limits of Empiricism'.[1] Believing that our empirical knowledge was obtained by induction, and deeply impressed by the force of Hume's criticism, Russell suggested that we had to assume some *principle of induction* which could not be based upon induction in its turn; a principle whose adoption marked the limits of empiricism. Russell's position was almost the same as one which, rightly or wrongly, I had ascribed to Kant in the first section of my *L.d.F.*.

Having been invited to participate in the discussion, I said that I did not believe in induction at all, although I did believe in empiricism—an empiricism which did not impose upon itself those Kantian limits which Russell was prepared to accept. This statement (which I formulated as pointedly as I could manage with the little English at my disposal) was taken for a joke by the audience, who graciously laughed and clapped. I then suggested that the whole trouble was due to the mistaken belief that *scientific knowledge* was an especially strict or certain or august kind of knowledge. This statement met with the same reception as the first. I concluded with an attempt to explain that, in the usual sense of 'know', whenever I *know* that it is raining, it must be *true* that it is raining; for if it is not true, then I simply cannot know that it is raining, however sincerely I may believe that I know it. In this sense of the word, 'knowledge' always means 'true and certain knowledge'; and 'to know' means, in addition, to be in possession of *sufficient reason* for holding that our knowledge is true and certain. But, I said, there was no such thing as scientific knowledge in this sense. If, nonetheless, we chose to label

---

[1] Published in the *Proceedings of the Arist. Soc.* **36**, 1936, pp. 131–150. My remarks alluded especially to pp. 146*ff.* [See Popper's *Unended Quest*, 1976, section 22. Ed.]

the results of our scientific endeavours with the customary name 'scientific knowledge', then we ought to be clear that scientific knowledge was not a species of knowledge; least of all, a species distinguished by a high degree of solidity or certainty. On the contrary, measured by the high standards of scientific criticism, 'scientific knowledge' always remained sheer guesswork—although guesswork controlled by criticism and experiment. (It could not even attain any positive degree of 'probability' if this term was used in the sense of the probability calculus, for example, in the form given to it by Keynes or Jeffreys.) I ended by saying that merely by recognizing that scientific knowledge consists of guesses or hypotheses, we can solve the problem of induction without having to assume a principle of induction, or any limits to empiricism.

My little speech was well received, which was gratifying, but for the wrong reasons, which was puzzling. For it had been taken as an attack upon science, and perhaps even as an expression of a somewhat superior attitude towards it.

Admittedly, I had attacked, by implication, Science with a capital 'S', and those of its devotees who were ready to take its pronouncements as gospel truth. But I knew, of course, that Russell, with his deep and critical understanding of science, and his love of truth, was no such devotee. Thus the dismissal of Science with a capital 'S', although implicit in what I had said, had not been my main point at all. Rather, what I had hoped to convey to the audience was this: if we assume that what is called 'scientific knowledge' consists only of guesses or conjectures, then this assumption is sufficient for solving the problem of induction—called by Kant 'the problem of Hume'— without sacrificing empiricism; that is to say, without adopting a principle of induction and ascribing to it *a priori* validity. For *guesses are not 'induced from observations'* (although they *may*, of course, be suggested to us by observations). This fact allows us to accept without reservation (and without Russell's limits of empiricism) Hume's logical criticism of induction and to give up the search for an inductive logic, for certainty, and even for probability, while continuing in our scientific search for truth.

This, my main point, was lost. And I realized that it could hardly have been otherwise. For if people think on inductive lines—and who does not?—then a remark like 'I do not believe in induction' can hardly be interpreted in any other sense than 'I do not believe in science'. Nor do I think that I should have conveyed my meaning better had I begun, say, with the words, 'I believe in the greatness of science, but I do not believe that the methods or procedures of science are inductive in any sense'. Had I said this, people would no doubt have heard 'Science' in place of 'science', and they might have concluded that I wanted to uphold some doctrine of intuition or intuitionism, or perhaps some form of scientific authoritarianism.

Having just used two 'isms', I may perhaps mention that I have often been reproached, even by some of my most sympathetic philosophic friends, for my bad habit of discussing philosophy in terms of 'isms'; and I am quite ready to admit that it might be more straightforward if, without any reference to 'isms', one could just explain one's tenets, state one's arguments, and be done with it. But my puzzling experience illustrates why this method does not always work. We never address ourselves to completely open minds. However open minded an audience may be, they cannot help harbouring, if only subconsciously, connected theories, views, and expectations about the world, and even about the ways in which we learn to know it. They have adopted positions; usually typical positions: 'isms'.

Most of us, especially most philosophers, hold a great number of theories consciously, and after critical examination; and we may be prepared both to defend these by argument and to give them up when good arguments are brought against them. But we all also hold theories which we take for granted more or less unconsciously and therefore uncritically; and these uncritically held theories often contain the strongest reason for continuing to hold those other theories consciously. That this is so has been known for a long time: Bacon described such unconscious assumptions as idols and as *prejudices*. In Plato's Dialogues, Socrates frequently makes his partners realize that certain positions taken up by them imply that they hold theories or views of which they are not fully aware and which sometimes are even mutually conflicting. Before him, Parmenides speaks of delusive opinions uncritically held by 'the mortals'.

One of the oldest, more interesting and perhaps more important tasks of philosophy is the critical examination of such 'positions' and the theories or views they involve—especially those which are uncritically taken for granted. In doing so it is often found that there are *clusters of related views* related by common assumptions, by common preferences, or by common dislikes. Obviously it is often convenient and even necessary to give names to these positions, or views, or clusters of views. Hence the 'isms'.

That 'isms' have gone out of fashion in modern philosophy, and that using them or similar names is felt to be a sign of bad taste, is due to the fact that the critical discussion of theories or positions or clusters of views has gone out of fashion. But fashions, especially in philosophy, should not be accepted. They should be examined critically because they are themselves nothing but 'isms'—'isms' adopted uncritically.

All this bears on induction. Many philosophers and even some scientists believe that induction is an undeniable fact of common sense: that the actual use of what is now often called 'inductive procedures' cannot be seriously denied. This may or may not be so (see section 3 below). But we should at any rate learn to listen to those who deny facts of common sense. Philosophers have been extremely patient in listening to people who asserted and others who denied the existence of tables and chairs, or meteorites, or ghosts (both in and out of machines), or analytic statements. But the discussions of all these assertions and 'isms' are not nearly so fundamental for philosophy as the one about the existence or non-existence of inductive procedures. So let us discuss the matter critically. Perhaps the assumption that inductive procedures—an 'inductive logic'—exist is, after all, a prejudice, so that all that does exist is merely a myth, a mistaken 'ism' ('inductivism').

If there is such a thing as inductivism, then this helps explain why 'isms' are unfashionable. For such things as uncritically held 'isms' are a danger to inductivism. Bacon saw this. But his remedy—purge your mind—was naive. So it has seemed better to inductivists to look away, or else to study 'isms' inductively.

These theories which, if held unconsciously, are obviously held uncritically, are often incorporated in our language; and not only in its vocabulary, but also in its grammatical structure. This was first seen, to my knowledge, by Bertrand Russell, when he pointed out that many philosophical theories depend on the mistaken assumption that 'all propositions are reducible to subject-predicate form', an assumption which is closely connected with the grammatical structure of Indo-European languages.[2] Later, a similar doctrine

[2]Bertrand Russell, *A Critical Examination of the Philosophy of Leibniz*, 1900. See for example p. 15 where Russell, I suppose for the first time in his writings, refers to what has later been called 'pseudo propositions'—that is to say, 'a

was developed by Benjamin Lee Whorf who stressed, more especially, the dependence of our idea of time upon our language.[3]

These facts about language are sometimes used to defend the following radical conclusion. We are, intellectually, it may be said, the prisoners of our language: we cannot think except in terms of theories (of substance, or of space and time, for example) which, unknown to us, are incorporated in our language; and we cannot escape by our own efforts—for example by means of a critical discussion—from our prison, for the critical discussion would have to be conducted with the help of our language; and it would therefore remain within it—within the prison. Only by learning a new language of a different structure—one which is essentially not fully translatable into our old language—and therefore only through the clash with a new culture, and a conversion to the new culture, could we be freed from our prison; or rather, could we enter another prison, possibly a bigger one.

It seems to me that there is a great deal in this doctrine of imprisonment, but that its consequences are exaggerated. Though the help rendered by culture clash may be immensely valuable, we may sometimes do without it: we may succeed by our own critical efforts in breaking down one or another of our prison walls. Russell did so when he discovered, or became aware of, one of these walls through a critical study of Leibniz's philosophy. Awareness of any of these walls amounts to its destruction, since the imprisonment is intellectual: it largely consists in our intellectual blindness to the prison walls.

It must be admitted that Russell's discovery and the consequent destruction of one of the prison walls did not free him—or us—completely. Some of the old walls still stand, just because we have not become aware of them; and even the destruction of the intangible wall for which our subject-predicate grammar was responsible does not mean that we can now escape into the open. We can merely escape into a wider prison (that of a language of relations). This fact, however, should not depress us. A life sentence confining us to an

meaningless form of words' which may be mistaken for a proposition. See also Russell's article 'Logical Atomism' in J. H. Muirhead, ed. *Contemporary British Philosophy*, First Series, 1924, pp. 360 and 367f., concerning 'the influence of language on philosophy' and more especially on 'the substance-attribute metaphysic'; and see also p. 61 of his book *My Philosophical Development*, 1959.

[3]See B. L. Whorf, *Language, Thought and Reality*, 1956.

intellectual prison from which we can, in principle, free ourselves by escaping into a wider one, and then on into another that is wider still, with no pre-assigned limits, is not only a bearable sentence but one that opens up a thrilling prospect of fighting for freedom: a worthy task for our intellectual life.

There are several completely analogous ways of being intellectually imprisoned. We may be imprisoned not merely in a language, but also in various systems of assumptions or theories or points of view (they have been called 'total ideologies'[4]) within a language; assumptions of which we may be unaware and which for this reason we may be unable to criticize or to transcend. Though all this is to be admitted, it should not be exaggerated. It is often said, for example, that a discussion cannot be fruitful unless the participants agree on fundamentals, or share some common background or 'conceptual framework'.[5] I deny the truth of any such assertion. Though a discussion may be very satisfactory for the participants if they agree on all important points, it will be more fruitful under less pleasant conditions.

It is one of the tasks of philosophical criticism to make conscious these various systems of beliefs, so that after a searching examination we may tentatively choose the best available. But this means that we must understand, examine, compare, and criticize coherent systems of assumptions, that is to say, 'approaches', or 'isms' (such as inductivism, or positivism, or intuitionism). In that meeting of the Aristotelian Society, my attempt to explain my solution was bound to fail because I had no opportunity to analyse, and to criticize, the approach or attitude which I have called 'inductivism', an attitude which was bound to put upon my words an interpretation which I had not intended.

Thinking people tend to develop some framework into which they try to fit whatever new idea they may come across; as a rule, they even translate any new idea which they meet into a language appropriate to their own framework. One of the most characteristic tasks of philosophy is to attack, if necessary, the framework itself.

[4]A criticism of the doctrine of 'total ideologies' and of similar sociologistic constructions will be found in chapter 23 of my *Open Society, op. cit.*; see for example the reference to Einstein's criticism of our 'categorial apparatus' on p. 220.

[5][See 'The Myth of the Framework', *op. cit.*, and 'Addendum: Facts, Standards, and Truth: A Further Criticism of Relativism', in *The Open Society, op. cit.* Ed.]

And in order to do so, it may become necessary to attack beliefs which, whether or not they are consciously held, are taken so much for granted that any criticism of them is felt to be perverse or insincere. Whenever the framework itself is attacked, its defenders will as a rule interpret, and attempt to refute, the attack within their own adopted framework. But in trying to translate critical arguments directed against the framework into a language appropriate to that framework, they are liable to produce fatal distortions and misunderstandings. A discussion in terms of 'isms' may diminish this to some extent by constantly stressing the fact that the framework itself is under fire.

2. The Critical Approach: Solution of the Problem of Induction.

I do not believe in Belief.
E. M. FORSTER

I

During the many years that have gone by since that meeting of the Aristotelian Society it never occurred to me that my own approach, my own framework, might be more liable to misunderstandings than others—that it might clash more seriously with certain widely held and unconsciously accepted views—and that as a consequence people might misinterpret my approach by identifying it with some form of irrationalism, scepticism, or relativism. It was only recently that I began to suspect this, and to suspect that my own approach to the theory of knowledge was more revolutionary, and for that reason more difficult to grasp, than I had thought. This suspicion arose from a new way of viewing my own approach, and its relation to the problem situation in philosophy; a way that was suggested to me by my friend W. W. Bartley, III. His views are striking in themselves.[1] But they also explain why certain misunderstandings of my position are almost bound to arise.

[1]Some of these views have now been published; see W. W. Bartley, III: *The Retreat to Commitment*, 1962, and 'Rationality versus the Theory of Rationality', in *The Critical Approach to Science and Philosophy*, ed. Mario Bunge, 1964. [See also now my 'The Philosophy of Karl Popper, Part III: Rationality, Criticism, and Logic', *Philosophia*, 1982. Ed.] My remarks in the text are based not upon these publications but upon conversations with Bartley prior to their publication. The present section was partly rewritten in 1979.

The central problem of the philosophy of knowledge, at least since the Reformation, has been this. How can we adjudicate or evaluate the far-reaching claims of competing theories and beliefs? I shall call this our *first* problem. This problem has led, historically, to a *second* problem: How can we *justify* our theories or beliefs? And this second problem is, in turn, bound up with a number of other questions: What does a justification consist of? and, more especially: Is it possible to justify our theories or beliefs *rationally*: that is to say, by giving reasons—'positive reasons' (as I shall call them), such as an appeal to observation; reasons, that is, for holding them to be true, or to be at least 'probable' (in the sense of the probability calculus)? Clearly there is an unstated, and apparently innocuous, assumption which sponsors the transition from the first to the second question: namely, that one adjudicates among competing claims by determining which of them can be *justified* by positive reasons, and which cannot.

Now Bartley suggests that my approach solves the first problem, yet in doing so changes its structure completely. For I reject the second problem as irrelevant, and the usual answers to it as incorrect. And I also reject as incorrect the assumption that leads from the first to the second problem. I assert (differing, Bartley contends, from all previous rationalists except perhaps those who were driven into scepticism) that we cannot give any positive justification or any positive reason for our theories and our beliefs. That is to say, we cannot give any positive reasons for holding our theories to be *true*. Moreover, I assert that the belief that we can give such reasons, and should seek for them is itself neither a rational nor a true belief, but one that can be shown to be without merit.

(I was just about to write the word 'baseless' where I have written 'without merit'. This provides a good example of just how much our language is influenced by the unconscious assumptions that are attacked within my own approach. It is assumed, without criticism, that only a view that lacks merit must be baseless—without basis, in the sense of being unfounded, or unjustified, or unsupported. Whereas, on my view, *all* views—good *and* bad—are in this important sense baseless, unfounded, unjustified, unsupported.)

In so far as my approach involves all this, my solution of the central problem of justification—as it has always been understood—is as *unambiguously negative* as that of any irrationalist or sceptic.

Yet I differ from both the sceptic and the irrationalist in offering an *unambiguously affirmative* solution of another, third, problem which, though similar to the problem of whether or not we can give valid positive reasons for holding a theory to be *true*, must be sharply distinguished from it. This third problem is the problem of whether one theory is *preferable* to another—and, if so, why. (I am speaking of a theory's being preferable in the sense that we think or conjecture that it is *a closer approximation to the truth*, and that we even have *reasons* to think or to conjecture that it is so.)

My answer to this question is unambiguously affirmative. We can often give reasons for regarding one theory as preferable to another. They consist in pointing out that, and how, one theory has hitherto withstood criticism better than another. I will call such reasons *critical reasons*, in order to distinguish them from those *positive reasons* which are offered with the intention of *justifying* a theory, or, in other words, of justifying the belief in its truth.

Critical reasons do not justify a theory, for the fact that one theory has so far withstood criticism better than another is no reason whatever for supposing that it is actually true. But although critical reasons can never justify a theory, they can be used to defend (but not to *justify*) our *preference* for it: that is, our deciding to use it, rather than some, or all, of the other theories so far proposed. Such critical reasons do not of course prove that our preference is more than conjectural: we ought to give up our preference should new critical reasons speak against it, or should a promising new theory be proposed, demanding a renewal of the critical discussion.

Giving reasons for one's preferences can of course be *called* a justification (in ordinary language). But it is not a justification in the sense criticized here. Our preferences are 'justified' only relative to the present state of our discussion.

Postponing until later the important question of the standards of preference for theories, I will now give Bartley's view of the new problem situation which has arisen. He describes the situation very strikingly by saying that, after having given a negative solution to the classical *problem of justification*, I have replaced it by the new *problem of criticism*, a problem for which I offer an affirmative solution.

This transition from the problem of justification to the problem of criticism, Bartley suggests, is fundamental; and it gives rise to misunderstandings because almost everybody takes it implicitly for granted that everybody else (I included) accepts the problem of justification as the central problem of the theory of knowledge.

For according to Bartley *all philosophies so far have been justificationist philosophies,* in the sense that all assumed that it was the *prima facie* task of the theory of knowledge to show that, and how, we can *justify* our theories or beliefs. Not only the rationalists and the empiricists and the Kantians shared this assumption but also the sceptics and the irrationalists. The sceptics, compelled to admit that we cannot justify our theories or beliefs, declare the bankruptcy of the search for knowledge; while the irrationalists (for example the fideists), owing to the same fundamental admission, declare the bankruptcy of the search for reasons—that is, for rationally valid arguments—and try to justify our knowledge, or rather, our beliefs, by appealing to authority, such as the authority of irrational sources. Both assume that the question of justification, or of the existence of positive reasons, is fundamental: both are classical justificationists.

Bartley observes that my approach has usually been mistaken for some form of justificationism, though in fact it is totally different from it. For even though I offer a negative solution to the classical problem of justification, resembling in this respect the sceptics and irrationalists, at the same time I dethrone the classical problem and replace it by a new central problem which allows of a solution that is neither sceptical nor irrationalist. For my proposed solution to the new problem is compatible with the view that our knowledge—our conjectural knowledge—may grow, and that it may do so by the use of reason: of critical argument.

My position, Bartley suggests, is liable to be misunderstood unless it is first grasped that the classical problem of justification has not only been removed from its central position, but that, seen from the new point of view, it must actually be dismissed as insignificant. Yet for the justificationist this is very difficult to see. For he argues like Hume: 'If I ask you why you believe in any particular matter of fact . . . , you must tell me some reason; . . . or you must allow that your belief is entirely without foundation.'[2]

Now like E. M. Forster I do not believe in belief: I am not

---

[2]David Hume, *Enquiry Concerning Human Understanding,* Section v, Part I; Selby-Bigge, p. 46. [*Cp.* Popper on methods of criticism in *The Self and its Brain, op. cit.,* pp. 172–3. Ed.]

interested in a philosophy of belief, and I do not believe that beliefs and their justification, or foundation, or rationality, are the subject-matter of the theory of knowledge. But if, in this passage from Hume, we replace the words 'believe in' by 'propose a theory or a conjecture about', and the words 'your belief' by 'your conjecture', then his pronouncement loses its force. For few will be shocked to hear that their conjecture is 'entirely without foundation'. To have some 'foundation', or justification, may be important for a belief; but it is not the kind of thing we should require for a conjecture or a hypothesis; at least not in the sense in which Hume uses the term 'foundation' (which corresponds to my phrase 'justification by positive reasons'). Admittedly, some people speak of 'the foundations of physical theory', for example; but this is either justificationist talk, or it means something quite different; and once we realize that physical theories are conjectures or hypotheses, and subject to revolutionary change, we might prefer not to speak about their 'foundations'—any more than about our belief in them.

Admittedly we can give some reasons for *proposing* a hypothesis, and for submitting it to critical discussion. But these are not justificatory reasons but are more in the nature of *explanations* of why—*in the light of our aims,* such as attaining more criticizable and more severely criticized theories—we offer one theory rather than another. These reasons and their logical role are utterly different from those that Hume had in mind. We may, for example, offer a perfectly good reason for proposing a hypothesis by pointing out that, if true, it would solve a problem which we want to solve (like Newton's theory which solved the problem of explaining Kepler's laws). A reason of this kind may be quite sufficient for proposing a hypothesis and recommending it as worthy of our critical attention. But it would not, of course, be a reason for supposing it to be true. It may not even be a reason for accepting it tentatively, or even for preferring it, for there may be other known hypotheses which solve the problem even better.

II

In this way we come to realize that Hume's epistemological problem—the problem of giving positive justifying reasons, or the problem of justification—might be replaced by the totally different problem of explaining, giving critical reasons, why we prefer one theory to another (or to all others known to us), and ultimately by the problem of *critically discussing* hypotheses in order to find out which of them is—comparatively—the one to be preferred.

A justificationist may, however, object that I have not really replaced one problem by another. He may argue, first, that instead of 'reasons why we prefer one theory to another' I could have said 'reasons why we *believe* that one theory is better than another'. To the extent that this point is verbal, I readily grant it; for although I do not want to philosophize about beliefs, I never quarrel over words. Secondly, he may point out that even if he were to admit that these 'reasons why we believe that one theory is better than another' are perhaps not of the same character as would be reasons for believing that, say, the first of these theories is *true*, he could still claim that they are 'positive reasons': that they are reasons for believing in the truth of *some* theory—that is, of the theory (the 'meta-theory' as it may be called) that the first theory is better than the second. In this way the justificationist might conclude that I have not really replaced the problem of justification by a different one.

Yet in saying this, the justificationist would merely fail to realize what he has admitted. First, there is a world of difference between a meta-theory that asserts that a theory A is better than a theory B, and another meta-theory that asserts that theory A is, in fact, true (or 'probable'). And there is a world of difference between arguments that might be considered as valid or weighty reasons in support of the one or the other of these two meta-theories. For example, in discussing competing explanatory theories or conjectures (about the structure of matter, say), we can often sum up the situation fairly by saying that, according to the present state of the critical debate, conjecture a is vastly superior to conjecture b, or even to all other conjectures so far proposed: that it appears to be a better approximation to the truth than any of these (and perhaps that it *may* even be actually true). But we shall not in general be able to say that, according to the present state of the critical debate, conjecture a is the best that will ever be produced in this field, or that it *appears* to be actually true. Thus one of the two meta-theories may do no more than sum up the present state of our *critical discussion* fairly, and may in that sense be merely negative, merely critical; while the other will not in general do this at all (even though it may sum up the present state of our belief, or of our intuitive conviction).

Second, there is, again, no attempt on my part to *justify* positively, or establish, in the traditional sense, that a preference for one theory rather than another is the correct one. The problem of justification is not simply shifted: it is done away with. The meta-theory is also not positively justified; it is conjectural—and open to criticism.

III

A more important objection seems to be the following. My claim to have replaced the problem of justification by another one is as groundless, a justificationist may say, as my claim to have given solutions differing from those of the sceptics and irrationalists. That the latter claim is groundless may be seen from the fact, the justificationist may argue, that my answer turns out to be identical with, or at best a variant of, relativism, pragmatism, and similar well-known views. For in saying that we should replace the question whether a theory is *true* by the question whether it is better or worse than some other theory, I clearly adopt, the justificationist may argue, a relativist position with respect to truth. And when I say that the latter question should be decided by appealing to the success of these theories, I reveal myself (he may argue) as a pragmatist, or perhaps even as a conventionalist.

But this objection implicitly attributes to me justificationist doctrines which I do not hold. For I do not say that we should *replace* the question whether a theory is true by the question whether it is better than another theory; nor do I say that a theory is better than another whenever it is more successful in some pragmatic sense. Both points are of great importance.

My position is this. I assert that the search for truth—or for a true theory which can solve our problem—is all-important: *all rational criticism is criticism of the claim of a theory to be true, and to be able to solve the problems which it was designed to solve.* Thus I do not *replace* the question whether a theory is true by the question whether it is better than another. Rather, I replace the question whether we can produce valid *reasons* (positive reasons) in favour of the truth of a theory by the question whether we can produce valid *reasons* (critical reasons) against its being true, or against the truth of its competitors. Moreover, to describe a theory as better than another, or superior, or what not, is, I hold, to indicate that it appears to come *nearer to the truth*.[3]

Truth—absolute truth—remains our aim; and it remains the implicit standard of our criticism: almost all criticism is an attempt to refute the theory criticized; that is to say, to show that it is *not true*. (An important exception is criticism attempting to show that a theory is not relevant—that it does not solve the problem which it was designed to solve.) Thus we are always searching for *a true theory* (a true and relevant theory), even though we can never give reasons (positive reasons) to show that we have actually found the true theory we have been searching for. At the same time we may have good reasons—that is, good *critical reasons*—for thinking that we have learned something important: that we have progressed towards the truth. For first, we may have learnt that a particular theory is not true according to the present state of the critical discussion; and secondly, we may have found some tentative reasons to believe (yes, even to believe) that a new theory comes nearer to the truth than its predecessors.

In order to be less abstract, I will give a historical example.

Einstein's theories have been much discussed by philosophers; but few of them have stressed the important fact that Einstein did not believe that special relativity was true: he pointed out from the start that it could at best only be an approximation (since it was valid only for non-accelerated motion).[4] So he proceeded to a further approximation, general relativity. And again, he pointed out that this theory could not be true either, but only an approximation. In fact, he searched for a better approximation for almost 40 years, until his death.[5]

[3]For a logical analysis of the term 'nearer to the truth', see my *Conjectures and Refutations: The Growth of Scientific Knowledge*, 1963, especially Chapter 10 and the Addendum. See also Section 4 below. [See also *Objective Knowledge*, 1972, especially Chapters 2, 3, and 9; and *The Self and Its Brain*, 1977, pp. 148–9. Ed.]

[4][See Popper's 'Preface 1982' to *Quantum Theory and the Schism in Physics*, Vol. III of the *Postscript*. Ed.]

[5]As it turned out after his death, it is possible that Einstein may have been

There is no trace of epistemological relativism in Einstein's attitude, in spite of the name 'Relativity Theory': he searched for truth, and he thought that he had reasons—critical reasons—indicating that he had not found it. At the same time, he (and many others) gave critical reasons indicating that he had made great progress towards it—that his theories solved problems that their predecessors could not solve, that they came nearer to the truth than their known competitors.[6]

This example may support my claim that in replacing the problem of justification by the problem of criticism we need give up neither the classical theory of truth as correspondence with the facts nor the acceptance of truth as one of our standards of criticism. (Other values are relevance to our problems, and explanatory power.)

Thus although I hold that more often than not we fail to find the truth, and do not know even when we have found it, I retain the classical idea of absolute or objective truth as a *regulative idea*; that is to say, *as a standard of which we may fall short*. The change made is not with respect to the idea of truth but with respect to any claims to *know* the truth; that is to say, to have at our disposal arguments or reasons which suffice, or even very nearly suffice, to establish the truth of any theory in question.

There is no need to be shocked by the discovery that we cannot justify or even support by arguments or reasons the claim that our theories are true. For critical reasoning still has a most important function with respect to the evaluation of theories: we can criticize and discriminate among our theories as a result of our critical

tragically mistaken in his doubts concerning general relativity; see for example, Charles W. Misner and John A. Wheeler, 'Classical Physics as Geometry', *Annals of Physics* **2**, 1957, pp. 525 *ff.*, where attention is drawn to G. Y. Rainich, *Transactions of the American Mathematical Society* **27**, 1925, pp. 106–136. See also Rainich, 'Electromagnetics in the General Relativity Theory', *Proceedings of the National Academy of Sciences of the U.S.A.* **10**, 1924, pp. 124 *ff.*; *The Mathematics of Relativity*, New York, 1950; and Wheeler's later papers, especially his report in *Logic, Methodology, and Philosophy of Science, Proceedings of the 1960 International Congress*, ed. E. Nagel, P. Suppes, and A. Tarski, 1962, pp. 361–374. *Since this footnote was written, the Misner-Wheeler theory, now called 'Geometrodynamics,' has greatly developed. See especially Wheeler's *Geometrodynamics*, 1962; and Charles W. Misner, Kip S. Thorne, and John A. Wheeler, *Gravitation*, 1973.

[6] Einstein's reasons for thinking so were criticized, in the most interesting fashion, by Alfred O'Rahilly, in his *Electromagnetics: A Discussion of Fundamentals*, 1938—a book which excels among the writings devoted to the attack on (and usually misrepresentation of) special relativity.

discussion. Although in such discussion we cannot as a rule distinguish (with certainty, or near certainty) between a true theory and a false theory, we can sometimes distinguish between a false theory and one which *may* be true. And we can often say of a particular theory that, in the light of the present state of our critical discussion, it appears to be much better than any other theory submitted; better, that is, from the point of view of our interest in truth; or better in the sense of getting nearer to the truth.

Thus I stress the critical (or, if you like, the negative) function of reason. Yet I also stress that reasoning is more important, more powerful, and less barren, than has usually been thought. Rational criticism is indeed the means by which we learn, grow in knowledge, and transcend ourselves.

IV

It seems to me that Bartley's simple formulation—that *justification* can be replaced by *non-justificational criticism*—and his emphasis on the change of focus involved in the transition from the various *justificationist philosophies* to a *critical philosophy which does not aim at justification* is most illuminating; at least I have found it so; and feeling that the reader might also benefit, I have decided to present Bartley's formulation here. It has helped me to see why the very idea of criticism is so often misunderstood by justificationist philosophers: they tend to whittle down the idea of valid criticism to the narrow task of proving the invalidity of certain attempts to *justify* certain beliefs. Bartley's formulation also helps to explain why I can agree with so much that has been said by various irrationalists against rationalism and rationalist attempts to justify our beliefs, though I make no concessions to irrationalism, but insist, on the contrary, that any theory or belief may, and should, be made subject to severe and searching *rational* criticism, and that we should search for reasons—for *rational* arguments— which might refute it. In fact, I have suggested that what distinguishes the attitude of rationality is simply openness to criticism.[7]

[7]See, for example, my *Open Society and Its Enemies* (fourth edition, 1962), Chapter 24; and my Academy lecture 'On the Sources of Knowledge and of Ignorance' (which is now the Introduction to my *Conjectures and Refutations*, 1963; see especially pp. 25–30). [See also the 'Addendum: Facts, Standards, and Truth: A Further Criticism of Relativism', *op. cit.* Ed.]

Irrationalists are quite right when they insist that we have 'sources of knowledge' other than reason and observation—for example, inspiration, or sympathetic understanding; or tradition, which is perhaps the most important 'source of knowledge', and which is so often ignored by rationalists because of its obvious fallibility.[8] But irrationalists are dangerously mistaken when they suggest that there is any knowledge, of whatever kind, or source, or origin, which is above or exempt from rational criticism.

V

Equally mistaken is the view that rational criticism is in the same boat as positive rational argument since it too must always be based upon some non-demonstrable presuppositions, so that its validity is essentially *relative* to these presuppositions; or in other words, that we are facing a situation in which position *A* is being criticized in terms of position *B* which, however, it is impossible to establish in its turn; so that no criticism of *A* in terms of *B* will be conclusive— and of course *vice versa*.

Most of the individual components of this argument are quite correct. But the conclusion does not follow at all. Thus I am prepared to admit that, in our criticisms, we often work with unjustifiable and non-demonstrable presuppositions. Thus our criticism is, indeed, never conclusive. But non-demonstrability of any kind never worries the critical rationalist. For his critical arguments—just like the theories which he is criticizing in terms of them—are conjectural. The difference is very simple. Justificational argument, leading back to positive reasons, eventually reaches reasons which themselves cannot be justified (otherwise the argument would lead to an infinite regress). And the justificationist usually concludes that such 'ultimate presuppositions' must in some sense be beyond argument, and cannot be criticized. But the criticisms, the critical reasons, offered in my approach are in *no* sense ultimate; *they too are open to criticism;* they are conjectural. One can continue to examine them infinitely; they are infinitely open to reexamina-

---

[8][See Popper's 'Towards a Rational Theory of Tradition', *Conjectures and Refutations.* Ed.]

tion and reconsideration. Yet no *infinite regress* is generated: for there is no question of proving or justifying or establishing anything; and there is no need for any *ultimate* presupposition. It is only the demand for proof or justification that generates an infinite regress, and creates a need for an ultimate *term* of the discussion. This is the heart of the difference between justification and criticism.[9]

Related to this objection is the widely held view that a purely critical method—that is, one which refrains from positively justifying anything—is impossible on the grounds that it would have to confine itself to criticism of the 'immanent' type. A piece of criticism is called 'immanent criticism' if it attacks a theory from within, by adopting all its assumptions or presuppositions, and only these; and it is called 'transcendent criticism' if it attacks a theory from without, proceeding from assumptions or presuppositions which are foreign to the theory criticized. Yet, so it is maintained, immanent criticism is relatively unimportant; for since it can do no more than point out logical inconsistencies *within* the theory criticized, it can never succeed against a consistent theory. (Moreover, it may even be said, pointing out inconsistencies may not be wholly immanent since it assumes or presupposes a logic: one that outlaws inconsistencies.) So the conclusion is reached that all criticism of a consistent theory, and thus the most important criticism, must be 'transcendent'. Whence it is contended that the defenders of a theory under attack could always reject any criticism as inconclusive, or as invalid—unless, indeed, the assumptions or presuppositions which underlie some transcendent criticism could be given a *positive justification*. So the methods of criticism and of positive justification are, it would seem, in the same boat.

This plausible view of the narrow scope and comparative insignificance of the critical method is mistaken. It seems to be connected with the correct observation that all, or most, critics of a system of thought adopt another system of thought which guides them in their critical attack. Though criticism may indeed tend to be 'transcendent' as regards its origin or guidance or inspiration, this does not mean that it has to be 'transcendent' in the logical sense of the term. In fact, no self-critical critic will as a rule be satisfied by his

[9][See Popper's *L.Sc.D.*, section 29. Ed.]

criticism unless he can shake off the traces of its transcendent origin: though perhaps guided by his own system of thought, he will transform his criticism until it becomes immanent—and thereby more effective against his opponent. For the theory under examination is not merely a system of assumptions, dogmas, conjectures, or what not; *it is also an attempt to solve a problem.* Therefore it *can* be immanently criticized as, for example, failing to solve its problems, or as succeeding no better than its competitors, or as merely shifting the problem to be solved, etc. In this way immanent criticism may point out serious weaknesses even in a consistent theory. As to the pointing out of inconsistencies, this will in most cases be accepted as immanent criticism precisely because the problem which a theory sets out to solve will be that of giving a *consistent* explanation of something (the result of an experiment, for example) that contradicted earlier theories. In general, the problem situation which stimulates the theory under criticism, and which alone gives point to it, always contains assumptions or presuppositions which include (and go far beyond) the acceptance of logical principles—such as the rule of rejecting contradictions—and which provide an ample basis for immanent criticism. (For example, if the problem situation which leads to the formulation of a theory involves the task of explaining certain observations or experiments, then other experiments may be used in order to criticize the theory *immanently*—provided, of course, that the defenders of the theory are prepared to admit the results of these experiments.[10]) Thus immanent criticism is both possible and important.

There is, however, no need to confine ourselves to immanent criticism. We are fully entitled to employ transcendent criticism. This consists in proceeding from a *competing theory;* and there is nothing wrong in trying to show that one theory exhibits weaknesses while another does not. On the contrary, this kind of mutual transcendent criticism may in the end allow us to say that, and why, a theory is preferable to its competitor.

---

[10]I owe the remark in parentheses to a comment of Alan E. Musgrave's on this passage.

VI

The critical approach which I have described here leads almost immediately to a straightforward solution of Hume's problem of induction (1739).[11]

Let us remember what Hume tried to show (in my opinion successfully, as far as logic goes).

(i) He indicated that there are countless (apparent) regularities in nature upon which everybody relies in practice, and many universal laws of nature, accepted by scientists, which are of the greatest theoretical importance.

(ii) He tried to show that any inductive inference—any reasoning from singular and observable cases (and their repeated occurrence) to anything like regularities or laws—must be *invalid*. Any such inference, he tried to show, could not even be approximately or partially valid. It could not even be a probable inference: it must, rather, be completely baseless, and must always remain so, however great the number of the observed instances might be. Thus he tried to show that we cannot validly reason from the known to the unknown, or from what has been experienced to what has not been experienced (and thus, for example, from the past to the future): no matter how often the sun has been observed regularly to rise and set, even the greatest number of observed instances does not constitute what I have called a positive reason for the regularity, or the law, of the sun's rising and setting. Thus it can neither establish this law nor make it probable.

[11]Hume's criticism of induction occurs first in his *Treatise of Human Nature*, Book i, Part iii, Section vi, where he explains that we must not, in attempting to justify our beliefs, appeal to experience 'beyond those particular instances, which have fallen under observation' (see p. 91 of Selby-Bigge's edition); that is to say, beyond what we actually know from observation. (Although he often mentions an inference from the past to the future, we should, I suggest, take this only as a special case of the inference from what is actually known from observation to what is not known.) Hume's argument against induction is further discussed in Part 2, section 16, of this volume, see note 4 to that section, and also Hume's *Enquiry Concerning Human Understanding*, Section iv, Part ii. For an extension, by Hume, of his criticism to all probabilistic theories of induction, see his *An Abstract of a Book Lately Published entitled A Treatise of Human Nature*, and my *L.Sc.D.*, section 81 and Appendix *ix. Hume's criticism of induction was unfortunately mixed up by Hume himself with his criticism of causality, of which it is, however, logically independent. With the most notable exception of Bertrand Russell (see below) commentators as a rule have failed to disentangle these two points.

(iii) He pointed out that there can be no valid reasons justifying the belief in a universal law other than those which are provided by *experience*.

The clash between (i) on the one side, and (ii) and (iii) on the other, constitutes Hume's problem, the *logical problem of induction*.

Points (ii) and (iii) may be re-formulated, a little more sharply and briefly, as follows.

(ii) There can be no valid reasoning from singular observation statements to universal laws of nature, and thus to scientific theories.

This is *the principle of the invalidity of induction*.

(iii) We demand that our adoption and our rejection of scientific theories should depend upon the results of observation and experiment, and thus upon singular observation statements.

This is *the principle of empiricism*.[12]

Now let us take (i) for granted. Then the logical problem of induction consists in the apparent clash between (ii), the principle of the invalidity of induction, and (iii), the principle of empiricism: empiricism appears to imply that without induction we cannot have scientific knowledge.

Hume realized that the clash between (ii) and (iii) was only apparent, for he accepted both (ii) and (iii) and dissolved the 'clash' by giving up rationalism. He decided that all our knowledge of laws is obtained from observation—in accordance with (iii)—by induction, and he concluded that, since induction is rationally invalid, this shows that we have to rely on association ('habit', which results from repetition) rather than on reason.

I too accept (ii) and (iii), but I do not draw any anti-rationalistic conclusion from them. Not only do I assert the compatibility of (ii) and (iii), but also that (ii) and (iii) are consistent with the following principle (iv):

(iv) We demand that our adoption and our rejection of scientific theories should depend upon our *critical reasoning* (combined with the results of observation and experiment, as demanded by (iii)).

This is the *principle of critical rationalism*.

In order to see that (i) to (iv) are consistent we merely have to

[12][See *Objective Knowledge*, p. 12. Ed.]

realize that our 'adoption' of scientific theories can only be tentative; that they always are and will remain *guesses or conjectures or hypotheses*. They are put forward, of course, in the hope of hitting upon the truth, even though they miss it more often than not. They may be true or false. They may be tested by observation (it is the main task of science to make these tests more and more severe), and rejected if they do not pass. Nothing in Hume's argument tells against the possibility of tests, or of rejecting a universal law because it is *contradicted* by observation statements. Indeed, we can do no more with a proposed law than test it: it is no use pretending that we have established universal theories, or justified them, or made them probable, by observation. We just have not done so, and cannot do so. We cannot give any positive reasons for them. They remain guesses or conjectures—though perhaps well-tested ones. Yet if we consider the problems they solve, and the criticisms and the tests they have withstood, we may have excellent critical reasons for preferring them to other theories—though only provisionally and tentatively.

What I have said here provides a complete solution to Hume's logical problem of induction. The key to this solution is the recognition that our theories, even the most important ones, and even those which are actually true, always remain guesses or conjectures. If they are true in fact, we cannot know this fact; neither from experience, nor from any other source.

The main points of my solution are:

(i) Acceptance of the view that theories are of supreme importance, both for practical and for theoretical science.

(ii) Acceptance of Hume's argument against induction: any hope that we may possess positive reasons for believing in our theories is destroyed by that argument. (But note that Hume's argument does not present any difficulty to those who hold that we may test our theories by trying to refute them.)

(iii) Acceptance of the principle of empiricism: scientific theories are rejected or adopted (though only temporarily and tentatively) in the light of the results of experimental or observational tests.

(iv) Acceptance of critical rationalism: scientific theories are rejected or adopted (though only temporarily and tentatively) as being better or worse than other known theories in the light of the results of rational criticism.

This, in brief outline, is my solution of Hume's problem—the logical problem of induction.[13] There have, of course, been other formulations of the problem, and there are aspects of the problem not yet sufficiently analysed in the present section. I shall therefore now proceed to follow the problem through a number of its aspects and phases or stages. The analysis of these will carry me beyond the logical problem of induction, as formulated here, to what I shall call the fourth, or metaphysical, phase or stage of the problem of induction.

Before pursuing these different aspects of the problem of induction, we first need to face some other issues which are neither logical nor methodological nor metaphysical: we need to consider some so-called 'facts' about induction and learning.

[13]For other formulations, see my *L.Sc.D.*, *passim,* and especially my letter to the Editor of *Erkenntnis,* reprinted in Appendix *i of *L.Sc.D.* [See later formulations in *Objective Knowledge,* Chapters 1 and 2. Ed.] One of the misunderstandings which my theory has encountered may be mentioned here. (It forms the basis of a review by G. J. Warnock, of my *L.Sc.D.* in *Mind* **69**, pp. 99–101.) In the introduction to Chapter X of *L.Sc.D.*, I wrote, 'we should try to assess how far it [that is, a hypothesis] has been able to prove its fitness to survive by standing up to tests'. As the context of my whole discussion shows, I did not, of course, mean to imply that a theory which has survived *until now,* and which thereby has proved its fitness to survive *until now,* has also proved its *fitness to survive future tests* (as the review assumes). On the contrary, I said again and again, emphatically, that, if we have to make a choice, we choose, for the time being, that theory which seems to be the best in the light of criticism, including tests; and that *this choice is perfectly reasonable, even though we cannot know whether the theory will survive future tests* (which may, or may not, be different from past tests), and even though we may fear—or hope—that it will *not* survive tests. (Incidentally, the reviewer is factually in error when he says that 'Popper says emphatically' that the problem of induction is 'insoluble'; for offering a solution to the problem—even one that is not a justification of induction—is not the same as saying, emphatically or otherwise, that the problem has no solution. This error is connected with the fact that what I call 'the problem of induction', or 'Hume's problem', is quite different from what the reviewer says I have so called.)

3. *On So-Called Inductive Procedures, with Notes on Learning, and on the Inductive Style*

I

It seems that almost everybody believes in induction; believes, that is, that we learn by the repetition of observations. Even Hume, in spite of his great discovery that a natural law can neither be established nor made 'probable' by induction, continued to believe firmly that animals and men do learn through repetition: through repeated observations as well as through the formation of habits, or the strengthening of habits, by repetition. And he upheld the theory that induction, though rationally indefensible and resulting in nothing better than unreasoned belief, was nevertheless reliable in the main—more reliable and useful at any rate than reason and the processes of reasoning; and that 'experience' was thus the unreasoned result of a (more or less passive) accumulation of observations.

As against all this, I happen to believe that in fact we *never* draw inductive inferences, or make use of what are now called 'inductive procedures'. Rather, we always discover regularities by the essentially different method of trial and error, of conjecture and refutation, or of learning from our mistakes; a method which makes the discovery of regularities much more interesting than Hume thought. The method of learning by trial and error has, wrongly, been taken for a method of learning by repetition. 'Experience' is gained by learning from our mistakes, rather than by the accumulation or association of observations. It is gained by an actively critical approach: by the critical use of experiments and observations designed to help us to find where we have gone astray.[1]

II

Thus while I agree with Hume's analysis of the logical problem of the *validity* of induction—that is, with his thesis of the invalidity of induction—I disagree with him and, I am afraid, with most people,

[1]See also this section, part IX, note 9 (and text) below.

about a purely *factual* question. I believe that *the allegation that we do in fact proceed by induction is a sheer myth*, and that the alleged evidence in favour of this alleged fact is partly non-existent, and partly obtained by misinterpreting the facts.

I hasten to add, however, that my factual thesis has no bearing whatsoever on my logical or my methodological or my epistemological doctrines. For the factual, psychological, and historical question, 'How do we come by our theories?', though it may be fascinating,[2] is irrelevant to the logical, methodological, and epistemological question of validity. Here again I follow Hume. Indeed, it was Hume's greatest achievement to separate these two problems sharply. By giving almost opposite answers to them he made it abundantly clear that they are quite distinct.

Some scientists find, or so it seems, that they get their best ideas when smoking; others by drinking coffee or whisky. Thus there is no reason why I should not admit that some may get their ideas by observing, or by repeating observations. And in this sense, I should be quite willing to mitigate my thesis that we never proceed by induction: let us replace 'never' by 'hardly ever'.

But having made this concession, I wish to explain that, whether a theory occurs to us first while smoking, reading or observing, or even in our sleep, the important question remains: What is its logical worth? Is it a good theory or a bad theory? Darwin, it has been said, got his theory of natural selection when reading Malthus. This is a point of considerable historical interest, but it has no bearing whatever on the question of the worth of Darwin's theory: even if Malthus's theory should be true, and well supported by the strongest evidence, Darwin's might be false and ill-supported; and even if Malthus's theory should be quite untenable, Darwin's might be excellent.

Of course, everybody will admit this. And yet it seems that there are few philosophers left who insist that we must distinguish sharply between questions of *validity* (such as whether we have any *reason* to rely on induction) and questions of *fact* (such as whether we actually rely on induction, or use 'inductive procedures'; or whether a theory was actually originated by way of induction, etc.).

[2]Jacques Hadamard has written a most interesting book on this matter: *The Psychology of Invention in the Mathematical Field*, 1945.

III

The most fashionable view of the matter seems to be this. When Hume found that induction is invalid he used the word 'invalid' in the sense of 'not in accordance with the canons of valid deductive reasoning'. But Hume's finding is trivial, according to the view I am reporting, for inductive reasoning is a species of reasoning which in some respects is similar to, and in other respects different from, deductive reasoning. It is therefore pretty obvious that its standards or canons will not in *every* respect conform to those of deductive reasoning. Inductive reasoning has its own standards, its own canons, its own 'procedures'. Thus an 'inductive procedure' or an 'inductive inference' will not in general be a 'valid inference'—that is, it will not be a valid *deductive* inference. But this is no reason why it should not be 'reasonable'—that is, conform to the appropriate standards of *inductive* reasoning: it is *inductively valid*.

But what are these alleged inductive standards? There are two answers, given by two different philosophical schools. One school teaches that the standards are those of 'probable' reasoning, in conformity with the laws of the *calculus of probability*. I have refuted this doctrine at length in my *L.Sc.D.* and elsewhere, and I shall add some further criticism later in this *Postscript*. The other school teaches that we should observe, and classify, the various 'inductive procedures' and the various usages of the word 'probable' which may occur when we speak about these 'inductive procedures' and that the results of these researches will allow us to lay down the laws or standards or canons of *inductive validity*.

Thus, according to this school, factual 'usages' are to establish standards. There is no longer a distinction between fact and standard (between the questions *'quid facti'* and *'quid juris'*[3]). There is no longer a logical problem of the validity or the 'justification' of induction. We cannot 'justify' the use of inductive procedures, according to this school, any more than we can 'justify' the use of deductive procedures, or cooking procedures, or any other procedures, except perhaps in some pragmatic sense of 'justify'; in other words, by their success. And inductive procedures are no more in need of a 'justification' than are deductive or any other procedures.

Before going on to criticize this fashionable view I wish to point

[3] *Cp. L.Sc.D.*, section 2, p. 31, where reference is made to Kant.

out that it is essentially an ancient view, translated into fashionable language. For the ancient expression 'the laws of thought', used to denote the principles of deductive inference, contained an allusion to the view that there was no way of 'justifying' the validity of deductive logic except by pointing to the psychological *fact* that we do, *in fact*, think in this way—or that we are, *in fact*, compelled to think in this way, or compelled to admit that certain deductive inferences are inescapable or 'necessary'.[4] Once the validity of deductive reasoning was thus reduced to *fact*, the way was open to the acceptance of some 'principle of induction' or of some 'canons of inductive reasoning' whose validity was likewise not in need of justification other than the fact that we did reason, or perhaps were compelled to reason, according to these inductive canons.

The now fashionable version of this ancient theory replaces a thoroughly unsound psychological argument for the identity of the question of validity and the question of fact by an equally unsound pragmatist or behaviorist or language-analytic argument.

All these devices are merely attempts to revive a theory long ago disposed of by Hume and Kant; and modern revivalists ride roughshod over the work of these great men without trying to counter their arguments or even to understand them. It is sheer dogmatism to assert that there is nothing in validity besides success. (No scientific theory was more successful than Newton's. If success were our only concern, no one would have criticized it, that is, reconsidered the question of its truth. Yet its criticism led to an important intellectual revolution; a revolution which ought to have deeply affected epistemological thought.)

The identification of valid modes of thought with actual thinking has been tried again and again, since time immemorial. One of the most influential and pernicious of these attempts was Hegel's philosophy of the identity of reason and reality. The latest theories of induction amount to a renewal of Hegel's attempt.

But the alleged facts which these theories of induction try to convert into standards are quite imaginary; and some people whose knowledge of science is by no means negligible—Albert Einstein, for example[5]—have denied that they are facts. But even if all of us who deny the existence of 'inductive procedures' are wrong, it

[4]Cp. *Conjectures and Refutations*, Chapter 9, especially pp. 207f.
[5]See especially Einstein's Herbert Spencer Lecture, *On the Method of Theoretical Physics*, Oxford 1933.

would be the height of dogmatism to assert that these disputed 'facts' create standards of reasoning whose validity is not open to further discussion. (And although the art of arguing critically about philosophical problems—and, with it, the great tradition of rational philosophical thought—is disappearing rapidly, I am unwilling to resign myself to this fact without trying to change it; and I am even less willing to follow the present fashion of elevating this fact into a new standard of philosophical excellence.)

IV

Philosophers are not of course the only ones who believe in induction and in the existence of 'inductive procedures'. As I said before, almost everybody does, including many psychologists—especially those interested in the theory of learning; many biologists; and quite a few physicists. To the physicists I shall say no more here, since I have already referred to Einstein. But I shall say something about 'learning theory', and also about what I have called 'the inductive style'—a manner of writing which is still very much the thing in some biological journals. (See subsections x and xi below.)

V

As to the theory of learning, no doubt *we can, and do, learn from experience*. I should even be prepared to say, with Hume and other classical empiricists, that *all learning is learning from experience*.

Yet when it comes to the interpretation of the thesis 'all learning is learning from experience', I differ from Hume and other classical empiricists. I differ from them radically in the assessment of the role which *repetition* plays in the process of *learning* and also in the assessment of the role which *observation* plays in the acquisition of *experience*.

As to learning and repetition, a serious source of confusion has been the failure to distinguish among three entirely different activities which are all called 'learning'. (I do not wish to imply that one could not with profit distinguish more than three.) I shall call them (1) learning by trial and error (or by conjecture and refutation); (2)

learning by habit formation (or learning by repetition proper); and (3) learning by imitation (or by absorbing a tradition). All three kinds can be found in animals as well as in human beings, playing their various characteristic parts in the acquisition of skills as well as of theoretical knowledge such as learning about new facts.

(1) Only the first of these three ways of learning, *learning by trial and error,* or by conjecture and refutation, is relevant to the growth of our knowledge; it alone is 'learning' in the sense of acquiring *new* information: of discovering *new* facts and *new* problems, practical as well as theoretical, and *new* solutions to our problems, old as well as new. This kind of learning includes the discovery of new skills and of new ways of doing things. In the processes of learning in this sense, mechanical repetition (like that of the drop that hollows the stone) plays no role whatever. It is not the repeated impact on our senses which leads to a new discovery, but something entirely different: our repeated and varied attempts to solve a problem which, unsolved, continues to irritate us. It is essential here that these 'repeated' attempts *differ* from each other, and that we repeat the same attempt only when it appears to us to be successful, and only in order to try it out again; that is, in order to test, if possible under varying conditions, the hypothesis that it leads invariably to a successful solution of our irritating problem.

Learning by trial and error comprises learning from systematic observation as well as learning from chance observation, though in different ways. *Systematic observation* always starts from a problem which we try to solve, or from conjectures which we try to test: this is what makes it systematic. Even where we try to determine some parameter by systematic measurement, there is an underlying hypothesis—the hypothesis that there is a parameter which is invariant with respect to certain changes in the conditions of our measurements. Without some such hypothesis, whether consciously proposed or unconsciously assumed, observation cannot be systematic. Yet even a so-called '*chance-observation*', although the least inventive way of making discoveries, is still a case of the trial and error method. For practically every example of a 'chance observation' is an example of the *refutation* of some conjecture or assumption or expectation, held either consciously or unconsciously. A 'chance observation' is like an unexpected stone in our path: we stumble over it just because we did not expect it—or more precisely because we *did* expect, though unconsciously, that the path would be smooth. Thus so-called 'chance observations' or 'accidental discoveries'—that is, stumbling-block discoveries—are not as accidental as one might think at a glance.

Moreover, most of the usual examples of 'chance discoveries' are based on inductivist misinterpretations: genuine examples even of stumbling-block discoveries seem to be rare. Perhaps Pasteur's first immunisation of chickens against chicken cholera (1880), or the discovery of the catalytic action of mercury salts on the conversion of naphthalene into phthalic acid (by heating with sulphuric acid)[6], are near to being genuine examples. But many of the others are not.

Oersted, for instance, was *searching* desperately for some electro-magnetic interactions. And Roentgen, when questioned about his discovery of X-rays, explained: 'I was *looking* for invisible rays'[7]—rays which he hoped to detect (as in the case of infra-red and ultra-violet rays) by means of a fluorescent screen. (This is why the screen was there.) Admittedly, unexposed photographic plates near Crookes tubes had many times shown—against expectation— signs of exposure, and this *might* easily have led to a genuine 'stumbling-block discovery' of X-rays; but all those involved, even Crookes himself, failed to gauge the significance of those signs of exposure. As to radioactivity, Becquerel consciously searched for new rays; and in his work on uranium salts, more particularly, he was guided by a (mistaken) hypothesis, due to Henri Poincaré. The discovery of penicillin was also not a chance discovery, for the kind of bacteriocidal (or bacteriostatic) effect observed by Fleming was well known to him and to others, and thus not even 'unexpected'. (See subsection x below.) Moreover, Fleming was very much alive,

[6]It seems that this story of this discovery, together with some variants of it, has become somewhat legendary among chemists. According to a private communication from Professor Alexander Findlay, the discovery was made in 1896 by Sapper, a young German chemist who accidentally broke a thermometer in the mixture and noticed a great acceleration in the rate of the reaction. It is important to add that this acceleration contained the solution of an urgent problem of industrial chemistry which he was trying to solve. Thus there was an accident; but it was preceded and prepared by the problem of which it was the solution, so that the observation was far from accidental. (See also A. Fleming's *Chemistry in the Service of Man*, 7th edition, 1947, pp. 318f.)

[7]See Otto Glasser, *Wilhelm Conrad Roentgen and the Early History of the Roentgen Rays*, 1933, p. 13.

even prior to his discovery, to the possible significance of this kind of effect for therapeutic purposes.

With animals as well as men, learning by trial and error originally results not so much in new 'knowledge' as in new skills. Yet all skills are linked to conscious or unconscious *expectations;* and the element of *error,* within the trial and error method, becomes manifest always in the disappointment of some expectation or other. (What we call our knowledge—'knowledge' in the subjective sense—may be said to consist of our conscious, perhaps verbally formulated, expectations.) As examples of new skills acquired by trial and error we may take piano playing, or cycling; or more precisely, finding the best fingering for a passage on the piano, or learning how to avoid a fall when riding a bicycle. In these cases we first try to solve a problem consciously by systematic trials—by the elimination or rejection (or falsification) of unsatisfactory solutions; later repetitions play a very different role. The cases are therefore helpful for contrasting the character and function of repetition in the kinds of learning here distinguished as (1) and (2).

(2) The second kind of learning—*habit formation through repetition proper* (or through 'mechanical' repetition)—should be clearly distinguished from the first. Here we do not look for any new solution of a problem, but try to become familiar with a solution previously discovered by trial and error (or learnt by imitation; see below under (3)).

To show how this second kind of learning differs from the first, it is instructive to consider our main examples, that of learning to play a certain passage on the piano, and that of learning to avoid a fall when riding a bicycle.

There are few human skills where constant 'practising'—that is, not only repetition but also more or less 'mechanical' repetition—is as important as in learning to play the piano. Yet we do not find anything new, such as a new fingering, through practising. Only after having discovered the new fingering by trial and error, that is, after comparing it with alternative solutions to the problem and rejecting less suitable solutions, can we begin to 'practise' it. Thus the function of mechanical repetition—of 'practising', or 'learning by rote'—is not to discover something new, but to establish familiarity with something previously discovered. Its function is not to make us conscious of a new problem (as is the function of testing repeatedly some tentative solutions) but to eliminate as far as possible the element of consciousness from our performance. And so we reach a state in which the original problem—for example that of co-ordinating the score and the movements of the fingers—vanishes completely, and we can give our whole attention to something more important—the musical idea, the phrasing of the passage. The function of 'practice', or repetition proper, in learning to ride a bicycle (or to drive a motor car) is the same: it does not produce a discovery, or even a *new* skill, though it may transform a discovery (a discovery of how to do things) into a new *skill*; and by making certain actions unconscious it leaves us free to give our attention to the traffic problems.

The inductivist doctrine that all learning is learning by rote, and that even the growth of our knowledge is the result of habit formation through mechanical repetition, is therefore mistaken. Repetition as such cannot attract our attention; rather, it tends to make our expectations unconscious. (We may not hear the clock ticking, but we may 'hear' it stop.)

The popularity of the idea that our knowledge grows through induction by repetition is doubtless due to mixing up learning in the two senses (1) and (2).

(3) Yet I may also briefly mention a third kind of learning—learning by imitation. It is one of the more primitive and important forms of learning; and here the highly complex instinctual basis of learning and the role played in it by suggestion and by the emotions are more obvious than in other ways of learning (though these are of course always present). What is important for our discussion is that from the point of view of the individual learner, learning by imitation is always a typical trial and error process: a child (or a young animal) *tries*, consciously or unconsciously, to imitate his parent and either corrects himself or is corrected by the parent. This trial and error process constitutes the first and fundamental stage of the imitative process. It is thus a stage of *discovery*: the child *discovers* how to walk by imitation; and this means, partly, by trial and error. It may be followed, of course, by a stage in which the newly discovered skill, as a result of 'practising', is executed unconsciously, and so becomes a habit.

43

VI

The inductivist interprets knowledge ('knowledge' in the subjective sense) as consisting of expectations; and so do I. Yet he further interprets an expectation as the memory of observations linked by associations which are the result of repetition. (He thinks that the dog expects food when the bell rings simply because on repeated previous occasions food has arrived after the bell rang.)

As opposed to this, I believe that new expectations are formed by trial and error: we form tentative expectations in fields of interest, fields in which we have problems, fields in which we are able to learn, that is, to correct our expectations. If the newly formed expectations are successful, they may become (by repetition) automatic, unconscious, and petrified, and we gradually cease to be able to learn in that particular field. Food-getting is a field in which the dog's behaviour is normally 'plastic': he can learn in this field, and he discovers regularities by trial and error, by anticipation and refutation. In this way he also first forms the theory that food will appear when the bell sounds; afterwards the anticipation or expectation may become habitual and petrified through repetition.

VII

The view presented here stands in sharp opposition to the theories of association and of conditioned reflex (see section 9 below). The first of these works by assuming that simple *terms of association* or 'data' are 'given' to us. This assumption is naïve and untenable, even if we forget about the complex stimulus situation, and assume that the stimuli are simple. The second of these theories is very different, and less objectionable, in so far as it makes no assumption about simple data or terms but tries to explain how stimuli (which may be highly complex) originally unrelated to an expectation (which also may be highly complex) may become signals able to release just this expectation. This is quite acceptable as a problem; yet the solution, the theory of the conditioned reflex, is unacceptable. It assumes the existence of elementary non-complex, non-conditioned reflexes out of which the conditioned reflex is built, and it assumes that all learning is to be explained as the conditioning of reflexes. Both assumptions are mistaken. Learning in the first and fundamental sense, that is, learning by trial and error, contains an element of invention or of creative action which goes far beyond any mere reflex; and to talk here, where invention and plasticity of action is paramount, of the 'conditioning of reflexes' wrongly suggests that all learning, even discovery and invention, can be explained by repetition. But we have seen that learning by repetition is less fundamental (and therefore also less elementary) than learning by trial and error.[8]

VIII

The inductivist's mistake is not confined to his failure to appreciate the difference between learning by trial and error and learning by rote, or to his consequent assumption that we can add to our knowledge by the formation of habits. He also believes that there is some raw material for knowledge in the form of perceptions or observations or sense-impressions or sense 'data' which are 'given' to us from the outside world, without our own intervention. This is an untenable psychological theory, amply refuted by the facts. In the cinema, what is 'given' to us is a sequence of stills, but what we see, or observe, or perceive, is movement; and we cannot help seeing the movement, even if we *know* that we are seeing only photographs of (say) an animated cartoon.

The simple fact is that seeing or perceiving or observing is a reaction, not simply to visual stimuli, but to certain complex situations, in which not only complexes and sequences of stimuli play a role but also our problems, our fears and hopes, our needs and satisfactions, our likes and our dislikes. Our reaction—that is, our immediate perceptual experience—is influenced by all this and also, largely, by our previous knowledge; by our expectations or anticipations, which provide a kind of schematic framework for our reactions. If we learn, in the sense of adding to our knowledge, by our observations or perceptions, then we do so because observing

[8][*I have published several criticisms of the so-called 'conditioned reflex'. In brief, I have asserted that Pavlov's dogs were not 'conditioned', but formed a theory, in a field (procuring food) in which theory formation is vitally important, that food comes when the bell rings. See especially my contribution to Karl R. Popper and John C. Eccles: *The Self and Its Brain*, 1977, pp. 91, 135–138; and Roger James's articles, referred to there, and now also James's excellent book, *Return to Reason*, Somerset, 1980.]

or perceiving consists in modifying, sharpening, correcting, and often falsifying, our anticipations. Thus inductivist theory is always superficial: a closer analysis shows that what inductivism naively takes as a 'datum' of our senses consists, in reality, of a complex give and take between the organism and its environment: the process of modifying or correcting our anticipations and refuting our conjectures, which is so characteristic of every sort of learning by which we add to our knowledge.[9]

IX

The classical empiricist view—that experience results from learning through the repetition of observation—is a closed system of prejudices whose critical examination is usually resisted and often resented. This system of prejudices is very popular, and has become part of 'common sense'; yet it also may be described as a highbrow and somewhat artificial *philosophical system* because there are many indications that 'experience' and 'learning from experience' are quite commonly and popularly used in the sense of the trial and error method: 'learning from experience' means, quite commonly, 'learning from our mistakes' (rather than 'learning by rote' or 'learning by the association of observations'—to say nothing of 'sense data'). As Oscar Wilde puts it: 'Experience is the name everyone gives to their mistakes.'[10]

In short, what is really the ordinary or common or popular usage of the word 'experience' entails a theory which is similar to my own theory, expounded at considerable length in my *L.Sc.D.* and elsewhere.[11] According to this theory, experience should not be

[9]With all this, compare Chapter 1 of my *Conjectures and Refutations*, especially pp. 43–52. [*Cf. Objective Knowledge*, Chapter 7. Ed.]

[10]Oscar Wilde, *Lady Windermere's Fan*, Act iii. *Cp.* section 15 of the *Addendum* (1961) to vol. ii of my *Open Society* (4th edition), pp. 388*f.* See also for example the article 'Experience' in the *Oxford English Dictionary* which, of course, also lists the *philosophical* or *epistemological* meaning, 'the observation of facts or events considered as a source of knowledge' (quoting Thomas Reid), but which otherwise supports the view that (apart from 'religious experience') 'experience' means, *commonly*, the result of learning from our mistakes.

[11]Apart from the section referred to in the preceding footnote see also the Introduction to my *Conjectures and Refutations* and my *Objective Knowledge*, Chapter 2.

taken as an ultimate 'source of knowledge', but rather as a system of fallible expectations or anticipations which each of us arrives at by trial and error. But my reference to the popular usages—or to the etymology—of 'experience' is not intended as an argument in favour of my own view, for such usages are often highly misleading (though we should not depart from common sense—whatever this may mean—without some fairly good reason). My intention is, rather, to point out that the analysts of ordinary linguistic usages fail to stick to their guns when, led by their inductivist philosophy of experience as a source of knowledge, they assume the existence of 'inductive procedures'.

X

The inductivism of many biologists is, I believe, traditional, going back to Bacon, Boyle, Leeuwenhoek, and to the early days of the Royal Society. I tried to sketch the philosophy behind this tradition—the belief that nature is an open book which must be read without prejudice—in the last section of my *L.Sc.D.* and, much more fully, in the Introduction to *Conjectures and Refutations*.

Nowhere is the power of the inductivist tradition as conspicuous as in what I have called *'the inductive style'*—a certain manner of reporting one's researches which is still the traditional way of writing in a number of biological journals, although by now it has almost disappeared from the journals of physics and chemistry.

The basic idea which inspires the inductive style is this: we must keep carefully to our actual observations, and must beware of theorizing; for this may make us acquire theoretical prejudices which may easily bias or taint our observations if we are not *very* careful.

For this reason a paper written in the inductive style has, essentially, the following structure:

(1) It first explains the preparations for our observation. To these belong, for example, the experimental arrangements, such as the apparatus used, its preparation for the experiment, and the preparation of the objects of observation.

(2) The main part of the paper consists of a theoretically unbi-

ased, pure description of the experimental results: the observations made, including measurements (if any).

(3) There follows a report of repetitions of the experiment, with an assessment of the reliability of the results, or of probable errors. (Lately this may include statistical work.)

(4) Optional: a comparison of the results with earlier ones, or with those of other workers in the field.

(5) Also optional: suggestions for future observations, for desirable improvements to the apparatus, and for new measurements.

(6) The paper is concluded (again optionally) by a brief epilogue, usually of a few lines only, and sometimes in smaller print, containing a formulation of a hypothesis suggested by the experimental results of the paper.

I do not, of course, suggest that these points are always rigidly adhered to. Some points may be omitted, others added. What I do suggest is that there is a tendency to make young biologists believe that this is the proper way to present results, and that even masters adhere to this way of presentation.

No doubt the idea which inspires the inductive style—the idea of adhering strictly to the observed facts and of excluding bias and prejudice—is laudable. And no doubt those trained to write in this way are unaware that this laudable and apparently safe idea is itself the mistaken result of a prejudice—worse still, of a philosophical prejudice—and of a mistaken theory of objectivity. (Objectivity is not the result of disinterested and unprejudiced observation. Objectivity, and also unbiased observation, are the result of criticism, including the criticism of observational reports. For we cannot avoid or suppress our theories, or prevent them from influencing our observations; yet we can try to recognize them as hypotheses, and to formulate them explicitly, so that they may be criticized.)

As an example of the inductive style I may mention here the classic paper in which Alexander Fleming reported the discovery of penicillin.[12] It describes his observation of the accidental invasion of a culture of bacteria by some agent which destroys them. It is a description of observations, of what happened; and although it does not *say*, of course, that this kind of thing was unexpected, or that it happened for the first time, its inductive style may leave the inno-

---

[12]A. Fleming, *British Journal of Experimental Pathology*, 1929, pp. 226 *ff.*

cent reader (for example a philosopher) with the impression that not only was the invasion of the culture of bacteria by penicillin accidental (this it was in a sense) but also that it was unexpected.

But the historical facts show that this impression would be mistaken. At least since Metchnikov (1845–1916), theories about 'antibodies'—that is cells, or molecules, or other microscopic agents which eat or destroy or inhibit the growth of dangerous bacteria—have been constantly discussed by bacteriologists. Nor was Alexander Fleming's bacteriocidal mould the first which had been observed to settle accidentally on a culture of microbes. Indeed, bacteriologists had long hoped that in this way they might one day find a powerful means of killing bacteria in man. In 1924 Sinclair Lewis had published *Arrowsmith*, a very good novel, in which an incident very much like that described in Fleming's paper plays a major role. (Its bacteriological parts were written in collaboration with Dr. Paul de Kruif, who later became well known for his *Microbe Hunters*, a popular and most readable history of bacteriology containing also a very good analysis of the methods of scientific discovery.)

Indeed, many similar incidents were known at the time, and many substances which were *prima facie* similar to penicillin; the main problem was whether any of them would be suitable for medical purposes. Fleming conjectured that penicillin would be suitable. But he failed for a decade to secure the much needed collaboration of a competent chemist. A decade after Fleming's discovery, Howard Florey and his collaborators discovered the surprising curative powers of penicillin, thus confirming Fleming's conjecture. Yet even these surprising powers were not wholly unexpected, for Paul Ehrlich (1854–1915) had hoped to find some such powerful substance, and the sulphur drugs (whose action seems to be somewhat similar to that of penicillin) were invented by workers brought up in the Ehrlich tradition.

Thus Fleming's discovery was not really accidental: it was the work of a great discoverer who knew very well what he was doing, and what was worth describing: and though it was an accident that the mould whose antibiotic properties he had observed turned out to be non-toxic, the existence of substances of this kind had been expected, and hoped for, for a long time. This expectation motivated the work both of Fleming and of Florey's team.

Yet we look in vain in the early papers for a statement of these motives, hopes, and expectations; of the problems which make the papers significant. Due to the inductive style of their publications, speculative hopes and anticipations are usually handed on among biologists by way of an oral tradition rather than in writing. At any rate, few biological journals would be prepared to accept a paper discussing such theoretical speculations, for they violate the accepted rules of the inductive style.[13]

XI

A criticism of the inductive style would be incomplete without suggesting something in its place. A scientist should of course be free to present his results as he sees fit. Yet a student of the logic of science should also be free to submit for criticism a kind of general structure which scientists might be well advised to adopt unless there are good reasons to deviate from it. A standard experimental paper, according to my plan, should be constructed as follows:

(1) A clear exposition of the problem—or, if the problem may be assumed to be well known, a clear reference to it and to an exposition of it. The author should also make it clear whether he accepts the problem situation as sketched by some predecessor or whether he sees the problem differently. This would give the author an opportunity to clarify for himself (and perhaps for others) the always shifting *problem situation.*

(2) A more detailed survey of the relevant hypotheses bearing on the problem (and of the experiments bearing on the hypotheses, indicating the degree to which these are able to contribute to the appraisal of the hypotheses).

(3) A more specific statement of the hypothesis (or hypotheses) which the author intends to propose, or to discuss, or to test experimentally.

(4) A description of the experiments and their results.

(5) An evaluation: whether the problem situation has changed; and if so, how.

[13][See Popper's 'Science: Problems, Aims, Responsibilities', *Federation Proceedings* 22, July–August 1963, pp. 961–972, esp. pp. 970–71. This paper, as well as Peter Medawar's 'Is the Scientific Paper a Fraud?' (BBC 3rd Programme, 1963–4), have now led to a change. Ed.]

(6) Suggestions for further work arising from the work reported.

These points seem to be fairly obvious, if my view of the procedure of science is adopted. Nevertheless, few authors adopt anything like this 'style': the inductivist tradition is too strong.

Yet it is becoming more and more urgent every day that every paper should explain its location within the problem situation in the various sciences. The staggering increase in published material, together with a decrease in personal contacts between workers in *neighbouring* fields, has led to an atomization of science that makes it more important than ever to stress the significance of theoretical problems and of theories. For theories constitute the network of co-ordinates for science.

XII

I have mentioned some reasons which, though they are not likely to convince an inductivist of the non-existence of inductive procedures, should at least show that there are alternative interpretations of the facts which he neglects at his peril; that the belief that inductive procedures exist may be challenged; and that, even if there should be some *prima facie* examples of inductive procedures, further analysis may show them to be examples of a method whose logical structure is entirely different: the method of trial and error.

Thus neither the appeal to the alleged *fact* of our using induction, or inductive procedures, nor the attempts to convert this *fact* into a standard, can be accepted in lieu of arguments: the inductivist who tries to justify our theories, or our beliefs, by an appeal to induction cannot avoid the task of giving a rational justification of induction.

XIII

The method of induction by repetition is intended to provide a standard of justification. (On the other hand, the method of trial and error—of learning from our mistakes—is purely critical, selective. It becomes justificationist or inductivist only if we mistakenly assume, with Bacon or Mill, that it is possible to justify a theory by the complete elimination of all its alternatives; but the number of untested alternatives is always infinite,[14] and there are always unthought-of possibilities.) The inductivist is, essentially, a justificationist (in the sense of section 2).

Hume thought that induction is *rationally* unjustifiable, but that it has its own kind of justification: it justifies itself in practice through its high degree of reliability in which we cannot but believe, though only irrationally. Modern inductivists also think that induction, though obviously different from deduction, has its own kind of justification: it justifies itself in practice, and sets its own standards: it is self-validating, self-authenticating. In induction, facts and standards are reconciled, as they are in God. The logical difference between Hume and the modern view is thus the difference between a somewhat reluctant and sceptical believer and a dogmatic theist.

But rationality—that is, criticism—and the dualism of facts and standards can easily be saved if we give up justificationism. If we do so, we also become aware of the logical gap between induction by repetition and the method of trial and error, or of learning from our mistakes.

Believers in inductive procedures have one thing in common with me: we both differ from Hume in thinking that, somehow, the facts or procedures of acquiring knowledge are closely related to some standards of rationality. But while they believe that the facts—or what they think are the facts—are self-authenticating and create their own standards of inductive rationality, I conjecture that, by and large, the rational standards (which I think are standards of criticism, that is, logical standards) are likely to determine our procedures. And since reason and logic tell us that, rationally, there is no induction and no justification but only criticism and elimination, it is a good idea to see whether those facts of scientific discovery cannot be interpreted—and better interpreted—as procedures of trial and error.

4. A Family of Four Problems of Induction.

The formulation of Hume's problem of induction given in section 2 above is, I believe, of fundamental importance. But there are other

[14]This was stated clearly by Jeffreys and Wrinch, *Phil. Mag.* **42**, 1921, pp. 369 ff.; see *L.Sc.D.*, p. 140 footnote *1; see also *Die beiden Grundprobleme*, esp. pp. xix f., footnote 11.

fomulations of the problem which bring out different aspects of it, and may be looked upon as different phases or stages of its discussion.

In the present section I intend to distinguish four such phases:

(1) A slight variant of the formulation given in section 2. It may be called Russell's challenge, and may be formulated as the question: '*What is the difference between the lunatic and the scientist?*' It is closely related to the 'problem of demarcation', that is, the problem of finding an adequate characterisation of the empirical character of scientific theories.

(2) The so-called '*problem of rational belief*'.

(3) The question whether we may draw inferences about the future or whether the future will be like the past; a question which Hume himself failed to distinguish sufficiently from the problem of induction. I shall call it '*Hume's problem of tomorrow*'.

These three questions are all of a logical or epistemological or methodological character, and I shall try to show in this section that no new ideas are needed for their solution.

But our discussion will lead us to distinguish a fourth phase or stage which, in spite of an apparent similarity with the third, is very different from it in its logical character. It may be called:

(4) 'The metaphysical phase of the problem of tomorrow', or '*the fourth or metaphysical phase of the problem of induction*'. The discussion of this fourth phase of the problem will be taken up in section 5. In the present section I will confine myself to a discussion of the first three phases or stages.

I

Bertrand Russell was the first philosopher since Kant to feel the whole force of Hume's problem of induction. Kant had seen clearly that, provided Hume was right, knowledge of universal character—and thus, he thought, scientific knowledge—could not exist; but since he believed that the example of mathematics and, more important still, of Newtonian mechanics, showed that we did in fact possess certain scientific knowledge, Kant felt that the central problem of philosophy was to explain how it was possible that it existed; that is, to explain why Hume was wrong.

Russell understood the problem similarly, though his detailed solution of it differed considerably from Kant's. (For example, by describing the laws of mechanics as mere probable knowledge, as opposed to the certain knowledge of mathematics, Russell further widened the gap between mathematics and physical science which Kant had discussed.)

Induction is discussed by Russell at length, in many places; first, I believe, in his incomparable book *The Problems of Philosophy* (1912). While in this slim but great volume he does not refer to Hume as the originator of the problem, he does so in his *History of Western Philosophy* (1946). There, in his chapter on Hume, he formulates the problem as follows.

If Hume is right that *we cannot draw any valid inference from observation to theory*, then our belief in science is no longer reasonable. For any allegedly scientific theory, however arbitrary, becomes as good—or as justifiable—as any other, because *none* is justifiable: the phrase 'My guess is as good as yours' would rule scientific method as its only principle. Thus if Hume were right there would be *'no difference between sanity and insanity'*,[1] and the obsessions and delusions of the insane would be as reasonable as the theories and discoveries of a great scientist.

To this challenge of Russell's, a simple and virtually complete answer is implicit in the discussions of section 2. Admittedly, we cannot justify the claim that the scientist's theory is true; no more, in fact, than the claim that delusions are true. Yet we may defend the claim that the scientist's theory is better—better even in the somewhat narrow sense of being better supported by observations. For observations *may* be crucial between two theories, in the sense that they may contradict one while being compatible with the other. Hume's argument does *not* establish that *we may not draw any inference from observation to theory:* it merely establishes that we may not draw *verifying* inferences from observations to theories, leaving open the possibility that we may draw *falsifying* inferences: an inference from the truth of an observation statement ('This is a black swan') to the falsity of a theory ('All swans are white') can be deductively perfectly valid.

I do not see how this solution of Russell's challenging formulation of the problem can be contested (except by confusing the first

[1]Bertrand Russell, *History of Western Philosophy*, 1945. See the penultimate paragraph of chapter xvii (p. 673 of the first English edition). The original passage is not italicized.

phase of the problem with the second or the third phase, still to be discussed). It seems that Russell, when suggesting that science would be impossible if Hume were right, simply overlooked the all-important fact that Hume's argument does not show the invalidity of *falsifying* inferences from observation to theory. Russell's suggestion fits into Kant's view of science as well-established knowledge (*scientia, epistēmē*), but it does not fit into Russell's own view of scientific theories as hypothetical or conjectural. In fact, in many passages Russell describes the method of science in a way which makes it unnecessary to speak of induction; for example, he writes: 'Logic, instead of being, as formerly, the bar to possibilities, has become the great liberator of the imagination, presenting innumerable alternatives which are closed to unreflective common sense, *and leaving to experience the task of deciding, where decision is possible, between the many worlds which logic offers for our choice.*'[2] The key words in this beautiful passage are perhaps, 'where decision is possible': I have little doubt that when Russell wrote this passage he saw that, where decision is possible, it may be possible only as a rejection of some of the 'innumerable alternatives', rather than as a positive decision for one of them. Yet apparently he did not see that this was the typical or even the only case; or that this fact makes it possible to solve Hume's logical problem of induction.

Only one important point has to be added to this solution of the problem: observation alone cannot always decide which of two competing theories is better, although it *may* do so, especially if the theories permit of a crucial experiment. In general, more than observation is needed: also needed is a critical discussion of the merits of the two theories. Such a discussion must consider whether they solve the problem which they are supposed to solve: whether they explain what they are supposed to explain; whether they do not merely shift the problem, for example, by an untestable *ad-hoc* assumption; whether they are testable, and how well they are testable.

These questions (closely related to what I call the 'problem of demarcation', to be discussed below in Chapter 2, sections 17 to 26) are very important. For it may well happen that we are faced with two theories of very different value, and that our observations are

---

[2]Bertrand Russell, *The Problems of Philosophy*, 1912, p. 148. (No italics in the original.)

equally compatible with both, although for very different reasons: with the one theory in spite of the fact that they test it severely, and with the other theory simply because it is not testable—because all observations whatever are compatible with it. (The first theory may be Newton's, say, or Kepler's theory that all planets move in ellipses, combined perhaps with his theory of World Harmony; and the second may be Plato's theory that all planets possess souls, and are gods.)

Only after questions about the explanatory value and testability of the two theories have been resolved may we say of them whether they are really competing with each other and whether they can be subjected to crucial observational tests which may decide against one of them and thereby show that the other is 'better'. In this way we may in the end come to say, after many trials and errors, that we have a theory which, according to the present state of our critical discussion, including observational tests, appears to come nearer to the truth than all the others considered.

II

Thus Russell's challenge has been met, and we have dealt with the first phase of the problem of induction.

The 'problem of rational belief', as the second phase of our problem may be called, is in my opinion less fundamental and interesting than the first. It arises as follows.

Even if we admit that there is no logical difficulty in showing that and how observations may sometimes help us to distinguish between 'good' and 'bad' theories, we must insist that no explanation has been given of the trustworthiness of science, or of the fact that *it is reasonable to believe* in its results—in theories which are well tested by observations. There is more to a good theory than that it has escaped falsification so far: even if we admit that we are always fallible and very prone to make mistakes, and that all scientific theories are conjectural, it is unreasonable to deny that there is a tremendous amount of positive knowledge in science. But how can we admit the reasonableness of this position—of this rather restricted kind of belief in science—and, at the same time, admit that Hume is right?

This is the new challenge. My view that it is less fundamental and interesting than the first is partly due to my limited interest in the philosophy of belief. But it is also due to the fact that no really new ideas are needed to meet it. Nonetheless, this new challenge, or the second stage of the problem, may help to clarify the situation.

I will assume that agreement has been reached on the conjectural character of scientific theories: that our scientific theories remain uncertain, however 'successful' and well 'supported' by evidence and the result of discussion, and that we may be unable to foresee what kind of change will become necessary. (Remember Newton's mechanics!) Accordingly, I will assume that if we speak here of 'rational belief' in science, and in scientific theory, we do *not* mean to say that it is rational to believe in the truth of any particular theory. The point is of greatest importance.

What, then, is the object of our 'rational belief'? It is, I submit, not the truth, but what we may call the *truthlikeness (or 'verisimilitude')* of the theories of science, so far as they have stood up to severe criticism, including tests. What we believe (rightly or wrongly) is not that Newton's theory or Einstein's theory is true, but that they are *good approximations* to the truth, though capable of being superseded by better ones.

But this belief, I assert, *is* rational. It is rational even if we assume that we shall find tomorrow that the laws of mechanics (or what we held to be the laws of mechanics) have suddenly changed (a possibility which will be discussed more fully in subsections III and IV below, devoted to the 'problem of tomorrow'). For in this case we should be faced with the problem of explaining not only the new observed regularities, but also the old ones. Our problem would be (a) to construct a theory from which the old theory could be obtained, under certain conditions, as a *good approximation*, and (b) to show what circumstances (initial conditions) brought about the change.[3] This approach, which ensures the survival of the superseded theory as an approximation, is demanded by realism and by the method of science. Simply to submit to the fact that the change has happened, and to record it, would amount to the acceptance of *miracles*, to the abandonment of the quest for rational explanation, and thus of the task of science—of rationality.

[3]See *L.Sc.D.*, section 79, p. 253.

These considerations show that the belief in the truthlikeness of well-corroborated results of science (such as the laws of mechanics) is indeed rational, and remains so even after these results have been superseded. Moreover, it is a belief capable of *degrees*.

We have to distinguish between two different dimensions or scales of degrees: the degree of the truthlikeness of a theory, and the degree of the rationality of our belief that a certain theory has achieved (a certain degree of) truthlikeness.

I have called the first of these two degrees 'degree of verisimilitude'[4] and the second 'degree of corroboration'.[5] They are 'comparative' in the sense that two theories can be compared with respect to verisimilitude or corroboration, without, however, leading in general (that is, with the possible exception of probabilistic theories) to numerical evaluations.

If two competing theories have been criticized and tested as thoroughly as we could manage, with the result that the degree of corroboration of one of them is greater than that of the other, we will, in general, have *reason to believe* that the first is a better approximation to the truth than the second. (It is also possible to say of a theory not yet corroborated that it is *potentially* better than another; that is to say, that it would be reasonable to accept it as a better approximation to the truth, provided it passes certain tests.[6])

According to this view, the rationality of science and of its results—and thus of the 'belief' in them—is essentially bound up with its progress, with the ever-renewed discussion of the relative merits of new theories; it is bound up with the *progressive overthrow* of theories, rather than with their alleged *progressive consolidation* (or increasing probability) resulting from the accumulation of supporting observations, as inductivists believe.

One example out of hundreds may illustrate this.

Most of us today strongly believe—and have reason to believe—in the Copernican model of the solar system (as revised by Kepler and Newton): in a certain arrangement of the planets, moving in near ellipses round the sun, accompanied by their moons. But what are our reasons for believing in the truthlikeness of this theory? (To

---

[4]See *Conjectures and Refutations*, Ch. 10, and *Addenda*. [See also *Objective Knowledge*, Chapters 2, 3, and 9; and 'A Note on Verisimilitude', in *British Journal for the Philosophy of Science* **27**, 1976, pp. 147–64. Ed.]
[5]See *L.Sc.D.*, Ch. X, and Appendix *ix.
[6]*Cp.* with this Ch. 10 of my *Conjectures and Refutations*, esp. pp. 215 *ff.*

be sure, we do not believe in its complete truth, since it is only a model and therefore bound to be an over-simplification and approximation—quite apart from the fact that it may be in need of an Einsteinian correction, and perhaps of some further revolutionary elaboration to account for the approximate validity of Bode's law.)

To tell *the story of the observational evidence* which has accumulated for many centuries, starting with the Egyptians and Babylonians, would do no justice to the powerful reasons we have for believing in the truthlikeness of the model. Quite apart from the fact that we should have to mention traditions (such as Homer's report of the sun's hesitancy on his path, or Joshua's of its standstill) which we have to explain away (by critical discussion) as myths, observation reports are highly selective, and this selection is influenced by preconceived ideas.

Our real reasons for believing in the truthlikeness of the Copernican model are much stronger. They consist in *the story of the critical discussion,* including the critical evaluation of observations, of all the theories of the solar system since Anaximander, not overlooking Heraclitus' hypothesis that a new sun was born every day, or the cosmologies of Democritus, Plato, Aristotle, Aristarchus, and Ptolemy. It was not so much the accumulation of observations by Tycho as the critical rejection of many conjectures by Kepler, Descartes, and others, culminating in Newton's mechanics and its subsequent critical examination, which ultimately persuaded everybody that a great step had been made towards the truth.

This persuasion, this belief, this preference, is reasonable because it is based upon the result of the present state of the critical discussion; and a preference for a theory may be called 'reasonable' if it is arguable, and if it withstands *searching critical argument*—ingenious attempts to show that it is not true, or not nearer to the truth than its competitors. Indeed, *this is the best sense of 'reasonable' known to me.*

The reasonableness of a belief, in the sense described here, changes with time and cultural tradition, and to a limited extent even with the group of people who are conducting a discussion; for new argument, new critical ideas, may alter the reasonableness of a belief. It goes without saying that new experiments may do the same.

However, *the present state of the critical discussion can be very*

*definite concerning the superiority of one theory over another. It can
also be very definite about the falsity of a theory: but not about its
truth.* When such definite appraisals have been the result of prolonged and thorough critical discussion, then they have, in the past,
usually been borne out by later discussions: reversals of the relative
appraisals of two theories, and revivals of theories which have been
definitely rejected, have been remarkably rare. (Newton's corpuscular theory of light was in no sense revived by Einstein's photon
theory, as is often asserted: it was a theory of transmission or
propagation; the photon theory, designed to meet emission and
absorption problems, succumbs to the wave theory in problems of
propagation.) As opposed to this, reversals of claims based on
'inductive evidence' have been surprisingly frequent, so that there is
something like inductive evidence against induction. (Not only the
false statement that all swans are white, but also most superstitions,
including medical superstitions, are supported by vast amounts of
inductive evidence.) The fact that reversals of critical appraisals are
rare can easily be explained: good critical arguments retain their
power—unless, indeed, they operate with undetected prejudices or
with fictitious or misinterpreted observational evidence.

To say that many of our critical appraisals have survived, and that,
even when our best theories have been superseded, we have rarely
reversed the judgment that they were the best available at the time
they were so judged, is to say that our critical method has been
surprisingly successful in the past. *But we must not conclude that it
will be so in the future.* Our problems may become too difficult for
us, or our intellects may decline. After all, only a very few among
thousands of well-trained scientists succeed in making contributions to the more difficult and fundamental problems of science; and
if these few should no longer be available science might stagnate. Or
some prejudices may become our undoing: the cult of impressive
technicalities or the cult of precision may get the better of us, and
interfere with our search for clarity, simplicity, and truth. There is
no royal road to science; there is no method which guarantees
success; and any theory of knowledge which, in explaining why we
are successful, allows us to predict that we shall continue to be
successful, explains and predicts too much.

This choice of the better theory is like that of the better witness.
When we are faced with witnesses who contradict each other, we try

to cross-question them, to analyse critically what they say, to check and counter-check relevant details. And we may decide—rationally decide—to prefer one of them; even though we assume that all witnesses, not excluding the best, are somewhat biased, since all testimony, even when confined to observation, is selective (like all thought), so that the ideal of 'the whole truth and nothing but the truth' is strictly speaking unattainable; though no doubt we are prepared to revise our preference for one of the witnesses in the light of new critical arguments, or new evidence.

This, in outline, is the 'positive' contribution which my theory makes to the second phase of the problem of induction. It is, however, necessary to stress the negative side also.

Though we may reasonably believe that the Copernican model as revised by Newton is nearer to the truth than Ptolemy's, there is no means of saying *how* near it is: even if we could *define* a metric for verisimilitude (which we can do only in cases which seem to be of little interest) we should be unable to *apply* it unless we knew the truth—which we don't. We may think that our present ideas about the solar system are near to the truth, and so they may be; but we cannot know it. Nor should we think that our discussion has furnished us with reasons to believe so: all it has done is to furnish us with good reasons to believe that we have progressed towards the truth: that is, we have good reasons to believe that some of our present ideas are *more truthlike* than some alternatives. And although the Copernican model was *the* great breakthrough, we no longer think that our sun is the center of the universe, or even of our galaxy.

We cannot justify our theories, or the belief that they are true; nor can we justify the belief that they are near to the truth. We can, however, rationally defend a preference—sometimes a very strong one—for a certain theory, in the light of the present results of our discussion.

The method of science is rational:[7] it is the best we have. It is therefore rational to accept its results; but not in the sense of pinning

---

[7] R. A. Wollheim, in a review of *L.Sc.D.* (*The Observer*, 15 February 1959), puts this point very clearly: '. . . as long as we think of scientists as trying always to establish laws, their activity must seem irrational; but once we conceive of them as trying always to falsify hypotheses, then the task in which they are involved seems consistent and comprehensible.'

our faith on them: we never know in advance where we may be let down.

Yet it *is* reasonable, or rational, to rely on the results of science *for all practical purposes.* For practice always means a choice: we may act in this way, or in that way. (Inaction is, of course, just one possible way of action.) And in so far as we accept, or reject, a scientific theory as a basis of practical action, this means choosing one theory rather than another. Where we are in the position to make such a choice, it will be rational to choose, of two competing theories, that which has survived prolonged critical discussion, including tests.

A last point should be made here. Belief seems to be something much needed in practical actions: man is a believing animal because he is an acting animal. The theoretician, *qua* theoretician, can do without it. For him, the theory which appears to have the greatest verisimilitude is not one to believe in, but one which is important for further progress. It is also the one he will single out as worthy of further criticism.

Of course, even the theoretician, *qua* theoretician, must act; for example, he must choose his problems. And in so far as he does so, he too may be guided by beliefs—and by doubts.

III

The third phase of the problem of induction is, in my opinion, even less fundamental than the second. The first phase was an urgent practical problem of method—how to distinguish between good and bad theories? The second was less important, but still of some urgency. For we do believe in the results of science, in some ordinary sense of 'believe', and this belief is reasonable, in some ordinary sense of 'reasonable'. Since it was not at once obvious how these facts can be accounted for within the logical framework of our solution, there was a problem here of some significance.

But the third phase of the problem, the 'problem of tomorrow' as I shall call it, seems to me, once we have reached clarity about the first two phases, no more than a typical philosophical muddle (unless, indeed, it is confused with the metaphysical fourth stage). Admittedly, an inductivist like Hume or Russell may think that it is indistinguishable from what I have called the problem of induction, and he may even think that it is a superior formulation of the same problem. But an essential part of my non-inductivist solution lies precisely in recognizing the fundamental character of the first phase of the problem and the inferior character of its third phase. We may formulate the third phase of the problem as 'How do you know that the future will be like the past?' or perhaps: 'How do you know that the laws of nature will continue to hold tomorrow?'.

The most simple and straightforward answer to the first of these two questions is: 'I do *not* know that the future will be like the past; on the contrary, I have good reason to expect that it will be different in many ways—indeed, in almost all those aspects which are mentioned by inductivists as examples of the "uniformity of nature". Thus our accustomed daily bread may turn into poison (remember the French case of mass poisoning from ergot); air may choke those who breathe it (remember the air poisonings in Hamburg); and our best and most trusted friends may turn into deadly enemies (remember the totalitarian societies).'[8]

If it is said that this answer to the first question is not to the point because what was meant by it was more clearly stated by the second question ('How do you know that the laws of nature will continue to hold tomorrow?'), then my answer is again, and emphatically: 'I do *not* know that what we regard today as a law of nature will not be regarded tomorrow as a refuted conjecture. In fact, this seems to be a more frequent occurrence than ergot poisoning.'

But it may be said that this answer is again not to the point; that the question asked was not about hypotheses which may be refuted, but about true genuine laws of nature, genuine natural regularities, and the possibility that these may change. To this question my answer is trivial. There are all kinds of changes in nature; but what we call a law of nature is a statement of something that remains *invariant* during changes; and if we find that what we had thought to be an invariant does change, then we have made a mistaken conjecture: it just was not a law of nature.

I have found, however, that inductivists are not satisfied by these

---

[8]My examples allude to those given by Russell with the opposite purpose—that is, in defence of induction—in his *Problems of Philosophy*, p. 69. [See *Objective Knowledge*, Chapter 1. Ed.]

answers, and that they do not feel that the third phase of the problem, the problem of tomorrow, is solved.

Their misgivings are connected with some of my views concerning 'degree of corroboration' (see Chapter IV below). I have often described it as *nothing but* a summarized report, or an appraisal of the way a theory has stood up so far to criticism and to tests, as well as of the thoroughness of the critical discussion, and the severity of the tests to which it has been submitted.

Sometimes I have also described it as the degree to which the theory in question 'has been able to prove its fitness to survive by standing up to tests'.[9] But as the context of such passages shows, I meant by this no more than a report about the past fitness of the theory to survive severe tests: like Darwin, I did not assume that something (whether an animal or a theory) that has shown its fitness to survive tests by surviving them has shown its fitness to survive all, or most, or any, future tests.[10] In fact, I believe that a theory, however well tested, may be refuted tomorrow—especially if somebody tries hard to refute it, and especially if he has a new idea about testing it.

But if the degree of corroboration is nothing but an evaluation of the past performance of the theory, does not the problem of induction, in the form of the problem of tomorrow, arise again? For does not the degree of corroboration of a theory—that is, its past performance—determine our expectation concerning its future performances? Do I not myself, in spite of my denial, mistakenly attribute to a theory a disposition to survive future tests on the basis of its past performance?

I agree that such an attribution on my part would amount to a breakdown of my theory: it would be an inductive inference. But I need not go in any way beyond the arguments already advanced[11] in order to clear up the issue.

The point is this. I do not believe that a highly corroborated theory is particularly likely, or probable, or what not, to survive future tests; or that it is more likely to do so than a less highly

[9]*L.Sc.D.*, Ch. X, before section 79 (p. 251).
[10]The point about 'fitness to survive' was misunderstood in the way here indicated by my reviewer in *Mind; cp.* note 12 to section 12, above.
[11]This discussion elaborates various passages of *L.Sc.D.*, especially section 79 and Appendix *ix (sub-section *14), and of 'Philosophy of Science: A Personal Report' (1957; now Chapter 1 of *Conjectures and Refutations*).

corroborated theory. On the contrary, the likelihood of a theory to survive will largely depend, among other factors, upon the rate of progress in that particular branch of science, or in other words, upon the interest research workers take in this particular field, and upon their efforts to design new tests. But the rate of progress may be very great just in a field in which the standards of criticism and of testing are very high—that is in fields in which we have highly corroborated theories.

Accordingly, I should *not* expect that a more highly corroborated theory will as a rule outlive a less well corroborated theory. The life expectancy of a theory does not, I think, grow with its degree of corroboration, or with its past power to survive tests.

But do I not (I will be asked) expect the sun to rise tomorrow, or do I not base my predictions on the laws of motion? Of course I do, because they are the best laws available, as discussed at length above. Even where I have theoretical doubts, I shall base my actions (if I have to act—that is, to choose) on the choice of the best theory available. Thus I should be prepared to bet on the sun's rising tomorrow (betting is a practical action), but not on the laws of Newtonian (or Einsteinian) mechanics to survive future criticism, or to survive it longer than, say, the best available theory of synaptic transmission, even though the latter has (or so it seems) a lesser degree of corroboration. As to practical actions (such as betting on predictions made by these theories), I should be ready to base them in both cases on the best theory in its field, provided it has been well tested.

The matter can also be put like this. The question of survival of a theory is a matter pertaining to its historical fate, and thus to the history of science. On the other hand, its use for prediction is a matter connected with its application. These two questions are related, but not intimately. For we often apply theories without any hesitation even if they are dead—that is falsified—as long as they are sufficiently good approximations for the purpose in hand. Thus there is nothing paradoxical in my readiness to bet on applications of a theory combined with a refusal to bet on the survival of the same theory.

My refusal to bet on the survival of a well corroborated theory shows that I do not draw any inductive conclusion from past survival to future survival.

IV

But do I really not draw inductive conclusions from past performance to future performance? Is Russell not right to stress that the '*only* reason for believing that the laws of motion will remain in operation is that they have operated hitherto, so far as our knowledge of the past enables us to judge'?[12] This inductivist view, in spite of its persuasiveness, is mistaken.

First, we should remember that nobody has ever observed the laws of motion operating: the laws of motion are our own invention, and are invented to solve certain problems—to explain certain events. If they have done so successfully, then we shall rely on them rather than on less successful competitors. But tomorrow somebody may present us with a new theory, a new set of laws of motion, which not only solves all the problems and passes all the tests which the old set solved and passed, but also suggests new crucial tests. They may suggest to us deviations from the old laws which hitherto have remained unnoticed. (All this has happened more than once since—in fact, once within a year of the publication of Russell's book. I am assuming here that Russell meant by 'the laws of motion' a Lorentz invariant set, since otherwise it would have happened even before he wrote.) Thus those laws of motion which 'have operated hitherto so far as *our knowledge* of the past enables us to judge' did *not* 'remain in operation' and there was *no* reason to expect them to do so, because so far as our present knowledge enables us to judge, they were never 'in operation'.

If, however, what Russell had in mind were those other laws of motion—the *real* laws of motion which, we may assume, operated in the past, then we have to stress that they were (and still are) *unknown* to us, and that no inductive inference—no inference from their past operation to their future operation—can be based on these *unknown* laws.

Of course, if any *real* laws have actually operated in the past (as I believe they have), they will continue to operate in the future. But *this* assertion is not based upon induction: it is based on the fact that we explain changes with the help of invariant laws, and that we should refuse to call anything a true (or real) law that does not

[12]Bertrand Russell, *The Problems of Philosophy*, p. 61. The italics are Russell's.

'operate'—that is, which does not hold—everywhere and at any time.

It might be suggested that we should interpret Russell's words more freely to mean something like 'The *only* reason for believing that the sun will be observed to rise tomorrow is that it has been observed to rise in the past; and for believing that the earth will continue to rotate is that, if our interpretation of our observation is correct, it has done so in the past. (And analogous remarks will hold for the laws of motion in general.)'. But this clearly does not work either (quite apart from the fact that observations which, as we have seen, are selective, are never the *only* reason for any reasonable or rational belief). For to say that our interpretation of past motions is *correct* is to say that certain theories are true.

It is not Russell's formulation which is mistaken, but the sentiments which he expresses (including his various references to probability). There simply is *no* reason to believe in the *truth* (or the probability) of any particular set of conjectures which we call a physical theory; though there may be reasons for preferring one theory to others as a better *approximation to the truth* (which is not a probability[13]). This makes all the difference.

V

Inductivists apply a fair and reasonable method of argumentation if they try to show that those well-known difficulties with which they have struggled so long, and with so little success, must arise, perhaps with some appropriate changes, within my theory also.

One difficulty for inductivists, so I gather, is to say why, having observed only black crows up to, say, 1950, one should prefer the law 'All crows are black' to the law 'All crows before 1970 are black, and after 1970 crows will be white'. For both these laws seem to account equally well for the available observational evidence. Nevertheless, it is said, we 'obviously' prefer the first to the second. The problem is to explain why we do so.[14]

[13]That is to say, verisimilitude does not satisfy the rules (for example, Keynes's rules) of the probability calculus. See *Conjectures and Refutations*, Ch. 10 and *Addenda*.

[14]I am uncertain when the above passage was written, but it was almost certainly

Now it may be thought that the same problem must arise within my theory of problem solving by conjectures, as follows. Somebody may be struck by the fact that two or three crows he has seen are black; he wants an explanation, and hits on the hypothesis that all crows are black. But somebody else suggests the competing hypothesis that all crows are black until 1970; or, to avoid the objection that our laws should contain only universals (while '1970' is an individual name[15]), he suggests a third hypothesis—that all crows are black except those photographed during a total eclipse of the sun (which are white).

Why, I am asked, do we prefer the first hypothesis to the second and the third, since all three seem to be related to the evidence equally well?

The answer is that the second and the third theories assert a connection, a dependence, which *would have to be explained*: for to accept an unexplained and inexplicable change at a certain date would amount to accepting a miracle.[16] (See sub-section II above.) This alone makes them inferior to the first. Besides, our present scientific theories of the colouring of birds are of a character which makes us suspect that any explanation of the change in question which would not be *ad hoc* would conflict with some well-corroborated theories (of genetics). For these reasons alone, I do not see any difficulty whatever in explaining my preference in a case like this.

written before my attention had been drawn by Professor Nelson Goodman to his *Fact, Fiction and Forecast*, 1955; see note 72a to Chapter 11 of my *Conjectures and Refutations*, p. 284. However this may be, my discussion shows that the problem of inductive support raised by Goodman with the help of his new and peculiar predicates ('grue', etc.) can be easily stated without such predicates. Thus the problem of these new predicates, and the problem of their exclusion, appears to me not so much a new riddle, to be solved by a theory of entrenchment, but rather like a red (or perhaps a reen) herring. [See my 'Goodman's Paradox: A Simple-Minded Solution', in *Philosophical Studies*, December 1968, pp. 85–8; 'Theories of Demarcation between Science and Metaphysics', in I. Lakatos and A. Musgrave, eds.: *Problems in the Philosophy of Science* (Amsterdam: North-Holland Publishing Company; 1968), pp. 54–57; 'Eine Lösung des Goodman-Paradoxons', in Gerard Radnitzky and Gunnar Andersson, eds.: *Voraussetzungen und Grenzen der Wissenschaft* (Tübingen: J. C. B. Mohr (Paul Siebeck) Verlag, 1981); and 'The Philosophy of Karl Popper: Part III, Rationality, Criticism and Logic', *Philosophia*, 1982. Ed.]

[15]Cp. *L. Sc. D.*, sections 13 to 15 and Appendix *x.
[16][Cp. *L. Sc. D.*, section 79. Ed.]

However, my chief objection to the argument is quite different.

All our hypotheses are conjectures, and anybody is free to offer conjectures—even conjectures that may appear quite silly to the majority of us. Only thus can we make way for bold, unconventional, new ideas. We have to pay for this freedom by often being confronted by ideas that seem to be silly. Few of these will be taken seriously; but some may; and some may sometimes, contrary to first impressions, turn out to be moves in the right direction. So, although every scientist who dismisses a theory as silly *a priori* takes a risk, there is no way to avoid such risks; not only is every proposal of a new conjecture risky, but so also is the decision whether to take it seriously or to dismiss it out of hand. As opposed to inductivists, I do not assert that there is *one* (inductively reached) theory which best accounts for or explains any given evidence. On the contrary, the idea of a *plurality of competing conjectures*—which, admittedly, we *try* to reduce by criticism—is essential to my methodology.

While such a plurality *may* baffle the inductivist, it creates no problem at all for me.

In practice, many of the more obviously 'silly' conjectures may be eliminated through criticism; as being untestable, or less testable; as being arbitrary, or *ad hoc*; as creating, without excuse, unnecessary new problems; and as conflicting with our most general ideas of what a satisfactory explanation should be like: ideas which are somewhat vague, but which, like scientific theories, develop by trial and error: ideas of the 'style' (mechanical, electrical, statistical, etc.) of a good explanation. Of course such considerations may make us sometimes reject a *good* theory. This too is one of the risks we run; it is part of the conjectural character of science.

Thus Galileo wrongly rejected the lunar theory of the tides because (as I have suggested elsewhere[17]) it was part and parcel of the astrological theory of stellar 'influences', which he rightly felt was in bad taste, although nobody seems ever to have taken the trouble to refute astrology observationally, or even to examine critically the vast amount of inductive evidence on which it is 'based'. Other examples are the early rejection on the Continent of Newton's theory of gravitation, and the early rejection in England, under the

[17]See my *Conjectures and Refutations*, p. 188, note 4 (and the reference there to p. 38, note 4).

influence of Newton's authority, of Huygens' wave theory of light, until the corpuscular theory was refuted by Young.

Thus the belief that the duty of the methodologist is to account for the silliness of silly theories which fit the facts, and to give reasons for their *a priori* exclusion, is naive: we should leave it to the scientists to struggle for their theories' (and their own) recognition and survival. Moreover, it is an inductivist belief. The inductivist who thinks that the only sufficient reason for accepting a theory is its support by past observations is, obviously, baffled if he finds that many highly unattractive theories can all claim to be just as well supported by past observations as the most attractive one. But for me the problem does not arise.

On the contrary, if a theory of high explanatory power looks attractive or promising to somebody, for whatever reason (such as his non-lunar theory looked to Galileo), then he is right to stick to it, and not to lose hope too soon, even in the face of internal difficulties, and even in the face of apparent empirical refutations: he may get it right in the end, in spite of everything. If not, he will have learned the more from his mistakes the greater his intellectual effort was in his attempt to meet criticism. A certain amount of dogmatism and pigheadedness is necessary in science if we are not to lose brilliant ideas which we do not at once know how to handle or to modify.

There is a place, and a function, within the critical method of science, even for the lunatic fringe. I once wrote that our universities should not try to produce scholars or scientists, but be satisfied with a more modest and a more liberal aim—that of producing men who can distinguish between a charlatan and a scholar or a scientist.[18] I was quickly set right by L.E.J. Brouwer who told me that even this formula was not liberal enough since it could be interpreted as encouraging that illiberal superiority with which the academic often looks down upon an outsider. He indicated that there was a place in science even for the charlatan, and rightly rejected anything that could be interpreted as supporting distinctions of this kind.

Thus if we were to give methodological reasons for condemning

[18]See note 6 to Chapter 11 of my *Open Society*. (In this note—an attack upon the idea of producing experts—the term 'expert' is used where I speak here of 'scholar'.)

all theories like the one that all crows are black except those photographed during a total eclipse of the sun, we might easily condemn a most important theory. We might give *a priori* reasons against Einstein's law that groups of fixed stars in the region of the Zodiac always exhibit the same relative angular distances except those taken during a total eclipse of the sun, when some of their relative angular distances will be found to be slightly different. Had Galileo known what photography is, he might easily have felt this Einsteinian law to have an unpleasant astrological flavour.

VI

Our discussion so far has been logical, or methodological, or epistemological. It may be summed up by saying that I have replaced the problem 'How do you know? What is the reason, or the justification, for your assertion?' by the problem: 'Why do you prefer this conjecture to competing conjectures? What is the reason for your preference?'

While my answer to the first problem is 'I do *not* know', my answer to the second problem is that, as a rule our *preference* for a better corroborated theory will be defended rationally by those arguments which have been used in our critical discussion, including of course our discussion of the results of tests. These are the arguments of which the degree of corroboration is intended to provide a summary report.

In this way the logical problem of induction is solved.

Of course many problems remain which may be said to be aspects or phases of the problem of induction; and some of these may be logical problems. The main problem remaining—I call it the fourth phase of the problem of induction—is, however, of a different character, in spite of its close connection with the problems already discussed.

5. *Why the Fourth Stage of the Problem is Metaphysical.*

The fourth phase of the problem is metaphysical. The challenge contained in it may be formulated as follows.

There are true natural laws. This we know, and we know it from experience. Hume says we do not; yet, in spite of what he says, we do: our belief that there are true natural laws is undoubtedly based, in some way or other, on observed regularities: on the change of day and night, the change of the seasons, and on similar experiences. Thus Hume must be wrong. Can you show *why* he is wrong? If not, you have not solved your problem.

I sympathize wholeheartedly with the spirit of this challenge, although I do not agree entirely with its formulation.

At the outset, we should note that there are a number of different ways to interpret the claim that there are true natural laws. There are, for example:

(1) There is (at present) at least one true universal statement, describing invariable regularities of nature.

(2) Some possible universal statement describing invariable regularities of nature (whether yet expressed or not) is true.

(3) There exist regularities in nature (whether ever expressed, or expressible, or not).

While all these claims are *related* to Hume's problem of induction, the issues connected with them are nonetheless quite different—and in a number of different respects.

First, what Hume showed was that we cannot derive a universal law like 'All swans are white' from howsoever many observations (or observation statements) of white swans. But the statements (1–3) just listed are not universal: they are singular existential statements. So the question arises how such a singular existential statement relates deductively to observation or perhaps to reflections upon our experience.

Our assertion—'There are true natural laws'—being existential, does not even refer to any particular physical law, but merely asserts that at least one such law is true. This is of considerable importance. I, for one, would not be prepared to point to any particular law of physics and say: 'This law is true, in its present formulation and interpretation: I feel certain that it will never be falsified, or modified, or recognized as merely conditionally valid, or as valid merely within certain limits.' At the same time, I do believe that at least *some* of the laws of our present system of physics are true in this sense; I should even say many of them are, if we include those on lower levels of universality.

Secondly, our assertion, at least in its first two interpretations, does not belong to physics. Rather, it speaks *about* the theories of physics (or perhaps of science in general). This may be expressed by saying that it belongs to the meta-theory of physics. (It belongs to what Tarski calls the 'semantics' of physical science.) While the statements of science are about non-linguistic objects, our assertion is thus about linguistic objects. So it belongs to some language (a 'meta-language') in which we can speak about some other language (the 'object language'), which in turn refers to the world. The statement 'There is some true natural law' is a conjecture about the world *and* also a commentary about natural laws. While Hume's original problem is concerned with the logical relationship between a natural law and some observational experience, our new problem is concerned with the relationship, logical or otherwise, between commentaries about natural laws and commentaries (or reflections) about observational experiences.

It might be objected here that our assertion that there is at least one true law of nature does, despite what I have just said, belong to science. The argument might go that laws of nature belong to science, and if *a* is a law of nature, then '*a* is true' follows from *a* (by Tarski's definition of 'true') and from '*a* is true' and '*a* is a law of nature' we obtain, of course, 'There is a true law of nature'. I admit the correctness of the derivation: 'There is a true law of nature' does indeed follow from any scientific law. Yet, since all scientific laws are conjectural, the statement 'There is a true law of nature' need not be true in its turn; it inherits the conjectural character of the law *a*. At the same time, it does *not* acquire scientific character simply by following from a scientific law: untestable, hence unscientific, statements follow trivially from any testable statement. And especially the statement 'There is a true scientific (or testable) universal law' is, in its turn, untestable. Of course, it follows from any (conjectural) assertion of a scientific law, as we have just seen.

Another problem which also weakens the interest of this derivation is that the arguments used in our scientific discussion in appraisal of scientific conjectures cannot be used in support of the conjecture that 'There is a true law of nature'. For these arguments merely support our *preference* for one law or another, and do not establish, or support, the view that any one of them is *true*.

Thirdly, in its third interpretation (the one in which I am chiefly interested here), our assertion takes on a *metaphysical* character—in

73

several of the many customary senses of the term 'metaphysical', and in a sense in which it can be used in contradistinction to 'logical', 'methodological', or 'epistemological'.

Unlike purely methodological assertions, and also *unlike* purely metalinguistic assertions, but *like* the conjectures of science itself, our assertion may be interpreted as a conjecture *about the world*. To assert that there exists a true law of nature may be interpreted to mean that the world is not completely chaotic but has certain structural regularities 'built-in', as it were. Hence it belongs to a theory of the structure of the world, to a kind of general cosmology: it is a conjecture of a metaphysical cosmology.

Clearly our assertion, being existential, cannot be empirically tested; it is not falsifiable; and it is not verifiable either, since no laws are verifiable. As our assertion is irrefutable, we may certainly describe it as 'metaphysical', in the technical sense in which this term is used in *L.Sc.D.* (Compare there section 6.)[1] And being neither falsifiable nor verifiable, it is presumably 'metaphysical' also in the sense of the positivists.

It is 'metaphysical' also in the traditional sense of the word, since it deals with subject matters which are regarded as characteristic of metaphysics. It deals with the same kind of subject matter as, for example, *the principle of universal causation,* one possible formulation of which is: 'For every event in this world, there exist true universal laws and true initial conditions from which a statement describing the event in question can be deduced.' This too is an assertion about the world and its structure.

It may be objected that, whatever the possible metaphysical interpretation of our assertion, it still belongs chiefly to methodology or the theory of knowledge. In order to search for true laws (as we do), we must *presuppose,* so it may be contended, the existence of such laws in our search. So 'There are true laws of nature' is a methodological presupposition.

But this objection is inconclusive. For one can very well search for something that does not exist, and without assuming or presupposing or postulating its existence. For example, when we test a law,

[1]See *L. Sc. D.*, passim; *Conjectures and Refutations*, especially Chapter 11 and Appendix to Chapter 10. See also my paper 'Indeterminism in Quantum Physics and in Classical Physics', in *The British Journal for the Philosophy of Science* 1, 1950, pp. 117*ff.*

we search for a counterexample to it. But we neither assume nor presuppose nor postulate the existence of such a counterexample. Indeed, no counterexample may exist: the law we are testing may be true.

Even though, then, we do not presuppose or assume 'There are true laws of nature', we may—and indeed, undoubtedly do—believe this. And perhaps this belief is psychologically important in our search for true laws. But even this would not make it a *methodological* presupposition; it would merely make it a *psychological* one.

Incidentally, I share this belief, and think it more reasonable than any alternative of which I know. The best way to understand—and evaluate—this belief is to regard it as a metaphysical conjecture about the structure of the world.

Before turning to this metaphysical issue, however, there is another more or less methodological objection; one which also exploits the connection, mentioned above, to the law of universal causation.

Many philosophers have held that the problem of the truth of the law of universal causation (or of the 'Uniformity of Nature', which is perhaps an even vaguer formula) is equivalent to Hume's problem. That is, it may be contended that the law of universal causation may be used as a principle of induction whose validity would make inductive inferences valid.

This suggestion is, however, wholly mistaken. It was perhaps excusable in the days before Einstein, but hardly since. Since Einstein, it should be clear that an inductive principle—a principle which would render inductive inference valid—cannot exist. For if a theory as well confirmed as Newton's may be found to be false, then clearly even the very best inductive evidence can never guarantee the truth of a theory. Consequently, no inductive principle which would allow us to draw inductive inferences will be valid: it would be refuted by the first refutation of a theory which was induced in accordance with the inductive principle in question.

But if a positive solution to Hume's problem in the form of a valid principle of induction cannot exist, then the law of universal causation—*whatever it may be*—cannot be a valid principle of induction. The same result may be argued more directly: the law of universal causation might be true and we might nevertheless fail to make any scientific progress—perhaps because the intitial conditions vary so radically that they practically never repeat themselves even approximately, or because of the complexity of the laws, or for other reasons. Thus the law of universal causation, even if true, would have no methodological significance. The significant and important methodological rule, 'Search for natural laws', does not follow from it. Nor is success promised to those who act in accordance with this imperative.

If the law of universal causation has no methodological significance, it is not surprising that the much weaker assertion that we are concerned with here, 'There exist regularities in nature'—the assertion whose validity is questioned in the fourth stage of our problem—is also without any direct methodological significance. Even if we knew for certain that there were regularities in nature, Hume's arguments against induction would hold. Millions of observations of men who speak English would not establish that all men speak English: no sequence of observations of the elements of a sample can tell us that we are faced with a *fair* sample. On the other hand, even if we knew that there were *no* invariable regularities— even if there were counterexamples to *all* apparent laws—there would still be much sense in *trying* to rationalize such an ultimately irrational world as far as possible by the critical method of trial and error.

Thus our metaphysical problem is largely academic, and quite different in character from the logical and methodological problems which we have solved. The fact that the fourth or metaphysical stage of our problem remains to be solved is hence perfectly compatible with my claim to have solved Hume's problem of induction completely on a logical, methodological, and epistemological basis. The solution of the fourth stage is not needed to establish my claim.

Hume's logical argument against induction simply does not immediately bear upon our metaphysical assertion that there exist regularities in nature. Nevertheless, it is perfectly true that we shall have to defend this metaphysical assertion against Hume—but not against his *logic*; rather, against his *metaphysics*.

We are now in a position to reformulate the fourth, or metaphysical, phase of the problem of induction more clearly. We can do this by stressing that it is an aspect of the *problem of tomorrow*.

I have (I may be told) agreed that the metaphysical conjecture that there are true natural laws, in the sense of 'There are regularities', is better than its known alternatives, and therefore one which is reasonable to believe. But if there are natural laws, then it can be argued as follows: if *a* is such a law, *a* will continue to apply, or operate, in the future—say *tomorrow*. But how can I say that it is reasonable to believe this if I agree with Hume? Admittedly, one scientific conjecture may be more reasonable than another, in the light of the present state of our critical discussion; but why, if we admit that Hume is right, should we accept it to be more reasonable to believe in this metaphysical conjecture than in its alternatives? Why should not, for example, *all* the apparent regularities slowly change? My earlier argument appears not to be applicable here because it was *methodological*: it showed why, if we wish to explain the world by the methods of science, any change in known regularities would have to be explained with the help of new (conjectured) laws. But now we are faced with different questions. Our questions now are: Why should not science and its method fail tomorrow completely because *all* regularities, whether previously thought of or not, fail? And why should it be reasonable to believe that this will not happen, and that even if there should be such changes tomorrow that science and its method should fail, there will be, unknown to us, at least one regularity which will continue to operate because it is truly invariable?

The discussion of the metaphysical problem will occupy us for some time. But as a first step towards a solution I wish to point out how naive any formulation of the problem is which uses such temporal terms as 'tomorrow' or 'the future'. For any such formulation takes one regularity naively for granted: *the order of time*.

In fact, all formulations like 'Will the future resemble the past?', are naively based upon the uncritical and unconscious acceptance of an intuitively 'natural' yet highly suspect *theory of time*. It is a theory which was held and expressed, for example, by St. Augustine, and which Newton was one of the first to formulate explicitly, in the words: 'Absolute, true, and mathematical time flows equally of itself, and from its own nature, without relation to anything external.'[2]

[2]Isaac Newton, *Philosophiae Naturalis Principia Mathematica*, 1687.

Those who formulate the problem of induction in temporal terms (like 'the future' or 'tomorrow') unconsciously presuppose this or an essentially similar theory of time, without, it appears, being aware of its problematical character. For they consider whether the laws of nature may change in the sense that in the future, or tomorrow, there will be regularities different from those in force until now. But this entails the theory that *the future, or tomorrow, will come, independently of a change in the laws of nature*. So they assume a flow of time which is independent of whatever happens, and independent of any change in the laws of nature. They thus assume precisely what Newton tries to describe. But they are more naive than Newton, and are unaware that, in the very formulation of their problem, they unconsciously assume that certain laws of nature—the laws of the flow of absolute time—are exempt from Hume's doubt. In other words, although they believe themselves to be empiricists and inductivists, they assume with Kant that these laws of time are valid *a priori*.

This view is equivalent to a metaphysical cosmology along the following lines. There is time (and presumably space), and everything in nature happens *in* time. The world is a totality of events, not of things (as Heraclitus first realized[3]), and events are essentially *in* time (and presumably *in* space).

This cosmology was challenged by Leibniz and by Berkeley who, independently, proposed a *relational theory of time and space*. Time was considered as a system of ordering relations (such as *before; after; simultaneous*) holding between *events;* and space as a system of ordering relations holding between things. The world is here again a totality of events. But these events are not *in* a time whose existence is a condition for an event to exist. Rather, only the totality of events exist, together with their temporal relations; and *'time'* is merely a word, a name for the abstract system of these temporal relations. This more sophisticated view may perhaps not be correct; yet it has been fairly generally accepted in science, even before Einstein.

---

[3]A defence of this somewhat controversial interpretation will be found in my *Open Society*, note 2 to Ch. 2, and *Conjectures and Refutations*, Ch. 5 (and the Appendix to Ch. 5).

6. *The Metaphysical Problem.*

The metaphysical principle of universal causation is one in which I do not believe. (*A fortiori*, I do not believe in the stronger principle of 'scientific' determinism which will be discussed in Volume II of this *Postscript*.) But I *do* believe, as already indicated, in the much weaker principle, 'There exists at least one true law of nature'. I shall outline a variety of arguments in its favour. Afterwards, I shall draw attention to some of the difficulties in the metaphysical position which I propose to adopt.

Let us first reconsider my answer to the first stage of the problem, given in sections 2 and 4. I emphasized there that scientific theories are guesses or conjectures *which may or may not be true*, and that we can never know of a theory that it is true, even if it is true. What I wish to emphasize now is this: the fact that we do not and cannot know that a theory is true is not in itself a reason why it should not be true. It may be a reason for suspending belief, but it certainly is not a reason for disbelief; that is to say, for believing that the theory is false.[1]

We may now reconsider the answer to the third stage, given in section 4. I said there that it is reasonable to act upon (and thus to believe in) a thoroughly discussed and well tested scientific theory, provided we are ready to change our minds in the light of new arguments; of new empirical evidence, for example.

Up to a point, this remark also solves the fourth stage of the problem. For to believe in a statement and to believe in the truth of a statement is the same. (This accords with Tarski's theory.) It may therefore be thought reasonable to believe that there exists a true law of nature, provided there exists a thoroughly discussed and well tested law of nature. Since in fact we have a considerable number of thoroughly discussed and well tested laws of nature, there are indeed empirical reasons for the belief that there exists at least one true law of nature.

It may, however, be felt that this reply is not yet entirely satisfactory. And the mentioning of Hume, in our original formulation of

---

[1] With Hume, knowledge is a kind of justified true belief. This whole approach clashes with mine. If I speak of 'belief' here, it is in a different sense—the sense, rather, of my *Objective Knowledge*. For me knowledge—that is, conjectural knowledge—is objective: it is outside, a product of our minds rather than a state of our minds. I do not take the 'problem' of belief seriously.

the problem, contains perhaps a clue to the reason for this dissatisfaction: Hume, if not an avowed idealist, was at least a sceptic as to
the reality of the physical world. His scepticism was closely connected with his views about induction. He admitted the strength of
our belief in a physical world ordered by laws, but asserted that this
belief was unfounded. This suggests that the fourth stage of the
question should have been: 'I believe that we live in a real world, and
in one exhibiting some kind of structural order which presents itself
to us in the form of laws. Can you show that this belief is reasonable?'

The issue raised here is that of metaphysical realism, in a form
which does not so much stress the existence of physical bodies as the
existence of laws. For physical bodies are only an aspect of the law-
like structure of the world which alone guarantees their (relative)
permanence; which means, on the other hand, that the existence of
physical bodies (about which Hume is so sceptical) entails that of
objective physical regularities. (*Cf.* section 16.)

7. *Metaphysical Realism.*

> A disastrous fear of metaphysics . . . [is the] malady of contemporary
> empiricist philosophizing. . . . This fear seems to be the motive of
> interpreting, for example, a 'thing' as a 'bundle of qualities'—'qualities'
> which may be discovered, it is assumed, among the raw material of our
> senses . . . I, on the contrary, do not think that any dangerous kind of
> metaphysics is involved in admitting the idea of a physical thing (or a
> physical object) as an autonomous notion into the system, together with
> the spatio-temporal structure appropriate to it.
>
> —ALBERT EINSTEIN

Fortunately, or perhaps unfortunately, the *L.Sc.D.* was not a
book on metaphysics—at least not metaphysics of 'the dangerous
kind' to which Einstein refers.[1] Nor is this *Postscript.* Yet I stated in
L.Sc.D. that I believed in metaphysical realism. (*Cf.* the second
paragraph of section 79, and the end of sections 4 and 28.) And I
believe in metaphysical realism still.

---

[1]At least not metaphysics of the 'dangerous kind' to which Einstein refers in the
motto translated from his contribution to *The Philosophy of Bertrand Russell,*
edited by P. A. Schilpp, 1944, p. 230.

Metaphysical realism is nowhere used to support any of the solutions proposed in *L.Sc.D.* (In this my method differs from the usual practice of the idealists who, from Berkeley and Hume to, say, Reichenbach, use their metaphysical views to support their epistemological theories.) It is not one of the theses of *L.Sc.D.*, nor does it anywhere play the part of a presupposition. And yet, it is very much there. It forms a kind of background that gives point to our search for truth. Rational discussion, that is, criticial argument in the interest of getting nearer to the truth, would be pointless without an objective reality, a world which we make it our task to discover: unknown, or largely unknown: a challenge to our intellectual ingenuity, courage, and integrity. There is no compromise in the *L.Sc.D.* with idealism, not even with the view that we know the world only through our observations—a view which so easily leads to the doctrine that all we know, or can know, are our own observational experiences. (*Cf. L.Sc.D.*, Chapter V.)

This robust if mainly implicit realism which permeates the *L.Sc.D.* is one of its aspects in which I take some pride. It is also one of its aspects which links it with this *Postscript*, each volume of which attacks one or another of the subjectivist, or idealist, approaches to knowledge. It may not therefore be out of place to discuss here and in the following nine sections, if only sketchily, some metaphysical problems as such, especially since they are connected, in several ways, with the problem of the structure and status of science (or of 'scientific knowledge' in the sense explained in section 1). This discussion will engage us down to section 16.

The intention of the empiricist philosophers, from Bacon to Hume, Mill, and Russell, was practical and realistic. With the possible exception of Berkeley, they all wanted to be down-to-earth realists. But their subjectivist epistemologies conflicted with their realist intentions. Instead of attributing to sense experience the important but limited power to test, or to check, our theories about the world, these epistemologists upheld 'the theory that all knowledge is derived from sense experience'.[2] And they equated 'is derived' either with 'is inductively derived', or, even more often, with 'originates'. They never saw clearly that it is not the origin of ideas

---

[2]This is an *Encyclopaedia Britannica* definition of empiricism quoted by Russell in the beginning of his paper on 'The Limits of Empiricism'; see note 1 to section 1.

which should interest epistemologists, but the truth of theories; and that the problem of the truth or falsity of a theory can, obviously, only arise *after* the theory has been put before us—that is to say, *after* it has originated with somebody, in some way or other—and that the history of its origin has hardly any bearing upon the question of its truth. (I well remember an old peasant up in the Tyrolese mountains who took it for granted that thunder was the noise made by the collision of heavy clouds, and lightning a very hot spark, due to their friction. I have little doubt that the historical *origin* of this straightforward theory must be less suspect—that is, nearer to the inductive model—than that of the more sophisticated theory which modern meteorologists have adopted.)

The empiricist philosopher's belief 'that all knowledge is derived from sense experience' leads with necessity to the view that all knowledge must be knowledge either of our present sense experience (Hume's 'ideas of impressions') or of our past sense experience (Hume's 'ideas of reflection'). Thus all knowledge becomes knowledge of what is going on in our minds. *On this subjective basis, no objective theory can be built*: the world becomes the totality of my ideas, of my dreams.

The doctrine that the world is my dream—that is, the doctrine of idealism—is irrefutable. It can deal with every refutation by interpreting it as a dream (just as psycho-analysis can deal with every criticism by psycho-analysing it). But the widespread belief that the irrefutability of a theory is a point in its favour is mistaken. *Irrefutability is not a virtue but a vice*. This also applies to realism, unfortunately: for realism is also irrefutable. (The refutation of realism is only an idealist's dream. Death, he dreams, may be the awakening which will finally demonstrate to us that while we were alive we were only dreaming. But as an argument this would not even tend to refute realism: if we were to realize, upon waking up, that we had been dreaming, we should do so because we were able to distinguish dream from reality. But this is just what the idealist says we cannot do.) From the irrefutability of idealism follows the non-demonstrability of realism, and *vice versa*. Both theories are non-demonstrable (and therefore synthetic) and also irrefutable: they are 'metaphysical'.

But there is an all-important difference between them. Metaphysical idealism is false, and metaphysical realism is true. We do not, of course, 'know' this, in the sense in which we may know that $2 + 3 = 5$; that is to say, we do not know it in the sense of demonstrable knowledge. We also do not know it in the sense of testable 'scientific knowledge'. But this does not mean that our knowledge is unreasoned, or unreasonable. On the contrary, there is no factual knowledge which is supported by more or by stronger (even though inconclusive) arguments.

Before considering the positive arguments in support of metaphysical realism more fully, I will first discuss some negative arguments: they support realism by way of a criticism of idealism.

From the point of view of a subjective or idealistic epistemology, the strongest form of idealism is solipsism. The epistemological argument in favour of idealism is that all I know are my own experiences, my own ideas. About other minds, I cannot know anything direct. In fact, my knowledge about other minds would have to depend upon my knowledge about bodies: we have no empirical knowledge of disembodied spirits. If bodies are merely parts of my dream, other minds must be even more so.

The problem of other minds has been endlessly discussed in recent years, largely in epistemological terms. I confess that I have not read all these discussions, and it is therefore not impossible that my simple argument for the existence of other minds has been used by others before (although I do not think it has). It satisfies me completely—perhaps because I always remember that in this kind of inquiry no arguments can be conclusive.

My argument is this. I know that I have not created Bach's music, or Mozart's; that I have not created Rembrandt's pictures, or Botticelli's. I am quite certain that I never could do anything like it: I just do not have it in me. (I know this particularly well since I made many attempts to copy Bach; it made me more appreciative of his inventive power.) I know that I do not have the imagination to write anything like the *Iliad* or the *Inferno* or *The Tempest*. If possible, I am even less able to draw an average comic strip, or to invent a television advertisement, or to write some of the books on the justification of induction which I am compelled to read. But on the solipsistic hypothesis, all these creations would be those of my own dreams. They would be creatures of my own imagination. For there

83

would be no other minds: there would be nothing but my mind. I know that this cannot be true.

The argument is of course inconclusive. I may perhaps underrate myself (and at the same time overrate myself) in my dream. Or the category of creation may not be applicable. All this is understood. Nevertheless, the argument satisfies me completely.

My argument is, no doubt, a little similar to Descartes's argument that a finite and imperfect mind cannot create out of itself the idea of God, but I find my own argument more convincing. The analogy with Descartes's argument suggests, however, a simple extension to the physical world. I know that I am incapable of creating, out of my own imagination, anything as beautiful as the mountains and glaciers of Switzerland, or even as some of the flowers and trees in my own garden. I know that ours is a world I never made.

I can only repeat that this argument satisfies me; perhaps because I never really needed it: I do not pretend that I ever doubted the reality of other minds, or of physical bodies. Indeed, when I think of this argument, I cannot but feel that solipsism (or, more generally, the doubt in the existence of other minds) is not so much a form of epistemology as a form of megalomania.

So much for solipsism, and other forms of idealist epistemology which question the existence of 'other minds'. It seems to me quite possible that arguments like the one proposed here prevented Berkeley from becoming a solipsist: being a Christian, he knew that he was not God. So he arrived at the view that there were other minds besides his own, and that it was God who made us perceive that many-splendoured thing, the world of our experience.

Berkeley's version of idealism is as irrefutable as any other, and has little to recommend itself. Even assuming that the epistemological argument favours the solipsistic thesis, it is clear that an appeal to the epistemological argument can no longer be convincing once realities are admitted which are not perceived, such as God and other human minds. Berkeley's attempt to reconcile epistemological idealism and Christianity leads to an apparent compromise which in fact damages both. (Christianity is damaged because Christ's physical suffering is no longer inflicted upon Him by men but by the immediate action of the deity.[3])

[3]See also section 11, below, text to footnote 7.

None of these arguments should be needed. Realism is so obviously true that even a straightforward argument such as the one presented here is just a little distasteful. There is a certain triteness and staleness about it that reminds me a little of a habit which I dislike: that of philosophizing without a real problem. 'It has been said that all sensible men are of the same religion and that no sensible man ever says what that religion is.'[4] I feel that to continue my argument would be to disregard the second half of this wise saying.

It would be unjust, admittedly, to say that the idealists had in fact no real problem. Their problem was the (positive) 'justification' of our knowledge and they were caught in a trap: in their own discovery that it was impossible to 'justify' realism. It has been pointed out to me by W. W. Bartley, III that it is unfair to judge them from the point of view that discards the whole programme of positive 'justification' as futile and replaces it by a programme of criticism. (*Cf.* section 2, above.) I accept this warning. Yet who amongst the idealist philosophers has ever stressed the point that even if realism is true we cannot justify it in their sense, no more than we can justify idealism if realism is false? And that consequently the impossibility of 'justifying' realism does not speak against its truth? And which of them made it clear that since this situation characterizes the logical structure of the problem, it is obviously quite futile to use, as an argument against realism, the fact that it cannot be 'justified'—or, indeed, any similar argument?

The exasperating staleness of the arguments of idealists and sensationalists results from their failure to see the inherent logical limitations of their justificationist programmes. They do not see, quite simply, that even a logical proof of the impossibility of justifying realism would not constitute a justification of its negation.[5]

My arguments apply not only to solipsism and Berkeleyan idealism, but to *all* other forms of this malady (so far as I know of them), especially to the various forms of positivism and phenomenalism, and also to the so-called 'neutral monism' of William James, Ernst Mach, and Bertrand Russell, as I shall show in the next section.

[4]*The Note-Books of Samuel Butler* (Shrewsbury Edition, 1926), p. 229.

[5]Bartley has drawn my attention to the fact that a similar point was raised by Ralph Barton Perry in 'The Ego-Centric Predicament', *Journal of Philosophy, Psychology, and Scientific Methods* 7 (1910), pp. 5–14. Perry's point was that if we assume, or admit, the fact of the ego-centric predicament, then *nothing follows* from this about the truth or falsity of realism or of idealism.

How little Russell *wanted* to be an idealist may be seen from the beautiful passage in which he describes his feelings after his conversion to realism: 'we . . . thought that *everything* is real that common sense, uninfluenced by philosophy or theology, supposes real. With a sense of escaping from prison, we allowed ourselves to think that grass is green, that the sun and stars would exist if no one was aware of them. . . .'[6]

But, being a believer in induction, Russell found that his epistemology did not actually deliver the goods he wanted. 'Theory of knowledge', he tells us, 'has a certain essential subjectivity; it asks, "How do *I* know what I know?" and starts inevitably from personal experience. Its data are egocentric, and so are the earlier stages of its argumentation. I have not, so far, got beyond the early stages, and have therefore seemed more subjective in outlook than in fact I am.'[7] The passage is interesting because of the frankness with which it reveals that the realist aim had not been attained yet, and also because of the clarity with which it locates the fundamental mistake: if we admit that our knowledge is guesswork, then Russell's fundamental question: 'How do I know what I know?' turns out to be badly put, for this question, in terms of knowledge, is very much like asking: 'Have you stopped beating your wife yet?' It assumes that I do know, and consequently that induction is valid. The apparently analogous question in terms of guessing, such as 'How (or why) do I guess what I guess?' is not really analogous at all: this question is psychological: it has no epistemological impact. Thus the proper answer to Russell's question is: 'I do *not* know; and as to guesses, never mind how or why I guess what I guess. I am not trying to prove that my guesses are correct, but I am most anxious to have them criticized, in order to replace them if possible by better guesses. And if *you* feel as doubtful about my guesses as I do, I hope you will help me by criticizing them ruthlessly.'[8]

The moment we replace the idea of knowledge by that of guesswork, the apparently 'essential subjectivity' of the theory of knowledge disappears. Perhaps *some* knowledge (*epistēmē?*) would have to be explained, essentially, on a subjective basis—on the basis of what *I* know securely. But guesses, as opposed to this, are pro-

[6]*The Philosophy of Bertrand Russell*, edited by P. A. Schilpp, 1944, p. 12.
[7]*Op. cit.*, p. 16.
[8]See also the end of section 4 above.

posals, and as such may be met by anybody's counter-proposals. The problem of their subjective basis in our senses ('there is nothing in my mind that was not first in my senses') need not be raised. *We move, from the very start, in the field of intersubjectivity, of the give-and-take of proposals and of rational criticism.*

Thus Russell's fundamental problem needs to be reformulated in terms of guesses; in terms of the hypothetical character of knowledge which (in another context) he would be the first to emphasize. I was therefore right, I think, to put this as my central point when I replied to Russell's paper at the Aristotelian Society, eight years before he published the passages quoted here. (*Cp.* section 1.)

Following the passage already quoted, Russell writes: 'If I ever have the leisure to undertake another serious investigation of a philosophical problem, I shall attempt to analyze the inferences from experience to the world of physics, *assuming them capable of validity,* and seeking to discover what principles of inference, if true, would make them valid.'[9] Thus Russell was prepared to adopt what Kant called a 'transcendental' method: the method of taking scientific knowledge as a fact, and of asking for the principles which would explain how this fact was possible. The result (given in Russell's *Human Knowledge, Its Scope and Its Limits,* 1948) could have been predicted—in fact, I had correctly diagnosed it by my remark at the Aristotelian Society. It was a theory of induction which accepted an inductive principle—or some rules of inductive inference—as valid *a priori.* The difference between Russell's *apriorism* and Kant's mainly lies in Russell's formulation of his inductive principle as a set of rules of *probable* inference.

The (transcendental) method, described by Russell in the passage just quoted, clearly amounts to a renunciation of his subjectivist approach. For here he accepts 'the world of physics' as the objective fact which epistemology ought to explain. Thus, even for Russell, the subjective method is not as essential as might be supposed. And there is no reason why it should govern the first steps if it is abandoned later. Russell's own analysis shows that the subjective basis cannot support the metaphysical realism which he himself wishes to establish, and that other—non-subjective—methods are needed for this purpose.

[9] *Op. cit.,* p. 16. (The italics are not in the original.)

These other methods, however, need not involve us in either Russell's or Kant's *apriorism*. Although they involve a break with traditional empiricist philosophy, and especially with Berkeley's and Hume's metaphysics which questions the reality of the physical world, they do not force us to break with empiricism—with the doctrine that no synthetic principle can be established as *a priori* valid. We can combine the two, empiricism and metaphysical realism, if only we take seriously the hypothetical character of all 'scientific knowledge', and the critical character of all rational discussion.

8. *Hume's Metaphysics. 'Neutral' Monism.*

Hume, like Russell, was a convinced realist whose subjective theory of knowledge led him to metaphysical results which, though he felt compelled to accept them on grounds of logic, he was constitutionally unable to believe in, even for an hour.[1] He seems to have despised his own firm belief in real things as irrational, even though practically unavoidable. He attempted to make use of this very contradiction—which he observed in his own mind—to solve his problem; but this led him nowhere: 'The perplexity arising from this contradiction', he writes, 'produces a propension to unite these broken appearances'—that is to say, his interrupted perceptions of a body—'by the fiction of a continu'd existence. . . . '[2] Nobody who reads his tortuous argument (Book i, Part iv, section ii of the *Treatise*) can help sharing his disappointment with the final results of what he first so confidently announces as 'my system'.[3] Having found that even by taking the bull by the horns he could not make him move another step—that even the contradictions did not stimulate 'the mind' to transcend them—he frankly states at the end of the

[1] '. . . whatever may be the reader's opinion at this present moment, . . . an hour hence he will be persuaded [that] there is both an external and an internal world'; *cp.* Hume's *Treatise*, end of Section ii of Book i, Part iv. (Selby-Bigge, p. 218.)

[2] *Treatise*, Selby-Bigge, p. 205. See also the footnote on p. 204 *f.*: 'This reasoning, it must be confest, is somewhat abstruse . . .; but . . . this very difficulty may be converted into a proof of the reasoning . . .'

[3] *Treatise*, Selby-Bigge, p. 199, lines 7-12. It seems that Hume, when he first wrote this passage, had no intention of adding to his system 'the second part' (p. 201); a 'third part of that hypothesis I propos'd to explain' (p. 205); and ultimately even a 'fourth member of this system' (p. 208).

section: 'Carelessness and inattention alone can afford us any rem-edy.' It is strange that this clear and candid thinker never suspected that, by showing that it did not deliver the goods, he had given what amounted to an annihilating blow to his own subjectivist theory of knowledge.

Hume's reaction to his findings after he had in effect destroyed his own theory is reminiscent of Frege's reaction to Russell's letter informing him of the paradoxes which Russell had discovered in his and Frege's theory. 'I wrote to Frege', Russell reports, 'who replied with the utmost gravity that *"die Arithmetik ist in's Schwanken geraten"*.' Thus it was not Frege's *theory* of arithmetic but arithmetic itself which, he thought, was tottering and threatening to collapse.[4]

Hume's 'system' committed him to idealism: the belief in physi-cal bodies—or in 'objects' which 'continue to exist'—was for him the result of an inescapable 'fiction of the imagination'.[5] Thus he agreed with Berkeley and with 'the most judicious philosophers', as he puts it, 'that our ideas of bodies are nothing but collections [of ideas] form'd by the mind . . .'[6] Yet not only were physical bodies thus reduced to nothing but bundles of ideas, but minds also, including Hume's own. For he boldly transcended Berkeley's views by teaching that '. . . what we call *mind* is nothing but a heap or collection of different perceptions, . . . suppos'd, tho' falsely, to be endow'd with a perfect simplicity and identity.'[7]

Hume's ontology has been much imitated, but none of the imita-tors has quite achieved the beautiful simplicity of his own system. It may be conveniently illustrated by the following diagram of a portion of the universe—say, of a symposium in which five philoso-phers are present (see next page):

The little circles may be interpreted as glimpses of bodies, or as elementary impressions, or perceptions, or as 'sense-data', or as their traces or reflections in our memory. They constitute the raw material of our universe, according to Hume's theory.[8]

[4]See *The Philosophy of Bertrand Russell, Library of Living Philosophers*, ed. by P. A. Schilpp, p. 13.
[5]*Op. cit.*, especially pp. 198 and 201.
[6]*Op. cit., p. 219.*
[7]*Op. cit.*, p. 207; *cp.* also section vi (p. 252, bottom).
[8][Compare this discussion with that in *The Self and Its Brain*, 1977, section 53, pp. 196-199. Ed.]

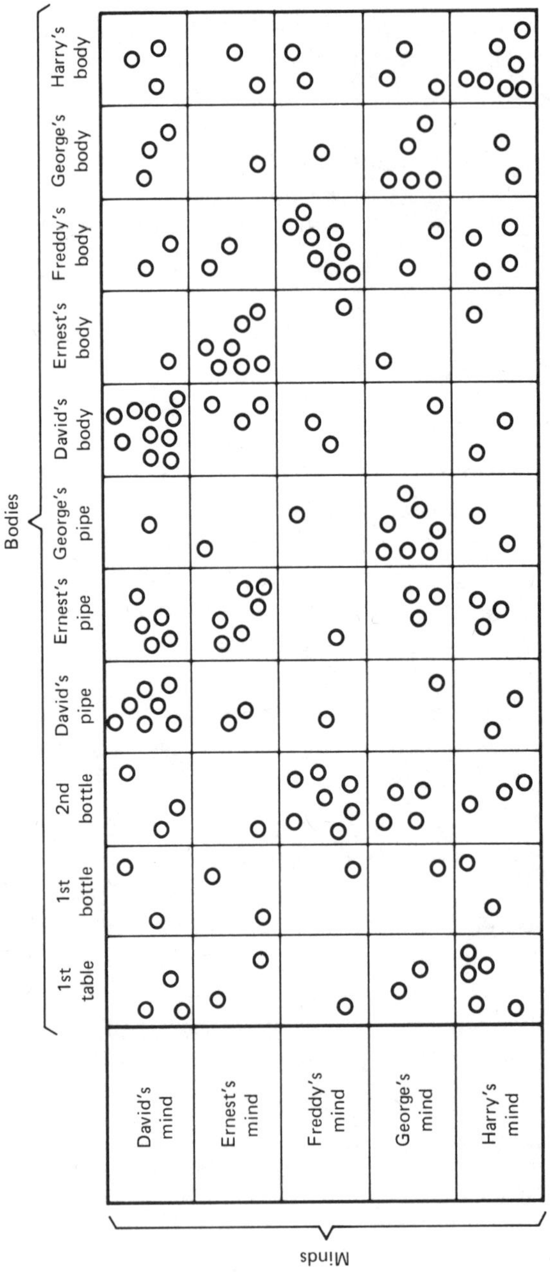
Bodies
1st table
1st bottle
2nd bottle
David's pipe
Ernest's pipe
George's pipe
David's body
Ernest's body
Freddy's body
George's body
Harry's body
David's mind
Ernest's mind
Freddy's mind
George's mind
Harry's mind
Minds

Each of the bundles of perception constituting each of the physical *bodies* in view is here represented by a vertical column containing its various aspects as they appear to the various minds. Each of the bundles of perception constituting each of the *minds* of the symposiasts is represented by a horizontal line containing the aspects of the various bodies as they appear to each mind.

Since the stuff, or the elements, of Hume's universe consist of perceptions of various bodies, belonging to various minds, this 'system' is undoubtedly idealistic. Yet systems which are almost identical with Hume's—such as Mach's, or Russell's, or Reichenbach's—have been described as systems of 'neutral monism', where the word 'neutral' is intended to mean 'neutral as between realism and idealism' (or perhaps as between materialism and spiritualism), and where 'monism' is intended to indicate that the stuff of the universe (or the character of its elements) is all of one kind: sensations, or impressions.[9]

As far as I can make out, the only difference between Hume's idealism and 'neutral' monism is this. If asked how the various bundles are bound up, Hume would have replied: by the *association of ideas*. The 'neutral monists' (Ernst Mach, or William James, or Bertrand Russell) would give a different answer. They would say (if I interpret them rightly): 'The bundles are bound up by *two kinds of natural law*: the horizontal bundles by psychological laws (including the law of association), and the vertical bundles by physical laws. In other words, minds are those bundles which obey the laws of psychology, while physical bodies are those bundles which observe the laws of physics. The elements themselves are at the same time elements of bodies and elements of minds; and they are therefore neutral.'[10]

As the reader will anticipate, I cannot accept this doctrine. The allegedly 'neutral' elements are simply perceptions—something that may be made to disappear by shutting our eyes, or our ears. Physical bodies, as characterized by the laws of physics, do not disappear

[9][See 'A Note on Berkeley as Precursor of Mach and Einstein', in *Conjectures and Refutations*, Chapter 6. Ed.]

[10]I may perhaps confess here that in winter 1926–7, I myself hit upon this form of neutral monism, without realizing at the time that it was fundamentally the same as Mach's and James's theory. Although I entertained the idea tentatively for a very short time, in order to see what could be done with it, I did not seriously believe in it for more than an hour—that is, until I found out its idealistic character.

in this way, while subjective visual sensations do. Thus 'neutral' monism is a form of idealism; and it is only to an idealist—to a philosopher thoroughly imbued with the subjectivist theory of knowledge—that this doctrine can appear to be 'neutral'.

Thus I reject 'neutral monism', together with idealism, as part of the subjectivist theory of knowledge.

9. *Why the Subjectivist Theory of Knowledge Fails.*

The subjectivist theory of knowledge fails for various reasons. First, it naively assumes that *all* knowledge is subjective—that we cannot speak of knowledge without a knower, a knowing subject. Secondly, what is traditionally its fundamental problem is misconceived. I have in mind the problem (in Russell's formulation; see note 7 to section 7) 'How do I know what I know?', with its implied naive empiricist answer 'From observation, or sense-experience'.[1]

Against all this, I contend that scientific knowledge is certainly not *my* knowledge. For I happen to know how little I know—how many thousand things there are which are 'known to science', but not to me (though I should love to know them). For me (and I should expect for any other subject) this fact alone should be enough to reject a subjectivist theory of scientific knowledge.

But even those bits of scientific and commonsense knowledge which I myself happen to possess do not conform to the preconceived scheme of the subjectivist theory of knowledge: few of them are entirely the results of *my own* experience. Rather, they are largely the results of my having absorbed certain traditions (for example by reading certain books), partly consciously, partly unconsciously. And they are no more closely linked with my own observational experience than are my metaphysical beliefs (religious or moral convictions, say) which also result from the absorption of certain traditions. In both cases, my own criticism of some of these traditions may play an important part in forming what I believe I know. But this criticism is almost always stimulated by the discov-

---

[1] [Later work considerably elaborates and extends the argument against the subjectivist theory of knowledge. See *Objective Knowledge*, Chapters 1–4; *Unended Quest*, sections 31, 38, 39; and *The Self and Its Brain* (with Sir John Eccles), Chapter P3. Ed.]

ery of inconsistencies within a tradition, or between different traditions. (It is hardly ever stimulated by discovering an inconsistency between a tradition and an observational experience of my own; for it is given to very few actually to falsify, by their own observations, a traditional theory.)

Thus scientific knowledge is not identical with *my* knowledge; and what is *my* knowledge—*my* commonsense knowledge or *my* scientific knowledge—is largely the result of my absorption of traditions, and (I hope) of some critical thinking.

There is, of course, a third kind of knowledge which also might be called 'mine': I know where I have to look for my ink bottle, or for the door of my room; I know the way to the railway station; I know that my shoe laces have a disposition to break if I am late. This kind of knowledge (which might be termed 'personal knowledge') is hardly traditional since it results from *my own experiences*; and it therefore comes closest to the kind of thing envisaged by the subjectivist theory. Yet even this 'personal knowledge' does not fit that theory; for it is embedded in the commonsense knowledge of traditional things—of ink bottles, shoe laces, railway stations; things about which we have to learn by absorbing a tradition. Admittedly, our observations, our eyes and ears, were immensely helpful in this process of absorption. Yet absorbing a tradition is a process fundamentally different from that envisaged by the subjectivist theory, which wants me to start from *my* knowledge and, moreover, from my observational experience.

The subjectivist is likely to reply that the processes he envisages are logically and genetically prior to those to which I am alluding. He will say that, before I can ever begin to absorb a tradition, I have to learn a lot about the world by observational experience.

At this stage I wish to ask, like a subjectivist: 'How do you know? What is the observational basis of your assertion?' For his reply is the result of sheer prejudice and lack of imagination. The biological sciences tell us a different tale: for example, we have very good reason to think that many insects have a great deal of 'inborn knowledge'—that some, in fact, are unable to learn, to modify their inborn reactions, in certain fields; that they use their senses, and 'recognize' certain things (food, or a potential mate), without having seen them before. But if this is true of some insects, it may well be true of mammals, to some degree at least, and of men: we may well be born with something like innate traditions or 'instincts' (though these may often mislead us). In fact the subjectivist's reply is a part of the empiricist mythology.

The next reply of the subjectivist is likely to be this: even if it should be true that insects and men have inborn knowledge of an instinctive kind, this knowledge must still be the result, in some way or other, of observational experience; the experience, that is, of previous generations.

For two different reasons, this reply is no more tenable than the previous one. First, even if true, it would shift the point at issue. Secondly, what it asserts seems to be untrue—at any rate, it conflicts with the views at present accepted by most geneticists.

As to the first of my two points, what is at issue is not empiricism in general, but subjectivist empiricism, or more precisely, the subjectivist theory of knowledge which starts from the problem 'How do I know?' and which assumes that all my knowledge can be traced back to my subjective observational experiences. But once it is admitted that *my* experiences are not enough—that we have to appeal to those of my ancestors (or anybody else), the subjectivist theory collapses. For 'knowledge' would then be admitted to be something inter-subjective, and this inter-subjective knowledge would then be admitted to be *genetically* prior to *my* subjective knowledge. (This incidentally proves, *a fortiori*, that the latter cannot be *logically* prior to the former.)

As to the second point, the retort is essentially a Lamarckian as opposed to a Darwinian one; for it is essentially intended as a defence of the view that, if we have inborn knowledge, it must, in the last resort, be still the result of *individually acquired observational experience*. I personally happen to have much sympathy with Lamarckism; but Lamarckism is not, at present, favoured by most biologists; and in any case, a theory of knowledge (whether empiricist or otherwise) should not uncritically adopt a conjecture such as Lamarckism on something like *a priori* grounds.[2]

Subjectivists are often misled by some of the ambiguities of the word 'knowledge'. 'Knowledge' is clearly derived from 'I know'.

[2][See *Objective Knowledge*, pp. 97, 149, 268–84. Ed.]

This suggests that 'knowledge' can only be that which is known—known to people.

But this view is clearly inadequate. Take a book containing a logarithmic table. There are people who know how to make such a table (how to calculate it, to arrange it, to print it), and others who know how to use it; but there is nobody who 'knows' the table (not even in the sense, say, in which some people 'know' the beginning of the multiplication table). Yet the table represents 'knowledge'—objective knowledge: answers, or partial answers, to countless important questions: most useful information. And this knowledge is not 'known' to anybody (not even to the compiler); it is only *available*: it is there, potentially, for everybody who is prepared to trust the editor and the publisher.

The case is similar with every scientific theory: a theory may contain, potentially or dispositionally, a wealth of information which nobody 'knows'—neither its discoverer nor its users: it can be tapped, and the information drawn from it; for example by applying it to sets of very special conditions.

Scientific knowledge in this objective sense can be studied, absorbed, used, applied. One and the same piece of it can be accepted or rejected, dogmatically or critically. It can be fervently believed in, or regarded as a rough guess: there are many divergent subjective attitudes, and many ways of reacting to traditions.

One use of objective knowledge, or of a piece of objective knowledge, is to help us to form some of our own subjective convictions. This way—the way leading from objective knowledge to subjective knowledge—is far more frequently taken than that leading from subjective knowledge to objective knowledge. Yet the subjectivist theory of knowledge assumes, as a matter of course, either that there is *only* subjective knowledge (that a book is merely a physical body of a kind that can arouse those associations in its user's mind which may constitute knowledge) or at least that objective knowledge—if we can speak of such a thing at all—can be no more than the record of, or a derivation from, some piece of subjective knowledge.

Although objective knowledge always results directly or indirectly from human actions, from steps taken in the light of subjective *and* objective knowledge, objective knowledge often emerges without having previously been known subjectively. This is invariably the case in all calculations (so far as the man who makes them is

95

concerned): here we wait for the result to emerge in some physical shape before we form the corresponding subjective conviction. (We should not forget that the inventors of the calculus did not really know how it worked, though they had a vague idea: thus they got their knowledge out of their calculations almost by magic. What they found was that the knowledge machine they had built worked extremely well—as a rule, though not always.)

It will be seen, from what I have said, that we can consider objective knowledge—science—as a *social institution,* or a set or structure of social institutions.[3] Like other social institutions, it is the result of human actions, largely unintended, and almost entirely unforeseen (*pace* Bacon). To be sure, it lives and grows largely through the institutionalized co-operation and competition of scientists who are not only inspired by curiosity—the wish to add to their subjective knowledge—but even more so by the wish *to make a contribution to the growth of knowledge*—that is, of objective knowledge. (Many great contributions that have been made to it consisted in errors, and in the detection of these errors.)

The study of how contributions are made, tested, accepted, rejected; of their hypothetical status; of the traditional standards applied to them, and of the refinement of these standards—this study is the most interesting and most fruitful part of the theory of knowledge—the theory of objective knowledge, that is.

Of course, I do not intend to suggest that there is no such thing as subjective knowledge, or that it is not important to the growth of objective knowledge: objective knowledge could not grow without it. The connection between the two is, as I have indicated, not at all simple. Thus there is every reason for having a theory of subjective knowledge also.

Yet any such theory would be part of an empirical science, not part of the logic of science, or epistemology. For its topic is the growth, the development of somebody's knowledge—of those subjective experiences which we express when we say: 'I knew it would be so!' They may become an object of psychological or—more interestingly, I think—of general biological study.

If we want to look upon scientific knowledge and activity as

[3]See my *Poverty of Historicism,* especially sections 32 and 21, and my *Open Society,* especially Chapters 23 and 14.

biological phenomena, we have to consider the part they play in the process of adaptation by the human animal to its environment and to environmental changes: to impending events. Animals, and even plants, prepare themselves for hibernation, say, on a purely instinctive basis. (How do I know this and other things which I am going to assert here? I do not know them: these are all guesses.) And some people do it, perhaps, by consulting the calendar, and placing an order with the coal merchant. A few others do it by studying oil combustion, and inventing a new, safer, and more economical type of oil burner.

The psychological or biological analogue of a hypothesis may be described as an *expectation*, or *anticipation*, of an event. This expectation or anticipation may be conscious or unconscious. It consists in the readiness of an organism to act, or react, in response to a situation of a certain specific kind. It consists in the (partial) activation of certain *dispositions*. (See section 27 of *L.Sc.D.*, especially p. 99, where the dispositional character of knowledge is briefly referred to.)

Classical examples of the way in which unconscious expectations may become conscious are: stumbling down a step ('I thought there was no step here'), or hearing the clock stop ('I was not aware that I heard it ticking, but I heard it when it stopped'). Our organism was anticipating, unconsciously, certain events, and we became conscious of the fact only after our expectations were disappointed, or falsified.

This dispositional preparedness for what is to come seems to be the true biological analogue of scientific knowledge.

In an animal organism, dispositions to react in a certain manner to certain kinds of stimuli are partly inborn. My thesis is that, so far as they are *acquired*, they are *modifications* of inborn dispositions which are 'plastic', and which develop, and change, upon being activated by stimuli, and especially also under the influence of failure and success (coupled perhaps with painful and pleasurable feelings); for the actions and reactions which are released by the stimuli are as a rule directed towards certain biological goals. In this way the organism develops its inborn dispositional knowledge: it learns by trial and error.

This is acquired knowledge which may be handed on by tradition though not (if we are to believe the Darwinian conjecture) by inheritance. Yet the inborn knowledge also changes; and the survival of the fittest inborn knowledge is again a trial-and-error process.[4]

The picture presented here should be contrasted with Hume's picture, which is still popular among epistemologists: the view that we in fact learn *by the repetition of observations*.

Repetition does play a part in the learning process. But this is 'learning' in a sense entirely different from learning by trial and error. (See section 3.) If we learn a skill (cycling, piano playing, speaking a new language), then the two processes may look like one, because learning by repetition takes over where learning by trial and error stops. Yet the difference is this: all *new* knowledge, all radical modifications of our dispositions, all discoveries, are the result of trial and error. Repetition merely makes us familiar with the newly acquired knowledge—it can make us forget what we are doing, and how we acquired our reactions, and especially that there were difficulties to overcome during the trial-and-error period.

So far as I am aware, theorists of the learning process are not yet clear about the gulf between these two senses of 'learning'. The theory of subjective knowledge—of the growth of our subjective knowledge—made its fatal mistake when it looked upon repetition rather than trial and error as the main instrument of learning. (See section 3 above. This mistake also plays a decisive role in all subjectivist theories of probability: see Part 2 of this volume.)

As to observation, and learning through the evidence of our senses, the theory sketched here is directly opposed to traditional empiricism.

The traditional theory is overly influenced by the undeniable truth that without our senses we could not have any knowledge of the world. I am prepared to admit even more: without our senses, we could not live. But I completely disagree with the traditional empiricist view that all our knowledge of the world is the 'result' of observation or sense experience, or that knowledge ever 'enters our intellect through our senses'. Apart from inborn dispositional knowledge, our knowledge does develop under the influence of

[4][See Donald T. Campbell: 'Evolutionary Epistemology', in Paul A. Schilpp, ed.: *The Philosophy of Karl Popper* (LaSalle, 1974), pp. 413–63. Ed.]

stimuli (transmitted by our senses). But stimuli, as a rule, act merely as *triggers*, though they may release cascades of new developments in our dispositional outfit. Similarly observations. They are significant only in the context of our expectations, our hypotheses, our theories.

The traditional view that our knowledge grows by the accumulation (or perhaps by the repetition) of perceptions or observations is sheer myth—probably the most widely believed myth of modern times. It is refuted by the fact that a blind and deaf man may know more, and may make greater contributions to knowledge, than one with sharp eyes and ears: surely, if our senses were intellectually as crucial as empiricists used to think, then the failure of these most important senses would induce the gravest intellectual deficiency. But it does nothing of the kind, as the great example of Helen Keller shows.

Traditional empiricism tries to describe the mind with the help of metaphors, as a *tabula rasa*—something like a well-wiped blackboard or an unexposed photographic plate—to be engraved by observations. This theory, which I have called '*the bucket theory of the mind*', views the mind as a bucket and the senses as funnels through which the bucket can slowly be filled by observations. The sum total of these observations (or perhaps the ordered or digested sum total) is 'our knowledge'. This view is radically mistaken.

There are two sides, as it were, to every higher organism: its inborn constitution, its dispositions to act and to react, its 'reactivity', is one, and its apparatus for receiving stimuli, its 'sensitivity' is the other. Whereas traditional empiricism views knowledge as located within the sphere of *sensitivity*, my view locates it within the sphere of the *activity and reactivity* of the organism.

This shift has far-reaching consequences. According to my view, observations (or 'sensations' or 'sense-data', etc.) are nothing like Bacon's 'grapes' out of which the 'wine of knowledge' flows: they are not the raw material of knowledge. On the contrary, observations always presuppose previous dispositional knowledge. An observation is the result of a stimulus that *rings a bell*. What does this mean? The stimulus must be *significant*, relative to our system of expectations or anticipations, in order to be able to ring a bell, and thus to be observed. The bell—or in other words, an interest, a disposition to respond to the stimulus, an expectation that the stimulus may prove significant—has to be there beforehand, or else the stimulus will pass unnoticed: it will not stimulate. Thus observations presuppose dispositional preparedness, that is to say, dispositional knowledge; and though observations may release dispositional changes—especially if they run counter to our expectations—they cannot possibly become parts or ingredients of the system of dispositions which constitute our subjective knowledge: they are not akin to subjective knowledge (though their *results* may be), but belong to a different realm of things.

Thus the usual way of regarding a 'conditioned reflex' is altogether mistaken. What the experiments do is not 'the linking of the sound of a bell with the saliva flow of the dog'; rather, they exploit the fact that the vitally important reactions connected with the dog's efforts to procure his food constitute, for obvious reasons, one of the most *plastic* dispositional systems of the animal: animals which must hunt for their food under very different conditions have to be able to adjust themselves to these conditions. The procuring of food is thus a field predisposed for learning: the plasticity itself is inborn. Hence the dog may easily form a new expectation—in fact, a new theory: he discovers that the bell announces his dinner. This is all. Where the dispositions are less plastic to start with, or where the animal's vital interests are not involved, attempts to set up a conditioned reflex generally fail.

Traditional empiricism assumes that we can derive or extract our knowledge of the world out of our sense experience by methods which can be seen to be reasonable. If this view—the bucket theory of the mind—is adhered to uncritically, then Hume's discovery that it is impossible to justify induction must lead to irrationalism. It must lead to the conclusion that, for example, our belief in physical bodies is an unjustifiable prejudice. And so indeed it is. Like all our knowledge it is, to start with, a prejudice—but one which may be examined critically. As a result of the critical examination of this conjecture, extending over the last hundred years, we now know more about bodies—for example, that they are processes (as Heraclitus foresaw). Yet we proceed perfectly rationally: we learn, we extend our knowledge, by *testing* our prejudices; by trial and error rather than by induction through repetition. Even our 'animal faith' in regularity is not the result of repetitions. It results from an inborn disposition (which may be activated by the stimulus of repetitions, or by that of one single event) to expect regularities.

Thus the problem of the subjectivist theory of knowledge, 'How do I learn, through my senses, to know about the world?', is wrongly put. And Hume's question, 'What causes induce us to believe in the existence of body?'[5], can be answered in a manner radically different from Hume's. We can say that what a dog growls at, or bites, or likes, he 'believes to be existent'. Existent, in the sense Hume has in mind, are, in the highest degree, those things which we strongly react to—against which we fight for our lives, or which we eat, or which we may be crushed by. The belief in their existence derives from an inborn disposition to treat them as potentially important. It is an inborn belief, although it is, undoubtedly, in need of many stimuli for its development. Such are the 'causes'—or so I conjecture—of this belief, of this prejudice. But they are not reasons. Reasons begin to appear only when we begin to criticize our prejudices—treat them as conjectures; for example when we begin to find out more about 'bodies'. And reasons of another kind may turn up when we criticize the subjectivist epistemology which led to Hume's question.

The dogmatic assumption that we should be able to answer a question of the theory of knowledge such as Hume's by an investigation of the way in which belief arises out of perceptions and associations, or of the way in which we 'immediately experience' colour patches or things, etc., lies at the root of most forms of idealism. It is surprising that, a hundred years after Darwin, philosophers naively continue to discuss the problem of epistemology in terms of the origin of our knowledge in sense-data or perceptions (or in terms of the kinds of words which we use if we discuss perceptions) or in terms of the number of 'repetitions' of the 'observation' of a black raven or a white swan.[6]

The deepest motive behind the subjectivist theory of knowledge is the realization that much of our alleged 'knowledge' is uncertain

[5]*Treatise*, Book i, Part iv, Section ii, first paragraph.

[6]See also my paper 'Philosophy of Science: A Personal Report', especially sections iv and v, and the second footnote to section ii. [This paper is now reprinted as Chapter 1 of *Conjectures and Refutations*. See also Chapters 1, 2, 3, and 7 of *Objective Knowledge*, 1972. Ed.]

(and therefore not really 'knowledge'), and the wish to start from *certainty*: from a *certain* basis, or at least from the most certain basis we have. And the experiences which are 'given to me' such as the experience of seeing something (or, with Descartes, the experience of doubting) seem to offer themselves as the natural starting points. Subjectivists uncritically assumed that upon the basis of these 'data' the edifice of knowledge—scientific knowledge—can be erected. But this assumption is incorrent. Nothing can be built on these 'data', even if we assume that they themselves exist. But they do not exist: there are no uninterpreted '*data*'; there is nothing simply 'given' to us uninterpreted; nothing to be taken as a basis. *All* our knowledge is interpretation in the light of our expectations, our theories, and is therefore *hypothetical* in some way or other.

This is exactly what we should expect if realism is true—if the world around us is, more or less, as common sense, refined by science, tells us that it is. If realism is true, if we are animals trying to adjust ourselves to our environment, then our knowledge can be only the trial-and-error affair which I have depicted. If realism is true, our belief in the reality of the world, and in physical laws, *cannot* be demonstrable, or shown to be certain or 'reasonable' by any valid reasoning. In other words, if realism is right, we cannot hope or expect to have *more* than conjectural knowledge: the miracle is, rather, that we have been so successful in our quest for conjectures. And in saying this I have in mind not only the miraculous successes of the last three hundred years.

Thus if realism is right, the aim of the subjectivist theory of finding a secure subjective basis upon which to erect our knowledge of the world—and sound reasons for a belief in the reality of the world—is an unrealizable and, indeed, an unreasonable aim.

Realism thus explains to us why our knowledge situation is *necessarily* precarious. If, on the other hand, some form of idealism is true, then *anything* may happen—and therefore, *possibly*, also that which does happen. Thus realism is the logically stronger of the two metaphysical theories. It is preferable for logical reasons: metaphysical idealism turns out to be void of *any* explanatory power.

In rejecting the subjective theory of knowledge, we undermine one of the strongest arguments or motives for idealism. But we still have to discuss Hume's sceptical arguments against the reality of matter.

10. *A World without Riddles.*

The subjective theory of knowledge—which tells us to construct the physical world out of my own 'egocentric' perceptual experience—has set itself a task which is both unnecessary and impossibly difficult. This is why it always relapses into some kind of idealism. And what makes idealism so unattractive is, precisely, the fatal ease with which it explains everything. For idealism does solve all problems—by emptying them.

One typical example is the idealist's (or the neutral monist's) solution of the body-mind problem—the problem of the immensely intricate physiological influences (of drugs, say) upon our mental state, and *vice versa*, of mental influences (of the realization of dangers, say) upon our physiological state. Idealism, in all its forms, including neutral monism allows us to solve this problem, in a split-second, so completely that it disappears. If all bodies, including my own, are just bundles of perceptions, then changes in the physical world will in general be also changes of the mental world, and *vice versa*: it thus can be predicted that *there will be interrelations* (whose detailed laws will have to be found by induction).

Another problem that vanishes in the light of subjectivism and idealism—or perhaps in the darkness they create—is the problem of matter. This is one of the oldest problems of philosophy. It is closely related to the problem of change. Change seems to presuppose something that remains unchanged during change. This led Heraclitus and Hegel to the theory of the identity of opposites; and it led Leucippus and Democritus to the more important theory that *all change consists in movement*—of indestructible material particles, of atoms which are not subject to any change apart from movement.

This was a clear and important theory of change, and of matter: matter was that entity which could move but which otherwise remained unchanged during change. Neither this theory of change nor this theory of matter was final, of course. Since the times of

Leucippus and Democritus, and more especially since Descartes—
with Leibniz, Boscovich and Faraday—the theory of matter has
developed into a theory of the *structure of matter itself*; a development which forms one of the most fascinating chapters in the
history of scientific and philosophical thought. (See Volume III of
the *Postscript*, section 20.)[1]

But what has the subjectivist epistemology to say to all this?
'When we gradually follow an object in its successive changes, the
smooth progress of the thought makes us ascribe an identity to the
succession', Hume writes; 'When we compare its situation after a
considerable change, the progress of the thought is broke; and
consequently we are presented with the idea of diversity: In order to
reconcile which contradictions the imagination is apt to feign something unknown and invisible, which it supposes to continue the
same under all these variations; and this unintelligible something it
calls a *substance, or . . . matter.*'[2]

The theory of substance or matter is intelligibly described in this
somewhat difficult passage of Hume's, and its connection with the
theory of change is clearly indicated; but it is dismissed as an
unintelligible fiction of '*the imagination*'. (As this *deus ex machina*
belongs to the mind, it ought to be 'nothing but' a bundle of ideas;
and ideas are not unintelligible. It is queer that Hume is never
bothered by *this* contradiction.) Moreover, the passage indicates,
quite consistently, that unobserved objects are likewise nothing but
fictions of the imagination; they are fictitious interpolations between the observed phenomena; they are '*interphenomena*', to use a
modern term due to Reichenbach (see section 13, below).

This 'unintelligible fiction' of substance or matter is not an idea,
according to Hume: it is neither a perception or impression, nor a
reflection; and it is therefore meaningless. For Hume syllogizes:
'We have no perfect idea of anything but of a perception. A substance is entirely different from a perception. We have, therefore, no
idea of a substance.'[3] And he further concludes from this that any

[1] [See *The Self and Its Brain*, Chapters P1 and P3. Ed.]

[2] *Treatise*, Book i, Part iv, section iii (Selby-Bigge, p. 220).

[3] *Treatise*, Book i, Part iv, section v (Selby-Bigge, p. 234). No doubt Hume's first
premise is meant to be continued 'and none of our ideas (whether perfect or not) is
entirely different from a perception'. Yet nothing, I think, can save the syllogism.
Changing the order of Hume's premises it may be put: 'A substance is something
entirely different from a perception. None of our ideas is something entirely

question concerning a substance must be meaningless: 'What possibility then of answering that question . . . when we do not . . . understand the meaning of that question?'[4]

Thus the problem of matter (and with it, that of the structure of matter) is dismissed as meaningless, on typical *a priori* grounds. Hume here followed Berkeley, who had dismissed the atomic theory as meaningless,[5] and he was in his turn followed by Mach.

This dismissal of what was destined to become one of the most important problems of science and philosophy is not an accident. In the world of Berkeley, Hume, and Mach, there can be no serious problems. According to positivism, 'our world is just surface—it has no depth'.[6] For it consists of nothing but our perceptions and their reflections in our memory. It is a world in which there is nothing to discover, since nothing is covered up. It is a world about which there is nothing to find out, nothing to learn. It is a world without any riddles.

Thus Hume's apparently so sceptical epistemology leads, like Bacon's, to the doctrine that nature is an open book, and that truth is manifest; since there is nothing but ideas, there can be no riddles. This explains why Hume's proof of the invalidity of inductive

different from a perception. Thus none of our ideas is the idea of a substance.' But the correct conclusion would read: 'Thus none of our ideas is a substance'; which would be trivial, considering that Hume has described substance as comprising, essentially, ideas *and* interphenomena. In order to obtain Hume's conclusion, his second premise—which is our first—would have to be replaced by, 'The idea of a substance is something entirely different from a perception'; but this would be untrue *a priori*, according to Hume, since in his view *no* idea is entirely different from a perception. The difficulty is fundamental since in a world containing nothing but ideas (and in which a belief is merely a 'vivacious idea') we cannot explain our belief in substance as anything like a 'fiction' which is *not an idea*. Berkeley and the modern language analysts would have avoided this difficulty by saying that the *word* 'substance' is meaningless (see also *Treatise*, Selby-Bigge, pp. 61-62); but then the difficulty arises in a new form: so far no criterion of the meaningfulness of words that makes sense has ever been produced. But, unlike the language analysts, Hume was not really interested in words but in worlds—in the worlds of our ideas and of our beliefs.

[4]*Treatise, loc. cit.* Yet Hume appeared to understand the meaning of the question whether or not a substance could be one of our ideas, and he thought he could answer it.

[5]See my 'Note on Berkeley as a Precursor of Mach', now in *Conjectures and Refutations*, Chapter 6.

[6]I am quoting here from memory, and probably not quite exactly, a brilliant conversational remark of Otto Neurath's when summing up his own philosophy of nature.

inference was so rarely taken seriously (as opposed to his criticism of causality). If nature was an open book, we would not need an inductive logic: we could induce, invalidly but fairly successfully, by association, or habit.

Hume might well have felt comments such as these to be a severe criticism of his work: 'For if truth be at all within the reach of human capacity, 'tis certain it must lie very deep and abstruse', he writes, and he says of his philosophy: 'I . . . would esteem it a strong presumption against it, were it so very easy and obvious.'[7] But this passage seems to refer merely to philosophical or metaphysical discussions in the narrow sense of the word; for in the immediate sequel, Hume clearly states his hope that his own philosophy of human nature may provide a short-cut to all (legitimate) knowledge: that it opens the book of nature for us. ' 'Tis evident', he writes, 'that all the sciences have a relation . . . to human nature. . . . Even *Mathematics, Natural Philosophy, and Natural Religion,* are in some measure dependent on the science of MAN. . . . Here then is the only expedient, from which we can hope for success . . . to leave the tedious lingring method, which we have hitherto followed, and instead of taking now and then a castle or village on the frontier, *to march up directly to the capital or centre of these sciences, to human nature itself;* which being once masters of, *we may everywhere else hope for an easy victory.* From this station we may extend our conquests over all those sciences. . . . . ' .[8]

This passage seems to indicate that Hume himself thought that truth would be manifest, once we had exorcised error with the help of the Delphian and Socratic principle 'Know thyself'. (Although it now became: Know that you cannot 'have . . . any idea of self'; for, as Hume says, 'from what impression cou'd this idea be deriv'd?'[9])

Hume's attack upon material things is part of the metaphysical aspect, as it were, of his attack upon universal laws, while his denial of the validity of induction represents the logical aspect of that attack. I do not think that his metaphysical and epistemological arguments are as strong as his logical arguments. In fact, they are not nearly as strong as Berkeley's epistemological and linguistic arguments against the realism (or materialism) of the scientists. I therefore turn now to Berkeley and his modern successors.

[7]*Treatise*, Introduction. (Selby-Bigge, pp. xviii f.)
[8]*Treatise, loc. cit. (pp. xix f.).* Italics mine.
[9]*Treatise*, Book i, Part iv, section vi. (Selby-Bigge, p. 251.)

11. *The Status of Theories and of Theoretical Concepts.*

Hume's theory, and that of the 'neutral' monists, is that the world consists of nothing but perceptions and their reflections, and that these are bundled up into minds and physical bodies. But what binds the bundles?

According to Hume, the law of association binds up the perceptions into minds; and in so far as physical causality is nothing but habit, it binds them up into bodies as well. In this sense, then, there is a kind of reality beyond the ideas; to wit, the law of their association, or their tendency to associate.

According to 'neutral' monism, the laws of psychology and of physics bind up ideas. These, in their turn, are of two kinds. There are, first, the laws of nature, obtained through observing phenomena, and through induction by simple generalization through habit or repetition. And there are, secondly, more abstract scientific theories, especially the physical theories of Newton. What is their status? Do they explain our observations? Both William James and Ernst Mach adopted a Berkeleyan answer to this question.

In Berkeley's view, scientific theories are *nothing but instruments* for the calculation and prediction of impending phenomena. They do not describe the world, or any aspect of it. They cannot do so because they are *completely void of meaning*. Newton's theory means nothing because such words as 'force', 'gravity', and 'attraction' mean nothing: they are *occult concepts*. His is not an explanatory theory, but only a mathematician's fiction, a mathematician's trick. Since it does not describe anything, it cannot be true or false— it can only be useful or useless, according to whether or not it serves its predictive purpose. Berkeley uses the term 'mathematical hypothesis' for this kind of meaningless but useful trick, designed for the mathematician's convenience: 'fabricated and assumed in order to abbreviate and ease the calculations', as Bacon puts it in his criticism of the Copernican theory.[1]

I completely disagree with Berkeley's *instrumentalist view of scientific theories,* and also with his theory of meaning: it relegates to the category of meaningless symbols words like 'matter' or 'substance', 'corpuscle' or 'atom', and in addition nearly all the words used in the theories of Copernicus, Galileo, Kepler, and Newton.

[1] *Novum Organum* ii, 36. Concerning Berkeley, see my 'Note on Berkeley as a Precursor of Mach', *Conjectures and Refutations*, Chapter 6, where the necessary references are given.

But nobody before Berkeley and hardly anybody after him came as close as he to seeing that according to an observationalist or phenomenalist theory of meaning, *all* scientific concepts must be completely void of meaning. For *all* scientific concepts are, in Berkeley's sense, 'occult'. They neither denote perceptions or observations or phenomena, nor can they be defined with the help of perceptions or observations or phenomena, or with the help of any concepts which denote perceptions or observations or phenomena. That is to say, they cannot be 'constituted', as I put it in my *L.Sc.D.*[2] (Nor can they be 'operationally defined'.[3])

This is Berkeley's truly great discovery, interpreted by him as a proof of the meaninglessness of scientific terms and theories. *I interpret it as a refutation of the observationalist or phenomenalist theory of meaning.*

Indeed, in Berkeley's sense of 'occult', all our scientific *concepts* are occult: they are used to describe unseen, and indeed invisible, structural properties of an unseen and invisible world.[4] But this does not mean that the *theories* formulated with the help of these concepts are 'occult' or 'metaphysical' or 'non-empirical': they may well have testable consequences. I am not afraid of 'occult' scientific concepts precisely because I have a criterion of demarcation which allows me to detect non-scientific theories by means that are more effective than any alleged criterion of the empirical or non-empirical character of the concepts or words involved.[5]

[2] See *L.Sc.D.*, the end of section 25. 'Constitution' is Carnap's name (in *Der logische Aufbau der Welt*) for a definition in phenomenal terms. See also section 20, below.

[3] Operational definitions of scientific concepts are as impossible as 'constitutions'. Take, for example, the operational definition of *length* sketched by Bridgman in his admirable book *The Logic of Modern Physics*. Length is defined with the help of a description of measurements involving the Paris metre (say). Now as can be seen from Bridgman's own analysis, corrections for *temperature* are involved in this description. But these would presuppose that we have defined *temperature measurements*, which in their turn presuppose *length*. Circles of this kind are essentially unavoidable. (See also the first footnote, and text, to section vi of my 'Three Views Concerning Human Knowledge', *Conjectures and Refutations*, Chapter 3, and note 26 to Appendix *x of the *L.Sc.D.*)

[4] I do not believe in ghosts (I mean, the objects studied by so-called 'psychical research'); not because they are occult, but rather because they are not occult enough. They are of a primitive kind of occultness, and they represent a naïve compromise between the ordinary and visible world, and the really hidden and invisible world which science tries to explore.

[5] When I wrote the *L.Sc.D.*—I am alluding to section 4; see footnote 1 to that

Turning now to criticize Berkeley's instrumentalist interpretation of scientific theories, I may perhaps say first that he did not go far enough in his criticism of theoretical concepts. Although he realized the non-existence of what he called 'abstract general ideas', he failed to realize that most predicates of ordinary language are also abstract in his sense. Berkeley himself, when *using* language, constantly uses abstract terms. Thus, where he speaks of 'infidelity' (in the sense of unbelief), as for example in that last thrust of *The Analyst* against 'the modern growth of infidelity', he clearly does not intend to designate by the word 'infidel' merely an increasing collection of his fellow men whom he dislikes (and who, say, behave in a certain observable manner), but he uses it as a name of a truly *abstract property* of these men. Universal terms in ordinary language are not merely general names, shared (as one might think in view of such words as 'table' or 'dog') by a concrete collection of concrete things, but they denote, like 'copper', 'crystal', 'apple', 'food', 'poison' or 'money', structural or relational or 'dispositional' properties of things which are 'abstract' precisely in Berkeley's sense. All universal concepts incorporate theories. And although some of these can be tested, they can never be exhaustively tested (and can never be verified). The statement 'Here is a glass of water' is open to an indefinite and inexhaustible number of tests—chemical tests, for example—because water, like anything else, is recognizable only by its *law-like behaviour* (cp. the end of section 25 of *L.Sc.D.*). Thus 'water' is dispositional, like every other universal concept. Even 'red' is dispositional, for 'This surface is red' asserts that this surface has a disposition to reflect red light. (The fact that there are no such things as non-dispositional universals invalidates, incidentally, the programme, proposed in Carnap's 'Testability and Meaning', of 'reducing' dispositions to non-dispositional observable proper-

section—I thought, optimistically, that positivists had become aware of the fact that statements and not concepts are important and that they had given up the attempt to demarcate metaphysics from science on the basis of the character of the concepts or words used. But I was mistaken. The operational analysis of the meaning of concepts is still very influential. And even members of the former Vienna Circle still continue in the ways of 'the older positivists', as I called them in 1934. For instance Carnap in his paper ('The Methodological Character of Theoretical Concepts', in *Minnesota Studies in the Philosophy of Science*, Vol. I, ed. Herbert Feigl and Michael Scriven, 1951), bases his definition of significant sentences (p. 60) upon that of significant concepts (p. 51).

ties.[6]) As a consequence, not only Newton's dynamics, but most statements of ordinary language would have to be described as meaningless in Berkeley's sense, since 'glass', and perhaps even more clearly 'water', although belonging to ordinary language, are genuine universals, and therefore abstract terms in Berkeley's sense.

In fact, the most common universal terms of ordinary language incorporate a great number of empirical as well as metaphysical or religious theories. Striking examples are the terms 'father' and 'son', which incorporate an empirical theory apparently unknown in certain primitive cultures—to say nothing about the metaphysical theory of causality which they also incorporate; and indeed, the quality of 'being a father' or that of 'being a son' cannot be perceived any better, in Berkeley's sense, than 'being gravitationally attracted by the sun'. (Again we find that Berkeley's idealism clashes with his Christianity. The reason is the same as before: the doctrine of incarnation is essential to Christianity.[7])

All this indicates that the difference between Newton's dynamics on the one hand and ordinary statements of ordinary language on the other—if there is such a difference—is only a matter of degree. But this means the end of Berkeley's theory. First, because the meaningfulness or meaninglessness of words can hardly be a matter of degree. (If a word has some meaning, however little, it has a meaning.) Secondly, because it is part of Berkeley's doctrine that statements of ordinary language, including laws of nature, are true or false because they describe something; from which we must now conclude that the same will hold for scientific theories such as Newton's. In other words, scientific theories are not *only* instruments, but genuine descriptive statements. *They are genuine conjectures about the world.*

This, of course, is a realist position: it makes it possible for us to say that a scientific theory, or a law of nature, can be true (even

[6]See my paper 'The Demarcation between Science and Metaphysics' (contributed in January 1955 to the Carnap volume of the *Library of Living Philosophers,* edited by P. A. Schilpp), especially section 4, from note 59 to the end of the section, reprinted in *Conjectures and Refutations,* Chapter 11. I am glad that Carnap has now accepted this view; for he writes on p. 65 of the paper referred to in the preceding note: 'There is actually no sharp line between observable properties and testable dispositions.' See also the last two pages of my 'Three Views Concerning Human Knowledge', *op. cit.*

[7]*Cp.* section 7 above, text to footnote 3.

though we can never be sure of it). And it further makes it possible for us to say—as we did in sections 5 and, especially, 6—that the metaphysical doctrine asserting the existence of a true law of nature is, somehow or other, backed by empirical arguments.

I have tried to show that a realist need not fear Hume's attack upon the reality of material things. But the weakness of Hume's idealist arguments certainly does not establish the truth of realism. Berkeley's attack upon the realistic interpretation of scientific theories seems to me more interesting and subtle than his and Hume's attack upon material things (although I am personally more attracted by Hume's manner of searching for the truth about the world than by Berkeley's linguistic arguments): a great deal can be learned from this attack of Berkeley's. And although it can easily be repulsed, by showing that Berkeley's arguments would prove too much, this does not refute his interpretation of scientific theories as instruments. Instrumentalism may well be true even if Berkeley's ingenious arguments in its favour are untenable. And as a certain form of instrumentalism has become a new kind of orthodoxy since 1927 (see Volume III of the *Postscript*), it will be necessary to criticize instrumentalism as such, quite independently of Berkeley's arguments. To do so, we may now leave our critical discussion of the subjectivist theory of knowledge, and return to our main topic: the theory of science.

12. *Criticism of Instrumentalism. Instrumentalism and the Problem of Induction.*

Although I have, I think, succeeded in turning the tables upon Berkeley, and in transforming his proof of instrumentalism into a disproof of his theory of meaning, I have not so far refuted the doctrine of instrumentalism as such; for there may be versions other than Berkeley's.

By instrumentalism I mean the doctrine that a scientific theory such as Newton's, or Einstein's, or Schrödinger's, should be interpreted as an instrument, *and nothing but an instrument,* for the deduction of predictions of future events (especially measurements) and for other practical applications; and more especially, that a scientific theory should not be interpreted as a genuine conjecture about the structure of the world, or as a genuine attempt to describe certain aspects of our world. The instrumentalist doctrine implies that scientific theories can be more or less useful, and more or less efficient; but it denies that they can, like descriptive statements, be true or false.

The whole issue of instrumentalism centres round the words 'nothing but'. For nobody who holds that scientific theories are genuine conjectures about the world would ever contest that they may *also* be looked upon as instruments for the deduction of predictions and for other applications.

I have already discussed, in *L.Sc.D.*, a certain form of instrumentalism, although not under this name. It is a view discussed by Schlick and attributed by him to Wittgenstein. (See *L.Sc.D.*, section 4, notes 4 and 7, and text.) This form of instrumentalism is similar to Berkeley's in so far as it agrees with his doctrine that scientific theories are meaningless. (According to Schlick, they are meaningless because they are not 'verifiable'. A doctrine somewhat related to instrumentalism is the conventionalism of Poincaré and Duhem which sees in scientific theories useful conventions rather than conjectures to be tested by experience. Since I criticized this doctrine too at some length in *L.Sc.D.*, I do not propose to consider it in the present chapter. But my present discussion would, with some alterations, apply to it.)

I have briefly sketched a criticism of instrumentalism elsewhere.[1] But since the topic is important in the present context, I am now going to develop this criticism a little more fully.

The idea of my criticism is this. In the applied sciences and in technology much use is made of *'computation rules'* (as I propose to call them). Examples of such computation rules are the rules and tables used in navigation, or the rules and tables used for calculating exposure times in photography (for those who do not have an exposure meter). These computation rules are indeed *nothing but*

---

[1]*Cp.* my paper 'Three Views Concerning Human Knowledge', *Conjectures and Refutations*, Chapter 5, section v. (This section contains an outline of an as yet unpublished paper, 'A Defence of Free Thinking in Quantum Theory', based on a lecture which I gave in Cambridge in 1953.) [See Volume III of the *Postscript*, and also the following papers by Popper: 'The Propensity Interpretation of Probability', in *The British Journal for the Philosophy of Science* 10, 1959, pp. 25-42; 'The Propensity Interpretation of the Calculus of Probability, and the Quantum Theory', in *Observation and Interpretation in the Philosophy of Physics*, The Colston Papers, Vol. IX, ed. S. Körner, 1957, pp. 65-70. Ed.]

instruments; that is to say they are intended, bought, and sold, as useful instruments, rather than as informative descriptions of the world.

Now if instrumentalism were true, then all scientific theories would be nothing but computation rules. Consequently, there could be no fundamental differences between the theories of the so-called pure sciences, such as Newton's dynamics, and those technological computation rules which we encounter everywhere in the applied sciences and in engineering.

But this is not so: *there are profound differences between theories and mere technological computation rules.* While instrumentalism can give a perfect description of these computation rules, it is quite unable to account for these differences. My arguments might be countered by establishing something like a translation rule which would allow the instrumentalist to translate into instrumentalist language anything the theorist might say to show the peculiar character of his approach. But such a defence of instrumentalism would not succeed. For this method would empty the instrumentalist thesis of any content, and thus rob it of the interest which it has at present. Moreover there really is a difference of attitude; for example, the instrumentalist does assume his range of ultimate practical aims to be fairly definitely circumscribed. He cannot very well accept the view that everything may possibly serve some purpose some day without emptying his thesis of any content. The theorist, on the other hand, may find a theory interesting in itself quite independently of any consideration of its future usefulness.

I will mention here ten points which throw some light on the differences between the theorist's and the instrumentalist's approach, and between theories and computation rules.

(1) The logical structure of theories is different from that of computation rules. The logical relations between two or more theoretical systems are different from those between two or more systems of computation rules. And the relations between theories on the one hand and computation rules on the other are not symmetrical.

Without discussing this point fully, it may be remarked that theories are deductive systems, conjecturally claimed to hold everywhere at all times. Computation rules may be presented in tabular form, and are drawn up with a limited practical purpose in mind.

Thus navigational rules useful for slow ships may be useless for fast aeroplanes. Computation rules (of navigation, for example) may be based upon a theory (Newton's, for example), in a sense in which no theory is based upon computation rules.

This argument is, however, not yet decisive: it might still be said that theories themselves are something like glorified computation rules—super-rules to be used for drawing up more special computation rules.

(2) Computation rules are chosen only because of their usefulness. Theories may be found to be false, but may still be useful for computation purposes. For example, we may assume that Newton's theory, or the conjunction of Newton's theory with Maxwell's (Herz's) theory of wireless waves, is falsified. Yet there is no reason whatever why computation rules used in navigation (including navigational radar) should not continue to be based upon both.

(3) In testing theories, we must attempt to falsify them. In trying out instruments, we only have to know the limits of their applicability. If we falsify a theory, we always look for a better one. But an instrument is not rejected because there are limits to its applicability: we expect to find such limits. If an airframe is 'tested to destruction' we do not reject the type afterwards because we succeeded in destroying it; we may reject it because its limits of applicability were narrower than those of another frame, but not merely because it had such limits (while we may reject a universal theory merely for this reason).

(4) Applications of a theory may be regarded as tests, and failure to yield the expected results may lead to the rejection of the theory. This is not so, in most cases, with computation rules and other instruments: any failure (say that of traditional navigation rules in air navigation) may lead us to reject our theoretical conjecture that they would be applicable in a certain field, but they will continue to be used in other fields. (The wheelbarrow may persist side by side with the tractor.)

(5) There is a definite tendency towards more and more general theories on the one side, and towards more and more specialized instruments (including computers) on the other side. The second tendency is explicable in terms of instrumentalism: from the practical point of view, we want instruments which are most convenient for the special purpose in hand. Thus it seems that the interests and

purposes of the theoretician may differ from those of the computer engineer.

(6) An interesting case would be one where we have before us two theories which at the moment are indistinguishable so far as foreseeable practical applications are concerned. For instruments and computation rules are designed for a given field of application. Thus the instrumentalist is bound to say that two theories are equally useful if, other things like 'ease of application' being equal, they lead to the same result within that field. The theoretician thinks differently: if the two theories are logically different, he will attempt to *find* a field of application in which they yield different results—even if nobody was interested in the field before. He will do so because he may then obtain a crucial experiment, falsifying one of the two. In this way new fields of experience may be opened up by the theories—fields of which nobody had thought before. This exploratory (and intellectual) function of theories may be explained away by saying that theories are instruments of exploration (like an explorer's ship or a microscope). *But then we should have to admit that there is a reality to be explored; and if so, it can be described, truly or falsely*—which is precisely what the Berkeleyan instrumentalist wishes to deny. (I say 'Berkeleyan instrumentalist' because it is clear that instrumentalism may also be combined with a metaphysical realism, and even with materialism. For example, we might deny that theories describe reality, but assert that there are things in themselves—even material things in themselves—which, though we may never describe them by our theories, may be operated upon, more or less successfully, by our instruments, including our theories.)

(7) This brings me to an important distinction between two types of prediction (or two senses of the term 'prediction') whose difference cannot, apparently, be appreciated by the instrumentalist. The one type is, say, the prediction of the next eclipse—its time, region of visibility, etc., or the prediction of the number of peas of different colour in Mendelian breeding experiments: generally speaking, the prediction of events of a known kind. The other type is the prediction of an event of a kind never seriously contemplated before the new theory was framed; an event of whose possibility we learn, as it were, from the theory. Examples of this second kind are Einstein's predictions concerning solar eclipses; or his predictions concerning

the conversion of mass into energy. Predictions of this kind often emerge from a new theory to the surprise of its author, whose only intention may well have been to remove some of the difficulties of the then existing theory. They open up a new and unsuspected world of facts—or perhaps a new aspect of our old world; and it would be difficult to fit this situation without strain into an instrumentalist frame of reference.

(8) The crucial question is of course whether or not the theory has, beyond its instrumental power, any informative content. In the light of the previous point this can hardly be denied, for if we can learn from the theory something about events of an unknown kind, the theory must be capable of *describing* these events to us (as indeed it does). But there are more direct arguments. Who would deny today that the Copernican theory describes the (approximate) structure of our solar system? Yet upon the suggestion of Osiander, the editor of Copernicus, who proposed an instrumentalist interpretation of Copernicus's system, this was denied by Cardinal Bellarmino, by Francis Bacon, and by Bishop Berkeley. What these men denied was, precisely, that the Copernican system had any descriptive or informative content.

(9) Theories, as opposed to computation rules, help us to *interpret* our experiences. An example is the interpretation, by Lise Meitner and Otto Frisch, of Hahn's observations as observations of uranium fission. It is of course no accident that Frisch was one of the authors of the drop theory of the nucleus which predicted that very heavy nuclei would split. (This fact may also explain why Frisch overlooked the problem of neutron emission and recapture: it was no part of the drop theory.) But what happened here on the frontier of science happens continually in ordinary life. We constantly interpret our experience with the help of theories. A bad taste or smell is interpreted as due to a rotten egg; and a skid—a completely theoretical term—*explains* or interprets an unusual and dangerous movement of a motor-car as due to insufficient friction between the tyres and the road. (This is an example of the way in which the dividing line—if there is such a thing—between ordinary language and theoretical language is constantly shifting.)

(10) Prediction has been considered by instrumentalists, ever since Berkeley, as one of the main practical tasks of science. The practical value of predictions is obvious and it does not require instrumentalism to justify it. Theorists, however, can explain the

importance of predictions for science in their own terms, without assuming prediction to be a task set to science from outside. To the theorist, predictions are important almost exclusively because of their bearing upon theory; because he is interested in searching for *true* theories, and because *predictions may serve as tests*, and provide an opportunity for the elimination of *false* theories.

The second kind of prediction mentioned above—prediction of an event hitherto unsuspected—gives us not only some measure of the newness of a theory but also a measure of its superiority over the old theory, and thereby of the advance made. Even if the new theory should be rejected one day, it will have led us to discover new kinds of empirical facts.

No doubt all ten points can be explained away by a determined instrumentalist. Yet to me it appears that they amply justify the rejection of instrumentalism.

For anybody who adopts an instrumentalist view, *the problem of induction disappears*. There is no question of the *truth* or *falsity* of instruments. Consequently, there is no question of the *validity* of the procedures or techniques used in designing or in improving instruments. But the problem of induction is only concerned with questions of truth, falsity, and validity.

Instrumentalism no doubt owes a great deal to the wish to solve, or to dissolve, the problem of induction. Berkeley believed in the induction of simple generalizations—say, that clouds of a certain kind usually bring rain; but he saw that an 'occult' theory—such as Newton's, or atomism—cannot be the result of induction. This did not create any difficulty for him, however, since he held that 'occult' theories, if successful, were nothing but useful instruments. Berkeley's solution of the problem was nearly forgotten, but it was rediscovered by Mach, who taught that theories were nothing but economical summaries and instruments for predictions; by Poincaré and by Duhem, who taught that they were nothing but conventions; and by J. S. Mill, Wittgenstein, and Schlick (and later by Ryle), who said that they were not genuine propositions, but pseudo-propositions whose function was to serve as rules of inference (or 'inference tickets')—as rules for transforming genuine observation statements into other genuine observation statements (that is, into predictions).

Berkeley, Mach, Poincaré and Duhem all believed that simple

generalizations of a low order of universality were based on induction. But they saw, at least, that more abstract theories could not be based on induction. Hume went further by pointing out that induction, even of simple generalizations, was invalid, and impossible to justify; yet he never doubted what he considered a fact of psychology or biology—that men and animals (influenced by association or habit) make inductions which prove very useful, however invalid any attempt to justify them must be.

I go further than Hume: I hold that inductive procedures simply do not exist (not even low-level ones) and that the story of their existence is a myth.[2] Berkeley, Poincaré, and Duhem were right to teach that it is impossible to obtain an abstract or higher-level theory such as Newton's, by induction from observation. But they were wrong to think that there is an essential difference, in this respect, between higher-level and lower-level laws; for all laws are useful inventions rather than inductive generalizations. (My own view, of course, is that they are not *merely* useful inventions, but genuine conjectures about the structure of the world that can be tested.) No doubt, lower-level hypotheses are not as abstract as higher-level explanatory theories. But they are theoretical and abstract nevertheless.

Let us now turn to the suggestion, due to Mill, Wittgenstein, and Schlick, that we should consider universal laws or theories as rules of inference rather than as genuine statements.[3]

The suggestion can be elaborated as follows. Take a universal statement such as 'All men are bipeds'. In its presence—that is, if we use it tacitly as a premise, or as a 'suppressed' premise—we may infer from the singular statement 'Socrates is a man' the conclusion 'Socrates is a biped'. Now let us make the assumption that only singular statements are 'meaningful'; or rather, let us make the

[2]See sections 3 and 9 above, and my paper 'Philosophy of Science: A Personal Report', especially sections iv to vi, reprinted as Chapter 1 to *Conjectures and Refutations*.

[3]It will be remembered that this suggestion was made by Wittgenstein and Schlick, in order to reconcile the non-verifiability of universal laws with the meaning-criterion of the *Tractatus;* see notes *4, 7, and 8 to section 4, and note 1 to appendix *i, of *L.Sc.D.* The reference to Mill alludes to his *Logic,* Book ii, Chapter iii, 3: 'All inference is from particulars to particulars.' See also the first note to section iv of my paper 'Three Views Concerning Human Knowledge', *op. cit.*

weaker assumption that, for some reason or other, we are interested only in singular statements (say because only singular statements describe empirical facts), but not in universal statements. We may then remain interested in the transition, or inference, from 'Socrates is a man' to 'Socrates is a biped'; and we may say that the whole interest of the universal statement 'All men are bipeds' rests in the fact that it allows us to make the transition from the one singular statement to the other. In other words, we may interpret the universal statement as giving us the right to make this transition, or as *validating certain inferences*, rather than as an assertion about a 'universal fact', whatever that may mean.

We certainly may look upon universal statements as validating certain inferences, or as equivalent to certain rules of inference: it is a simple fact of logic that, whenever we have a valid inference which proceeds from more premises than one, we may interpret any of the premises (if true) as a (valid) rule permitting us to draw the conclusion from the other premises. Thus we can say that 'Socrates is a man', if true, validates the inference from 'All men are bipeds' to 'Socrates is a biped'; and similarly, that 'All men are bipeds', if true, validates the inference from '*x* is a man' to '*x* is a biped'. Yet this fact must not be interpreted as establishing that a universal statement is, essentially, a rule of inference, or nothing but a rule of inference.

In order to discuss this issue, let us agree to say of statements that they can be *true or false*, and of inferences, or rules of inferences, that they can be *valid or invalid*. (We say that they are valid if they are truth-preserving; that is to say, if they are such that the truth of the premises ensures the truth of the conclusion.) We can then say that in all the usual languages, including the language of mathematics, of science, and of ordinary English, a universal statement like 'All men are bipeds' is factually true if, and only if, the transition from '*x* is a man' to '*x* is a biped' is *factually* truth preserving (or, as we may say, '*factually* valid'). Moreover, a universal statement will be *logically* true (or 'analytic') if, and only if, the corresponding inference, or rule of inference, is *logically* truth preserving, or *logically* valid.

Because of these equivalences, nothing is gained if we 'explain' universal statements as rules of inference—unless, indeed, we wish to make the proposal to *replace* universal statements by rules of inference, perhaps because we think that rules of inference might be less likely than universal statements to create philosophical confusions or puzzles. But this would amount to the proposal to *avoid* or to *discard* universal statements, and to replace them by rules of inference; or in other words, to the proposal to use a language in which universal statements do not occur, and are replaced by rules of inference.

A language of this kind could be constructed without difficulty. It would not be an ordinary language, but an artificial language, in so far as its grammar would forbid the use of universal statements. All its statements would be singular, and they could therefore be (in the sense of the *Tractatus*) truth-functions of atomic statements, and verifiable. Moreover, by expurgating all universal statements, the problem of induction might seem to disappear.[4]

An artificial language like this would no doubt conform to many positivist intentions. Yet neither the possibility of constructing it, nor its actual construction and adoption could help to solve any problem. Nor would it clarify anything. As a matter of fact, the proposal to replace universal statements by rules of inference has, by suggesting pseudo-solutions, created new philosophical confusions and intricacies which, judging from past experience, will not soon be dispelled.

Owing to the equivalence of universal statements and rules of inference, all the proposal can achieve is to replace the problem of the truth of universal statements by that of the validity of the corresponding rules of inference. There is no gain whatsoever in this replacement since the one problem is precisely equivalent to the

---

[4]One particular way of constructing a language L in which universal statements are forbidden (and replaced by an appropriate inference licence) is this. We lay down the general rule that whenever we decide to accept, in ordinary languages, the statement 'All A's are B's', we accept, in L, all singular statements obtainable by substitution from 'If x is an A, then x is a B'. This rule has the desired effect. If we formulate this rule for eliminating universal statements as one for eliminating universal theoretical systems, we may express it in the following form: Replace any universal axiom 'All A's are B's' of the theory by the *axiom-schema* 'If x is an A then x is a B'! Or in other words: Accept any singular statement that can be obtained by substitution from this schema as an axiom of the theory!

This rule will lead, in general, to theories in which infinitely many singular axioms will replace each axiom of the ordinary theory. The situation is closely analogous to that produced by applying W. Craig's theorem to the elimination of a certain class of predicates—say, to 'theoretical predicates'; and the philosophical confusion created in both cases is also closely analogous. (See also *L.Sc.D.*, note 4 to Appendix *x.)

other. Thus it seems that philosophers who proposed this replacement must have been confused about the issue. They believed in a gain, or in a solution, where there was neither.

But the situation is even worse.

It is customary to use only *logical*, or *analytic*, rules of inference. By the suggestion that all universal statements should be considered as rules of inference, we are thus implicitly encouraged to take all universal statements for analytic. (This would amount to the adoption of a radical form of conventionalism: see Chapters IV and VII of the *L.Sc.D.*) The custom of using only logical, or analytic, rules of inference is connected with the fact that in ordinary life as well as in mathematics or in science we are hardly ever conscious of the rules of inference we use. We use them unconsciously, implicitly, relying on their validity without question: there can hardly be such a thing as a *problematical* rule of inference.[5] As a consequence, we rarely question the validity of rules of inference, and we never think of putting them to experimental tests. Yet if we were, as proposed, to interpret all universal theories as rules of inference, then we should have to treat these rules of inference like universal theories: we should have to test them, to try to falsify them—unless we give up the critical method of science.

Philosophers who have advocated this proposal do not seem to have noticed these consequences. In regarding scientific laws as analytic, they came close to conventionalism. And yet, amongst them were some outstanding opponents of conventionalism.

All this may indicate the confusion created by the proposal that universal statements should be regarded as rules of inference.

More recently, a new attitude towards the problem of induction has become fashionable. The idea has spread among philosophers that induction may indeed be a myth, and that the problem of induction may indeed be solved by pointing out that there is no such thing as induction. Yet some philosophers remain impressed by the fact that many scientists assert that they are making inductions—that they obtain laws by generalizing their observational results.

From these two ingredients the following theory is derived.

---

[5]We can obtain from probability theory a system of almost valid rules of inference; yet although only almost valid, these rules are in no way problematic.

Scientists, it is said, do make what may be called 'inductions' in various fields of science—in each according to its peculiar method or procedure. But these 'inductions' should not be mistaken for inferences: making inductions is simply a skill, taught by example, and acquired in scientific practice. Since there is no problem of the validity of skills, the problem of induction—which is the problem of the validity of induction—disappears. In brief, inductive *inference* is a myth: only inductive *procedures* or inductive *techniques* exist (in fact, there is a great variety of them); but these cannot give rise to any 'problem of induction'.

This theory has nothing to offer to anybody but an instrumentalist. And for him it is superfluous. For if scientific theories are nothing but instruments, then the problem of induction does not arise anyway, as we have seen.

Those of us, on the other hand, who do not accept instrumentalism but look upon scientific theories as hypotheses or conjectures—as the tentative results of our search for true theories—will find the remark that there is *a multiplicity of inductive procedures* of little help: it can only multiply the difficulties of the problem of induction. By contrast, the remark that there are *no inductive procedures whatever* (even though there is a wealth of procedures by which we may *test* our theories—procedures which can all be fully analysed in terms of deductive logic) may help to eliminate the problem of induction, and thus to solve it.

13. *Instrumentalism Against Science.*

According to the instrumentalist, scientific theories cannot be real discoveries: they are gadgets. Science is an activity of gadget-making—glorified plumbing. (In the world of appearances, in the world without riddles, there is no place for real scientific discoveries.) The philosophy of instrumentalism on the other hand turns out to be a real philosophical discovery: it reveals, it unveils the true character of science—the reality behind its appearance. It reveals that scientific theories are not what they appear to be. They are not, as they appear to be, descriptions of a world behind the world of appearance: they stand revealed, by philosophical analysis, as mere instruments. (And the 'great scientists' stand revealed, by philosophical analysis, as glorified plumbers. What a relief!)

The tendency of instrumentalism is anti-rationalist. It implies that human reason cannot discover any secret of our world. Thus we do not know more about the world today than we did four hundred years ago. Our knowledge of facts has not increased: only our skill in handling them, and our knowledge of how to construct gadgets. There is no scientific revolution, according to instrumentalism: there is only an industrial revolution. There is no truth in science: there is only utility. Science is unable to enlighten our minds: it can only fill our bellies.

It is not surprising to find such views in Bishop Berkeley. Indeed, Berkeley made it clear that he published them largely in the hope of defending religion against the onslaught of science and of 'free-thinking'; against the claim that reason, unaided by divine revelation, can discover a world behind the world of appearance. But it is a little surprising to find support for these instrumentalist views in the camp of the admirers of science.

My first example of an admirer of science who supports instrumentalism is Reichenbach. There is no more doubt about Reichenbach's allegiance than about the bishop's: Reichenbach is in the opposite camp. He believes in the teaching of science. 'The philosopher must admit', he writes, 'that nature may very well be as the quantum physicist describes it; whether this description is true is a matter of empirical science.'[1] There can be as little doubt about Reichenbach's fundamental realism as about Hume's realistic convictions, or Russell's. Moreover, Reichenbach—like the 'neutral' monists—holds fast to the belief that his analysis of physics is philosophically neutral. 'Those who claim that causality is *a priori* and extends to unobservables, and those who refuse to speak of the causal behaviour of unobservables, are alike in that they commit the mistake of judging quantum mechanics from the viewpoint of certain philosophical doctrines. It is possible to speak about the logical status of quantum mechanics in neutral terms.'[2] This possibility he claims to have realized in his inquiry: 'The inquiry is carried through without the presupposition of any philosophical conception', he writes.[3] But he is in fact no more neutral than the neutral monists: he accepts implicitly the subjective theory of knowledge.

---

[1] The quotation is the concluding sentence of Reichenbach's paper, 'The Principle of Anomaly', *Dialectica* **2**, 1948, described by the author as a summary, with improvements, of his *Philosophical Foundations of Quantum Mechanics*, 1944.

[2] *Loc. cit.*

[3] *Cp.* the end of his *Summary, op. cit.*, p. 349.

He is a Berkeleyan or Humean idealist; and like a Berkeleyan instrumentalist, he is better informed about the objects treated by physics than is physics itself. For his whole analysis of quantum physics is based upon the distinction between the observed *phenomena*, such as this tree outside my window whilst I am looking at it, and the *interpolations* between these phenomena, called by him 'interphenomena', such as the same tree after I have dropped my eyes in order to write down this sentence. 'If we say that a tree exists while we do not look at it', Reichenbach writes, '. . . we interpolate an unobserved object between observables; and we select the interpolated object in such a way that it allows us to carry through the . . . *postulate of identical causality* for observed and unobserved objects . . . .'[4] And he continues: 'If we abandon this postulate, we arrive at different descriptions of the unobservables. These descriptions are not false. . . . The usual language, according to which the object persists when observation ceases, is constructed by singling out one description among this class; this normal description, or *normal system*, is determined by the postulate of identical causality for observed and unobserved objects. *The postulate is neither true nor false but a rule which we use to simplify our language.*'[5]

Now let us consider this doctrine for a moment. What Reichenbach says—and what he wants to say, as his book and the article show—is this. The objects which we interpolate between observations are more or less *arbitrarily chosen* or 'selected'. The 'normal system' consists in selecting the objects in such a way that the causal laws for observed and unobserved objects remain the same. But this 'normal' choice is merely the free adoption of a postulate which is neither true nor false; and we prefer it merely because we wish to simplify our language. (Thus the 'postulate' may be described as a convention.)

Reichenbach uses the following example: ' . . . we could assume', he writes, 'that the unobserved tree splits into two trees'.[6] But in this case, he explains, we may have to alter our causal laws, as far as unobserved objects are concerned; for we may observe the

---

[4]*Op. cit.*, p. 341. The dots in this quotation indicate a lengthy omission, of more than three sentences. Nevertheless, this quotation presents Reichenbach's doctrine accurately.

[5]*Loc. cit.* The last sentence of the quotation is italicized by me.

[6]*Loc. cit.*

tree's shadow, even when we do not observe the tree; and so we should have to assume that the two trees (into which the tree splits when unobserved) 'cause only one shadow'.[7]

Such assumptions would clearly be awkward; and this is the only reason, according to Reichenbach, why we usually adopt the 'normal system', that is to say, the 'postulate of identical causality for observed and unobserved objects': we adopt it in order 'to simplify our language'.

As against this view, I wish to point out that the splitting of the tree if we look away, and its re-assembling if we look back, would clearly violate the laws of physics. For it would involve (according to physics) very considerable forces, accelerations, energy expenditure, etc. But this is the same as saying that *physics informs us unambiguously that the tree does not split when we look away.* (Nor does it, according to physics, disappear: this would involve the destruction of matter and therefore something of the nature of an atomic explosion.) Thus when Reichenbach says that the 'normal' assumption—according to which the unobserved tree behaves as if it were observed—'is neither true nor false' but merely a device 'to simplify our language', he reveals to us that the clear and unambiguous pronouncements of physics are, in reality, nothing but simplifying devices. And when he asserts that we may choose to adopt other 'postulates' (of course, as long as they yield the same results as far as phenomena or observations are concerned) then he agrees implicitly with Berkeley and Mach that physical theories are only instruments for the description and prediction of phenomena.

Thus Reichenbach is led, by his subjectivist theory of knowledge, to a position which he abhors: to the belief that the philosopher may know more, or may know better, than the physicist; to the belief that philosophy can correct the pronouncements of physics.

This idealist position of Reichenbach's turns out to be essential to his solution of the problems of quantum theory. For his result is this. In quantum theory, it is 'impossible to give a definition of interphenomena in such a way that the postulates of causality are satisfied'.[8] Thus according to quantum theory, *'the class of all descriptions of interphenomena contains no normal system'*.[9] Rei-

[7]*Loc. cit.*
[8]*Philosophical Foundations of Quantum Mechanics,* 1944, pp. 32f.
[9]*Loc. cit.* The italics are Reichenbach's.

chenbach calls this result, for which he claims to have given a proof, 'the principle of anomaly' (or sometimes 'the principle of causal anomaly').

Now from the idealistic point of view, this is indeed a result. For it can be formulated as follows. Quantum theory delivers the goods. It allows us to calculate observations correctly. True, it does not allow us to introduce unobserved objects in the normal way. But since the normal way of introducing unobserved objects is in any case merely a matter of convenience and linguistic simplicity, this negative result need not worry us.

I intend to discuss in Volume III of this *Postscript* the problem of interpreting quantum physics from my own realistic point of view (incidentally, without further reference to Reichenbach). Here I only wish to say that Reichenbach's solution seems to me as unsatisfactory as all idealistic solutions, because it is too cheap. In spite of the impressive mathematical and logical apparatus used, it suffers from the old malady: it is fatally easy. For it amounts to saying: 'The formalism delivers the goods. What else do you want? Do you want to understand it? But this is easy: it is just an instrument for the prediction of phenomena; and understanding an instrument means knowing how it is constructed and how it works. If you want to know more—if you want, for example, to get an idea of how atoms or electrons or photons interact—then you must have forgotten that unobserved entities (such as atoms or electrons or photons) are nothing but interphenomena, introduced only for our own convenience; thus your question is really pointless; and you may even be asking a metaphysical pseudo-question.'

This is not, of course, a paraphrase of what Reichenbach actually says. But neither is it a parody. On the contrary, it is a fair rendering of his results and of his attitude (which is similar to that of Bohr and Heisenberg). In any case, the main point which my rendering is intended to bring into focus is Reichenbach's subjective idealism, and the fact that it is an essential ingredient of his results, not merely a slip, or just a way of putting things which might also be expressed in the 'language of realism'.

My second example is Carnap. As in the case of Reichenbach, nobody can doubt the sincerity of Carnap's allegiance to science. In his early days, especially in *Der logische Aufbau der Welt*, Carnap tried to build a system of the world of science upon the foundations of a person's subjective sense-experience; but a little later he was converted, by Otto Neurath, to 'physicalism'. In his big book— *Logical Foundations of Probability*—he develops a probabilistic theory of scientific induction. This theory leads to the result, correct in my opinion (*cp.L.Sc .D.*, section 80), that in an infinite universe, the logical probability of a universal law, even upon the most favourable empirical evidence, will always equal zero. But Carnap unfortunately identifies the degree of confirmation of a law with its logical probability. He therefore obtains the undesired result that the degree of confirmation of any law (whether accepted or rejected) must also equal zero. Thus *no law is confirmable*. This result makes him raise the question (p. 574): 'Are Laws Needed for Making Predictions?' His answer is that they are not needed, and that they are therefore 'not indispensable' for science. The argument thus goes one step beyond instrumentalism. Like the instrumentalists, Carnap assumes here that the sole function of science is to make predictions; like them, he believes that laws are 'efficient instruments' for this purpose; but he also holds that they are not needed. In fact, his argument would show that they are completely redundant. 'We see', Carnap writes, 'that the use of laws is not indispensable for making predictions. Nevertheless it is *expedient*, of course, to state universal laws in books on physics, biology, psychology, etc. . . . Although these laws stated by scientists do not have a high degree of confirmation,[10] they . . . serve as *efficient instruments* for finding those highly confirmed singular predictions which are needed in practical life.'[11]

Admittedly, universal laws are statements in Carnap's theory; but since they turn out not to be confirmable, and also not to be indispensable in science—even though they are 'efficient instruments for predictions'—it is clear that Carnap's theory of induction has led him to a view closely akin to Reichenbach's: only singular statements of possible observations really matter; for the status of

[10]This is an understatement of Carnap's: in an infinite universe, the degree is zero.

[11]*Logical Foundations of Probability*, 1950, p. 575. The italics are mine. My dots replace the words 'have a high qualified instance confirmation and thus'. I have omitted them because I do not wish to explain here, and to criticize, Carnap's concept of a 'qualified instance confirmation' whose inadequacy I have pointed out elsewhere; see *The British Journal for the Philosophy of Science* **6**, 1955, pp. 157–163; and **7**, 1956, pp. 249–256, especially section (7), pp. 251*ff.*

laws is merely that of non-confirmable statements whose sole function is that they are efficient but nevertheless *completely* redundant instruments of prediction.

Considering the very different attitude adopted by Carnap at an earlier place in the same book (I have in mind especially his reference to Einstein and to myself on p. 193, which shows that his intention was to do justice to theories), he doubtless neither intended nor expected these results. They are the unintended consequences of his implicit acceptance of an inductivist epistemology.[12]

In later volumes of this *Postscript*, we shall meet further examples of the unfortunate part played by subjectivist, inductivist, and instrumentalist interpretations of science and, more especially, of quantum physics.

14. *Science Against Instrumentalism.*

We may now bring to a conclusion our discussion of the subjective theory of knowledge. The question of the reasonableness of metaphysical realism was raised (at the end of section 6) in connection with Hume's sceptical attitude. Our examination of Hume's, and his successors', subjective theory of knowledge has led to the result that no serious arguments can be found in this theory against the reasonableness of metaphysical realism.

The reality of physical bodies is implied in almost all the common sense statements we ever make; and this, in turn, entails the existence of laws of nature: so all the pronouncements of science imply realism. These arguments make it reasonable to believe that there are true laws of nature, even though this view is neither verifiable nor falsifiable and is therefore metaphysical. Our discussion of

---

[12]I criticized Carnap's findings about univeral laws more fully in 'The Demarcation between Science and Metaphysics', contributed to the Carnap volume of the *Library of Living Philosophers*, edited by P. A. Schilpp; reprinted in *Conjectures and Refutations*, Chapter 11. Carnap's later paper, 'The Methodological Character of Scientific Concepts' (see note 5 to section 11, above), revises some of the findings which I criticized.

The aim of his new paper is to re-admit theories as significant, and to give a criterion of their significance. I believe, however, that his new criterion of significance can easily be shown to be both *too narrow and too wide*, exactly as can the older positivist criterion of meaning, criticized for this very same reason in section 8 of my *L.Sc.D.* (See section 19 below.)

Hume's scepticism and that of his followers has brought nothing to light that could bring any weight against these arguments. We found only a specious dogma—that whatever is in my mind (or in my language) can reflect only what has previously been in my senses. In its trail we even found a series of unintended and unacceptable consequences, including contradictions, frankly described as such by Hume.

Perhaps the most remarkable thing we found was the attitude of many members of this school of thought towards their results. Like Frege who, faced with contradictions in his theory of arithmetic, believed that arithmetic was tottering, the followers of Berkeley and of Hume are inclined to believe that reality is tottering. And while they admire science, their philosophy of science leads them with necessity to results amounting to the view that science is tottering. To them, what scientists regard as the greatest discoveries of science, the discoveries of laws, are nothing but tricks, or transformation rules, or else nothing but redundant though perhaps quite efficient modes of speech.

These unintended consequences ought to be regarded as so many refutations of the theories of knowledge which give rise to them.

Any criticism directed against competing theories should always be used in an attempt to find weak spots in one's own position. But I do not find any unintended consequences or absurd results arising from my own approach. According to my approach, it is reasonable to accept the views of common sense as long as they stand up to criticism: science arises from criticism *and* common sense *and* imagination. (For example, I believe, with common sense, in the reality of material things, and thus of matter. I am not a 'materialist', however; not only because I believe in minds, but because I do not believe that the doctrine that matter is ultimate and inexplicable has stood up to criticism.[1] It is, moreover, a dull and unimaginative doctrine. If one day we are able to explain matter satisfactorily— say, as a disturbance in a field of forces or of propensities—we should beware of thinking that we have *explained it away*, as some 'spiritualists' are inclined to think even today. By explaining an

[1][See Popper's discussion of the history of the idea of matter in his 'Metaphysical Epilogue', Volume III of the *Postscript;* in *The Self and Its Brain*, Chapters 1 and 3; and in 'Philosophy and Physics', in *Atti del XII Congresso Internazionale di Filosofia*, Venice, 1958 (Florence, 1960) **2**, pp. 367–74. Ed.]

eclipse, or a thunderstorm, or an atom bomb, we do not show that these are unreal, or mere appearances. By asserting that a motor car, or a piece of cheese, is a structure of parts which are not motor cars or bits of cheese, we do not imply that motor cars or bits of cheese are illusions. Yet strangely enough, even Russell speaks as if physics, by explaining the structure of matter, had shown that matter was an illusion.)[2]

Admittedly, there are people who consider some of my theories perverse, but these 'perverse' results of mine are all of the character of denials of commonly held *philosophical* beliefs. Thus it has been considered perverse of me to stress the importance of falsification rather than of verification. This, Schlick said, was perverse, because we want to be right rather than wrong. I see no reason to quarrel with Schlick's psychological argument. I too prefer to be right; and it is just because I like to be right that I prefer to correct myself or, if necessary, to be corrected by others. Thus I try hard to detect faults in my own arguments, that is to say, to criticize them, to refute them. More generally, our tests are attempted falsifications just because we want to get nearer to truth and to avoid falsity: because we prefer to be right. Others too have complained, like Schlick, that I fail to take account of positive or supporting arguments (as opposed to refutations). But this complaint betrays an inductivist prejudice. It is the main idea of induction and verification that we seek supporting instances. With this philosophical or methodological idea I disagree indeed: mere supporting instances are as a rule too cheap to be worth having; they can always be had for the asking; thus they cannot carry any weight; and any support capable of carrying weight can only rest upon ingenious tests, undertaken with the aim of refuting our hypothesis, if it can be refuted.

It has also been suggested that my attempt to deny the popular view that 'science starts from observation' is perverse—if not, indeed, an insincere attempt to impress people by saying something

[2]See, for example, *Portraits from Memory*, 1956, p. 145: 'The truth is, of course, that mind and matter are, alike, illusions. Physicists . . . discover this fact about matter . . .' I am inclined to assert the opposite. Never before have we had such strong scientific (as opposed to common sense) arguments for the reality of matter as now, when we have begun to understand it. It has been tranformed from a dubious metaphysical idea about which we could know nothing into a structure about which many clear and highly interesting questions can be asked, and perhaps even answered.

new and paradoxical. But this view is merely part of a popular *philosophy*—of an inductivist theory of knowledge; and its denial is far less perverse than the clash with common sense to which the consequences of this popular theory give rise. Moreover, the main tenets of popular inductivism have to be given up when it is realized that even Newton's theory is perhaps not true, in spite of the incredibly strong empirical evidence in its support.

I cannot find any consequences of my own theory that clash with common sense, or with science; nor have such clashes been found by my critics, although they have found plenty of clashes with accepted philosophical doctrines. But the subjectivist theory of knowledge, and with it the instrumentalist interpretation of scientific theories, clashes not only with common sense but also with science, and with the rationalist tradition. It rejects these—though without intending to do so. And just as it rejects them, it may be rejected by them.

With this I conclude my criticism of the arguments in favour of idealism. But I still have to say something about realism, both in favour of it and in criticism of it.

15. *The Aim of Science.*

> No fairer destiny could be allotted to any physical theory than that it should itself point out the way to introducing a more comprehensive theory in which it lives on as a limiting case.
>
> ALBERT EINSTEIN

So far I have argued in support of realism largely by way of criticizing idealism. I will now offer some positive arguments for realism, before proceeding, in the next section, to point out some of its difficulties. The positive arguments I have in mind rest on the relation between realism and the aim of science.

Since publishing the *Logik der Forschung* (that is, since 1934) I have developed a more systematic treatment of the problem of scientific method: I have tried to start with some suggestions about the aims of scientific activity, and to derive most of what I have to say about the methods of science—including many comments about its history—from this suggestion. Here I will confine myself to explaining the suggestion, and to pointing out its bearing on the problems of realism.

To speak of 'the aim' of scientific activity may perhaps sound a little naive; for clearly, different scientists have different aims, and science itself (whatever that may mean) has no aims. I admit all this. Yet when we speak of science, we do seem to feel, more or less clearly, that there is something characteristic of scientific activity; and since scientific activity looks pretty much like a rational activity, and since a rational activity must have some aim, the attempt to describe the aim of science may not be entirely futile.

I suggest that it is the aim of science to find *satisfactory explanations* of whatever strikes us as being in need of explanation. By an *explanation* (or a causal explanation) is meant a set of statements one of which describes the state of affairs to be explained (the *explicandum*) while the others, the explanatory statements, form the 'explanation' in the narrower sense of the word (the *explicans* of the *explicandum*).

We may take it, as a rule, that the explicandum is more or less well known to be true, or assumed to be so known. For there is little point in asking for an explanation of a state of affairs which may turn out to be entirely imaginary. (Flying saucers may represent such a case: the explanation needed may not be one of flying saucers, but of the reports of flying saucers; yet should flying saucers exist, then no further explanation of the *reports* would be required.) The *explicans*, on the one hand, which is the object of our search, will as a rule not be known: it will have to be discovered. Thus, scientific explanation, whenever it is a discovery, will be *the explanation of the known by the unknown.*[1]

The *explicans*, in order to be satisfactory (satisfactoriness may be a matter of degree), must satisfy a number of conditions. First it must logically entail the *explicandum*. Secondly, the *explicans* ought to be true although it will not, in general, be known to be true; in any case, it must not be known to be false—not even after the most critical examination. If it is not known to be true (as will usually be the case) there must be *independent* evidence in its favour. In other words, it must be *independently* testable; and it will be the more

[1]See the last paragraph of my text (before the final quotation) of my 'Note on Berkeley as a Precursor of Mach', reprinted in *Conjectures and Refutations*, Chapter 6, p. 174.

satisfactory the greater the severity of the independent tests it has survived.

I still have to elucidate my use of the expression 'independent', with its opposites, '*ad hoc*', and (in extreme cases) 'circular'.

Let *a* be an *explicandum*, known to be true. Since *a* trivially follows from *a* itself, we could always offer *a* as an explanation of itself. But this would be highly unsatisfactory, even though we knew in this case that the *explicans* was true, and that the *explicandum* followed from it. *Thus we must exclude explanations of this kind because of their circularity.*

Yet the kind of circularity I have in mind here is a matter of degree. Consider the following dialogue: 'Why is the sea so rough today?'—'Because Neptune is very angry.'—'By what evidence can you support your statement that Neptune is very angry?' 'Oh, don't you *see* how *very* rough the sea is? And is it not always rough when Neptune is angry?' This explanation is found unsatisfactory because (just as in the case of the fully circular explanation) the only evidence for the *explicans* is the *explicandum* itself.[2] The feeling that this kind of almost circular or *ad hoc* explanation is highly unsatisfactory, and the corresponding requirement that explanations of this kind should be avoided are, I believe, among the main forces in the development of science: dissatisfaction is among the first fruits of the critical or rational approach.

In order that the *explicans* should not be *ad hoc*, it must be rich in content: it must have a variety of testable consequences, and among them, especially, testable consequences which are different from the *explicandum*. It is these different testable consequences which I have in mind when I speak of *independent* tests, or of *independent* evidence.

Although these remarks may perhaps help to elucidate somewhat the intuitive idea of an independently testable *explicans,* they are still quite insufficient to characterize a satisfactory and independently testable explanation. For if *a* is our explicandum—let *a* be again 'The sea is rough today'—then we can always offer a highly unsatisfactory *explicans* which is completely *ad hoc* even though it has independently testable consequences. We can still choose these consequences as we like. For we may choose, say, 'These plums are

[2]This kind of reasoning survives in Thales (Diels-Kranz 10, vol. i, p. 456, line 35); Anaximander (D.-K. A11, A28); Anaximenes (D.-K. A17, B1).

juicy' and 'All ravens are black'. Let *b* be their conjunction. Then we can take as explicans simply the conjunction of *a* and *b*: it will satisfy all our requirements so far stated, but it will be *ad hoc*, and intuitively utterly unsatisfactory.

Only if we require that explanations shall use universal laws of nature (supplemented by initial conditions) can we make progress towards realizing the idea of independent, or non-*ad-hoc*, explanations. For universal laws of nature *may* be statements with a rich content, so that *they may be independently tested* everywhere, and at all times. Thus if they are used as explanations, they *may* not be *ad hoc* because they *may* allow us to interpret the *explicandum* as an instance of a reproducible effect. All this is true, however, only if we confine ourselves to universal laws that are testable, that is to say, falsifiable. It is here that the problem of demarcation, and the criterion of falsifiability, comes in.

The question 'what kind of explanation may be *satisfactory*?' thus leads to the reply: an explanation in terms of testable and falsifiable universal laws and initial conditions. An explanation of this kind will be the more satisfactory the more highly testable these laws are, and the better they have been tested. (This applies also to the initial conditions.)

In this way, the conjecture that it is the aim of science to find satisfactory explanations leads us further to the idea of improving the degree of satisfactoriness of our explanations by improving their degree of testability, i.e., by proceeding to better testable ones; which means—as shown in Chapters VI and VIII of *L.Sc.D.*—proceeding to theories of ever richer content, of a higher degree of universality, and a higher degree of precision. This, no doubt, is fully in keeping with the history and actual practice of the theoretical sciences.

We may arrive at fundamentally the same result in another way. If it is the aim of science to explain, then it will also be its aim to explain what so far has been accepted as an *explicans*, such as a law of nature. Thus the task of science constantly renews itself. We could go on for ever, proceeding to explanations of a higher and higher level of universality—unless, indeed, we were to arrive at an *ultimate explanation;* that is to say, at an explanation which is neither capable of any further explanation, nor in need of it.

Are there ultimate explanations? The doctrine which I have called 'essentialism' upholds the view that science must seek ultimate explanations in terms of essences:[3] if we can explain the behaviour of a thing in terms of its essence—of its essential properties—then no further question can be raised (except perhaps the theological question of the Creator of the essences). Thus Descartes believed that he had explained physics in terms of the *essence of a physical body* which, he taught, was extension; and some Newtonians, following Roger Cotes, believed that the *essence of matter* was its inertia and its power to attract other matter, and that Newton's theory could be derived from, and thus ultimately be explained by, these essential properties of all matter. Newton himself was of a different opinion. It was a hypothesis concerning the ultimate or essential causal explanation of gravity itself which he had in mind when he wrote in the *Scholium generale* at the end of the *Principia*: 'So far I have explained the phenomena . . . by the force of gravity, but I have not yet ascertained *the cause of gravity itself* . . . and I do not arbitrarily [or *ad hoc*] invent hypotheses.'[4]

I do not believe in the doctrine of ultimate explanation. In the past, critics of this doctrine have as a rule been instrumentalists: they interpreted scientific theories as *nothing but* instruments for prediction without any explanatory power. I do not agree with them either. But there is a third possibility, a 'third view', as I have called it. It has been well described as a 'modified essentialism'—with the emphasis upon the word 'modified'.[5]

This 'third view' which I uphold modifies essentialism in a radical manner. First of all, I reject the idea of ultimate explanation. I

---

[3]I have discussed and criticized essentialism more fully in my paper, 'Three Views Concerning Human Knowledge', *Conjectures and Refutations*, Chapter 3, where I also refer to my earlier discussions (in the last footnote to section ii). [Essentialism and the demand for ultimate explanations are also of course intimately connected with justificationism. See section 2 above. Ed.]

[4]See also Newton's letters to Richard Bentley of January 17th and especially February 25th, 1693 ('1692–3'). I have quoted from this letter in section iii of 'Three Views Concerning Human Knowledge', where the problem is discussed a little more fully.

[5]The term 'modified essentialiam' was used as a description of my own 'third view' by a reviewer of 'Three Views Concerning Human Knowledge', in *The Times Literary Supplement* 55, 1956, p. 527. To avoid misunderstandings, I wish to say here that my acceptance of this term should not be construed as a concession to the doctrines of 'ultimate reality', and 'ultimate explanation', and even less as a concession to the doctrine of essentialist definitions. I adhere to the criticism of this doctrine which I have given in my *Open Society*, Chapter 11, section ii (especially note 42), and in other places; see below, section 31; also note 2 to section 19 of the *L.Sc.D.*

maintain that every explanation may be further explained, by a theory of higher universality. There can be no explanation which is not in need of a further explanation, for none can be a self-explanatory description of an essence (such as an essentialist definition of body, as suggested by Descartes). Secondly, I reject all *what-is? questions*: questions asking what a thing is, what is its essence, or its true nature. For we must give up the view, characteristic of essentialism, that in every single thing there is an essence, an inherent nature or principle (such as the spirit of wine in wine) which necessarily causes it to be what it is, and thus to act as it does. This animistic view explains nothing; but it has led essentialists (like Newton) to shun relational properties, such as gravity, and to believe, on grounds felt to be *a priori* valid, that a satisfactory explanation must be in terms of inherent properties (as opposed to relational properties). The third and last modification of essentialism is this. We must give up the view, closely connected with animism (and characteristic of Aristotle, as opposed to Plato), that it is the essential properties inherent *in each individual or singular thing* which may be appealed to as explaining this thing's behaviour. For this view completely fails to throw any light on the question why different individual things should behave in like manner. If it is said, 'because their essences are alike' the question arises: *why should there not be as many different essences as there are different things?*

Plato tried to solve precisely this problem by saying that like individual things are offspring, and thus copies, of the same original 'Form', which is therefore something 'outside' and 'prior' and 'superior' to the various individual things; and indeed, we have as yet no better theory of likeness. Even today, we appeal to their common origin if we wish to explain the likeness of two men, or of a bird and a fish, or of two beds, or two motor cars, or two languages, or two legal procedures; that is to say, we explain similarity in the main genetically; and if we make a metaphysical system out of this, it is liable to become a historicist philosophy. Plato's solution was rejected by Aristotle; but since Aristotle's version of essentialism does not contain even a hint of a solution, it seems that he never grasped the problem.[6]

[6] As to Plato's theory of forms or ideas, it is 'one of its most important functions to explain the similarity of sensible things . . .'; *cp.* my *Open Society,* Chapter 3,

By choosing explanations in terms of universal laws of nature, we offer a solution to precisely this last (Platonic) problem. For we conceive all individual things, and all singular facts, to be subject to these laws. The laws (which in their turn *are* in need of further explanation) thus explain regularities or similarities of individual things or singular facts or events. And these laws are not inherent in the singular things. (Nor are they Platonic ideas outside the world.) Laws of nature are conceived, rather, as (conjectural) descriptions of the hidden structural properties of nature—of our world itself.

Here then is the similarity between my own view (the 'third view') and essentialism: although I do not think that we can ever describe, by our universal laws, an *ultimate* essence of the world, I do not doubt that we may seek to probe deeper and deeper into the structure of our world or, as we might say, into properties of the world that are more and more essential, or of greater and greater depth.

Every time we proceed to explain some conjectural law or theory by a new conjectural theory of a higher degree of universality, we are discovering more about the world: we are penetrating deeper into its secrets. And every time we succeed in falsifying a theory of this kind, we make an important new discovery. For these falsifications are most important. They teach us the unexpected. And they reassure us that, although our theories are made by ourselves, although they are our own inventions, they are none the less genuine assertions about the world; for they can *clash* with something we never made.

Our 'modified essentialism' is, I believe, helpful when the question of the logical form of natural laws is raised. It suggests that our laws or theories must be *universal*, that is to say, must make assertions about all spatio-temporal regions of the world. It suggests, moreover, that our theories make assertions about structural or relational properties of the world; and that the properties described by an explanatory theory must, in some sense or other, be deeper than those to be explained.

These two ideas—that of structural or relational properties of our world, and that of the depth of a theory—are in need of elucidation.

section v; see also notes 19 and 20, and text. The failure of Aristotle's theory to perform this function is mentioned there (in the third edition, 1957) at the end of note 54 to Chapter 11.

We often explain the law-like behaviour of certain individual things in terms of their *structure*. Thus we can explain, and understand, the working of a clock after taking it to pieces a few times, and putting it together again; for in this way, we can learn to understand its structure, and its way of working as a consequence of its structure. Now if we look a little more closely at this procedure, then we find that, in a structural explanation of this kind, we always presuppose some law-like behaviour other than the one to be explained (and 'deeper' than this). For example, what we wish to explain, in the case of the clock, is the regular motion of its wheels and hands. We do so by analyzing its structure; but we have also to assume that the various parts making up the structure are *rigid* (i.e., that they retain their geometrical shapes and 'extensions') and that they are *impenetrable* (i.e., that they push one another along—if one part gets into another's way—instead of one part moving, as it were, through the other). These two law-like properties, the rigidity and impenetrability of certain physical bodies, may in their turn again be structurally explained: for example, by *lattices of atoms* which, it has been conjectured, constitute the material structure of this type of body. But in this second explanation, we not only conjecture that certain parts—the atoms—are arranged in a lattice structure, but we assume in addition that certain laws of attraction and repulsion hold between the atoms. These in their turn may be further explained by the sub-atomic structure of the atoms, together with laws governing the behaviour of the sub-atomic particles, and so on. All this may be expressed by the admittedly vague metaphor that the laws of nature state 'structural properties of the world'. (The metaphor is vague just because, at any level, it is not only the *structure* which explains, but also the laws; but it is permissible because, at any level, the laws are partly explained by structures, and also because it is at least conceivable that at some level, structure and law may become indistinguishable—that the laws *impose* a certain kind of structure upon the world, and that they may be interpreted, alternatively, as *descriptions* of that structure.[7] This

[7]Certain problems concerning laws of nature, and their dual character of being at the same time necessary in some sense (here alluded to by the word 'impose') and contingent (here alluded to by the word 'description') are dealt with in appendix *x of *L.Sc.D.* For the problem of explaining rigidity ('extension') and impenetrability, see also the 'Metaphysical Epilogue' in Volume III of the *Postscript*, sections 21 and 27, where some remarks may be found on the field theory of matter.

seems to be aimed at, if not yet achieved, by the field theories of matter.) So much for the idea of structure.

The second idea in need of elucidation is that of '*depth*'. It defies, I think, any attempt at an exhaustive logical analysis; it is, nevertheless, a guide to our intuitions. (This is so in mathematics: in the presence of the axioms, all its theorems are logically equivalent, and yet there is a great difference in 'depth' which is hardly susceptible to logical analysis.[8]) The 'depth' of a scientific theory seems to be most closely related to its simplicity and so to the wealth of its content. (It is otherwise with the depth of a mathematical theorem whose content may be taken to be nil.) Two ingredients seem to be required: a rich content, and a certain coherence or compactness (or 'organicity') of the state of affairs described. It is this latter ingredient which, although it is intuitively fairly clear, is so difficult to analyse, and which the essentialists were trying to describe when they spoke of essences, in contradistinction to mere accumulations of accidental properties. I do not think that we can do much more than refer here to an intuitive idea, nor that we need do much more. For in the case of any particular theory proposed, it is the wealth of its content, and thus its degree of testability, which decides its interest, and the results of actual tests which decide its fate. From the point of view of method, we may look upon its depth, its coherence, and even its beauty, as mere guides or stimuli to our intuition and to our imagination.

Nevertheless, there does seem to be something like a *sufficient condition* for depth, or for degrees of depth, which can be analyzed logically. I shall try to explain this with the help of an example from the history of science.

It is well known that Newton's dynamics achieved a unification of Galileo's terrestrial and Kepler's celestial physics. It is often said

[8]It has been suggested that, in mathematics, the depth of a theorem can be measured by assuming that it (a) increases with the number of steps—i.e., the length—of the shortest proof, and that it (b) decreases with the length of the theorem itself. This would relativize the idea of depth since a certain theorem t which in the formalized language L_1 is short and which needs a very long proof in L_1 may be a long formula in the language L_2, and may yet serve as an axiom of L_2. But intuitively there seems to be something absolute about depth—a quality that is clearly missed by the proposed measure, and which has something to do with (a) the depth of the idea or the character of the (simplest) proof (within the simplest system) rather than its length, and (b) the fertility or general applicability of this idea, as a method of proof, rather than the brevity of the theorem.

that Newton's dynamics can be induced from Galileo's and Kepler's laws, and it has even been asserted that it can be strictly deduced from them.[9] But this is not so: from a logical point of view, Newton's theory, strictly speaking, contradicts both Galileo's and Kepler's (although these latter theories can of course be obtained as approximations, once we have Newton's theory to work with). For this reason it is impossible to derive Newton's theory from either Galileo's or Kepler's or both, whether by deduction or induction. For neither a deductive nor an inductive inference can ever proceed from consistent premises to a conclusion that formally contradicts these premises.

I regard this as a very strong argument against inductivism.

I shall now briefly indicate the contradictions between Newton's theory and those of his predecessors. Galileo asserts that a thrown stone or a projectile moves in a parabola, except in the case of a free vertical fall when it moves, with constant acceleration, in a straight line. (We neglect air resistance throughout this discussion.) From the point of view of Newton's theory, both these assertions are false, for two distinct reasons. The first is false because the path of a long-range projectile, such as an inter-continental missile (thrown in an upward or horizontal direction) will be not even approximately parabolic: it will be elliptic. It becomes approximately a parabola only if the total distance of the flight of the projectile is negligible, compared with the radius of the earth. This point was made by Newton himself, in the *Principia* as well as in his popular-

[9]What can be deduced from Kepler's laws (see Max Born, *Natural Philosophy of Cause and Chance*, 1949, pp. 129–133) is that, for all planets, the acceleration towards the sun equals at any moment k/r^2, where r is the distance at that moment between the planet and the sun, and k a constant, the same for all planets. Yet this very result formally contradicts Newton's theory (except on the assumption that the masses of the planets are all equal or, if unequal, then infinitely small as compared with the mass of the sun). This fact follows from what is said here, in the text following footnote 11, about Kepler's third law. But in addition, it should be remembered that neither Kepler's nor Galileo's theories contain Newton's concept of *force*, which is traditionally introduced in these deductions without further ado; as if this ('occult') concept could be read off from the facts, instead of being the result of a new interpretation of the facts (that is, of the 'phenomena' described by Kepler's and Galileo's laws) in the light of a completely new theory. Only *after* the concept of force (and even the proportionality of gravitational and inertial mass) has been introduced is it at all possible to link the above formula for the acceleration with Newton's inverse square law of attraction (by an assumption like the one that the planets' masses are negligible).

ized version, *The System of the World*, where he illustrates it with the help of the following figure.[10]

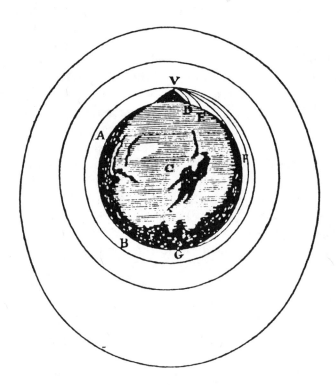

Newton's figure illustrates his statement that, if the velocity of the projectile is increased, and with it the distance of its flight, it will, 'at last, exceeding the limits of the earth, . . . pass into space without touching it'. And it will describe, approximately, a Keplerian ellipse throughout.

Thus a projectile on earth moves along an ellipse with finite eccentricity, rather than on a parabola. Of course, for sufficiently short throws, a parabola will be an excellent approximation; but the parabolic track is not strictly deducible from Newton's theory unless we add to the latter a factually *false* initial condition (and one which, incidentally, is unrealizable in Newton's theory since it leads

[10]See Newton's *Principia, the Scholium* at the end of section ii of Book i; p. 55 of the 1934 edition (Motte's translation revised by Cajori). The figure, from *The System of the World,* and the quotation given here, will be found on p. 551 of this edition.

to altogether absurd consequences) to the effect that the radius of the earth is infinite. If we do not admit this assumption, even though it is *known to be false*, then we always get an ellipse, in contradiction to Galileo's law according to which we should obtain a parabola.

A precisely analogous logical situation arises in connection with the second part of Galileo's law which asserts the existence of an acceleration *constant*. From the point of view of Newton's theory, the acceleration of free falling bodies is never constant: it always increases during the fall, owing to the fact that the body approaches nearer to the centre of attraction. This effect is very considerable if the body falls from a great height, although of course negligible if the height is negligible as compared with the radius of the earth. In this case, we can obtain Galileo's theory from Newton's if we again introduce the *false* assumption that the radius of the earth is infinite (or the height of the fall zero).

The contradictions which I have pointed out are far from negligible for long-distance missiles. To these we may apply Newton's theory (with corrections for air-resistance, of course) but not Galileo's: the latter simply leads to false results, as can be shown with the help of Newton's theory.

With respect to Kepler's laws, the situation is similar. It is obvious that, in Newton's theory, Kepler's laws are only approximately valid—that is, strictly invalid—if we take into account the mutual attraction between the planets.[11] But there are more fundamental contradictions between the two theories than this somewhat obvious one. For even if, as a concession to our opponents, we neglect the mutual attraction between the planets, Kepler's third law, considered from the point of view of Newton's dynamics, cannot be more than an approximation which is applicable to a very special case: to planets whose masses are all equal or, if unequal, negligible as compared with the mass of the sun. Since it does not even

[11]See, for example, P. Duhem, *The Aim and Structure of Physical Theory*, 1905; English translation by P. P. Wiener, 1945, Part ii, Ch. vi, section 4. Duhem says more explicitly what is implicit in Newton's own statement (*Principia*, book i, proposition lxv, theorem xxv); for Newton makes it quite clear that whenever more than two bodies interact, Kepler's first two laws will be at best only approximately valid, and even this in very special cases only, of which he analyses two in some detail. Incidentally, formula (1), below, follows immediately from book i, proposition lix, in view of book i, proposition xv. (See also book iii, proposition xv.) Thus my own analysis, like Duhem's, is implicit in Newton's.

approximately hold for two planets if one of them is very light while the other is very heavy, it is clear that Kepler's third law contradicts Newton's theory in precisely the same way as does Galileo's.

This can easily be shown as follows. Newton's theory yields for a two-body system—a binary star system—a law which astronomers often call 'Kepler's law' since it is closely related to Kepler's third law. This so-called 'Kepler's law' says that if m_0 is the mass of one of the two bodies—the sun, say—and if m_1 is the mass of the other body—a planet, say—then, choosing appropriate units of measurement, we can derive from Newton's theory

(1) $$a^3/T^2 = m_0 + m_1$$

where a is the mean distance between the two bodies, and T the time of a full revolution. Now Kepler's own third law asserts that

(2) $$a^3/T^2 = constant,$$

that is to say, the same constant for *all* planets of the solar system. It is clear that we obtain this law from (1) only under the assumption that $m_0 + m_1 =$ constant; and since $m_0 =$ constant for our solar system if we identify m_0 with the mass of the sun, we obtain (2) from (1), provided we assume that m_1 is the same for all planets; or, if this is factually *false* (as is indeed the case, since Jupiter is several thousand times larger than the smallest planets), that the masses of the planets are *all zero as compared with that of the sun*, so that we may put $m_1 = 0$, *for all planets*. This is quite a good approximation from the point of view of Newton's theory; but at the same time, putting $m_1 = 0$ is not only strictly speaking false, but unrealizable from the point of view of Newton's theory. (A body with zero mass would no longer obey Newton's laws of motion.) Thus, even if we forget all about the mutual attraction between the planets, Kepler's third law (2) formally contradicts Newton's theory which yields (1).

It is important to note that from Galileo's or Kepler's theories we do not obtain even the slightest hint of how these theories would have to be adjusted—what (false) premises would have to be adopted, or what conditions stipulated—in order to interpret these theories by another and more generally valid one such as Newton's. *Only after we possess Newton's theory can we find out whether, and in what sense, the older theories are approximations to it.* We may express this fact briefly by saying that, although from the point of view of Newton's theory, Galileo's and Kepler's are excellent approximations to certain special Newtonian results, Newton's theory cannot be said, from the point of view of the other two theories, to be an approximation to their results. All this shows that logic, whether deductive or inductive, cannot possibly make the step from these theories to Newton's dynamics.[12] Only ingenuity can make this step. Once it has been made, Galileo's and Kepler's results may be said to corroborate the new theory.

Here, however, I am not so much interested in the impossibility of induction as in *the problem of depth*. And regarding this problem, we can indeed learn something from our example. Newton's theory unifies Galileo's and Kepler's. But far from being a mere conjunction of these two theories—which play the part of *explicanda* for Newton's—*it corrects them while explaining them*. The original explanatory task was the deduction of the earlier results. Yet this task is discharged, not by deducing these earlier results but by deducing something better in their place: new results which, under the special conditions of the older results, come numerically very close to these older results and, at the same time, *correct them*. Thus the empirical success of the old theory may be said to corroborate the new theory; and in addition, the corrections may be tested in their turn—and perhaps refuted, or else corroborated. What is brought out strongly by the logical situation which I have sketched, is the fact that the new theory cannot possibly be *ad hoc* or circular. Far from repeating its *explicandum*, the new theory contradicts it and corrects it. In this way, even the evidence of the *explicandum* itself becomes independent evidence for the new theory. (Incidentally, this analysis allows us to *explain*—along lines similar to those of *L.Sc.D.*, section 57—*the value of metrical theories*, and of measurement; and it thus helps us to avoid the mistake of accepting measurement and precision as ultimate and irreducible values.)

I suggest that if in the empirical sciences a new theory of a higher level of universality successfully explains some older theory *by correcting it*, then this is a sure sign that the new theory has penetrated deeper than the old one. The demand that a new theory should

---

[12]The concepts of force (*cp.* footnote 9, above) and of action at a distance introduce further difficulties.

contain the old one approximately, for appropriate values of the parameters of the new theory, may be called (following Bohr) the *'principle of correspondence'*.

Fulfilment of this demand is a sufficient condition of depth, as I said before. That it is not a necessary condition may be seen from the fact that Maxwell's electromagnetic wave theory did not correct, in this sense, Fresnel's wave theory of light. It means an increase in depth, no doubt, but in a different sense: 'The old question of the direction of the vibrations of polarized light became pointless. The difficulties concerning the boundary conditions for the boundaries between two media were solved by the very foundations of the theory. No *ad hoc* hypotheses were needed any longer for eliminating longitudinal light waves. Light pressure, so important in the theory of radiation, and only lately determined experimentally, could be derived as one of the consequences of the theory.'[13] This brilliant passage, in which Einstein sketches some of the major achievements of Maxwell's theory and compares it with Fresnel's, may be taken as an indication that there are other sufficient conditions of depth which are not covered by my analysis.

The task of science, which, I have suggested, is to find satisfactory explanations, can hardly be understood if we are not realists. For a satisfactory explanation is one which is not *ad hoc*; and this idea—the idea of *independent evidence*—can hardly be understood without the idea of discovery, of progressing to deeper levels of explanation; without the idea, therefore, that there is something for us to discover; and something to discuss critically.

And yet it seems to me that within methodology we do not have to presuppose metaphysical realism. Nor can we derive any help from it, except of an intuitive kind. For once we have been told that the aim of science is to explain, and that the most satisfactory explanation will be the one that is most severely testable and most severely tested, we know all that we need to know as methodolo-

[13]A. Einstein, *Physikalische Zeitschrift* 10, 1909, pp. 817 *f*. The abandonment of the theory of a material ether (implicit in Maxwell's failure to construct a satisfactory material model of it) may be said to give depth, in the sense analyzed above, to Maxwell's theory as compared with Fresnel's; and this is, it seems to me, implicit in the quotation from Einstein's paper. Thus Maxwell's theory in Einstein's formulation is perhaps not really an example of *another* sense of 'depth'. But in Maxwell's own original form it is, I think.

gists. That the aim is realizable we cannot assert—neither with nor without the help of metaphysical realism, which can give us only some intuitive encouragement, some hope, but no assurance of any kind. And although a rational treatment of methodology may be said to depend upon an assumed, or conjectured, aim of science, it certainly does not depend upon the metaphysical and most likely false assumption that the true structural theory of the world (if any) can be stated in human language.

If the picture of the world which modern science draws comes anywhere near to the truth—in other words, if we have anything like 'scientific knowledge'—then the conditions obtaining almost everywhere in the universe make the discovery of structural laws of the kind we are seeking—and thus the attainment of 'scientific knowledge'—almost impossible. For almost all regions of the universe are filled by chaotic radiation, and almost all the rest by matter in a likewise chaotic state. In spite of this, science has been miraculously successful in proceeding toward what I think should be regarded as its aim. This strange fact cannot, I think, be explained without proving too much (*cp.* section 3, text to footnote 1). But it can encourage us to pursue that aim, even though we may not get any further encouragement to believe that we can actually attain it; neither from metaphysical realism nor from any other source.[14]

*Addendum 1980 to Section 15

(1)   The preceding section was, as noted above, first published in 1957. It contains, among other things, a refutation of the view, held

---

[14]Most of this section (15) was first published in *Ratio* **1**, 1957 (in both English and German editions). The material is republished here with the permission of the Editor, my friend the late Julius Kraft.

*[It has also now been reprinted in *Objective Knowledge* as Chapter 5. The idea discussed in this section that theories may *correct* an 'observational' or 'phenomenal' law which they are supposed to explain (such as, for example, Kepler's third law) was repeatedly expounded in my lectures from my New Zealand days on. One of these lectures stimulated the correction of a supposed phenomenal law (see the 1941 paper referred to in my *Poverty of Historicism*, 1957, 1960, footnote on pp. 134*f.*). Another of these lectures was published in Simon Moser's volume *Gesetz und Wirklichkeit* (1948), 1949 (see especially pp. 57*f.*), and reprinted in Hans Albert: *Theorie und Realität*, 1964, (see especially p. 100), an English translation of which is published as an Appendix, 'The Bucket and the Searchlight: Two Theories

by such men as Isaac Newton and Max Born, that Newton's theory can be derived from Kepler's laws, either by an inductive or by a deductive argument.

When I first wrote this section, I did not lay much stress upon the refutation of the historical myth that Newton's theory is the result of induction, because I thought that I had destroyed the theory of induction twenty years earlier; and I was enough of an optimist to believe that all the resistance still emanating from the defenders of induction would soon disappear. (Nevertheless, I did criticize in some detail Carnap's then current theory of probabilistic induction, with the result that he eventually gave it up; and the last form in which he defended induction was completely different from the famous theory which he developed in his large but, in my opinion, untenable book, *Logical Foundations of Probability*.)

Since then, inductivists have taken some heart; partly because I have no longer replied to their arguments, which were all clearly refuted in various parts of my earlier writings. I no longer replied to them because I thought, and still think, that the issue was long settled and therefore boring.

(2) Nevertheless, it may be a good thing to repeat here, very briefly, one of the more interesting arguments against induction, an argument which is implicit in the preceding section.

By induction I mean an argument which, given some empirical (singular or particular) premises, leads to a universal conclusion, a universal theory, either with logical certainty, or with 'probability' (in the sense in which this term is used in the calculus of probability).

The argument against induction that I wish to restate here is very simple:

Many theories, such as Newton's, which have been thought to be the result of induction, actually are *inconsistent* with their alleged (partial) inductive premises, as shown above.

But if this is so, then induction, in any important sense, collapses. So much about non-probabilistic induction.

of Knowledge', to *Objective Knowledge*. The same idea of mine was also the 'starting-point' (as he puts it on p. 92) of P. K. Feyerabend's paper 'Explanation, Reduction and Empiricism' (in Herbert Feigl and Grover Maxwell, editors, *Minnesota Studies in the Philosophy of Science* 3, 1962) whose reference [66] is to the present section (as first published in *Ratio*). Feyerabend's acknowledgement seems to have been overlooked by the authors of various papers on related subjects.]

As for a probabilistic inductive argument: according to the probability calculus, if we are given a number of consistent inductive premises, then any inferred conclusion which is inconsistent with them, can have, relative to these premises, only zero probability.

(3)   Newton's theory was no doubt greatly indebted to Galileo's and to Kepler's theories; so much so that these were regarded by Newton himself as (partial) inductive premises of his theory.

Galileo's theory of free falling bodies contained a *constant*, g, the constant of acceleration. From Newton's theory it follows that g is not a constant but a variable, dependent (a) upon the mass of the attracting body (in Galileo's case the Earth) and (b) upon the square of the distance from the center of mass.

Therefore, Galileo's theory is inconsistent with Newton's.

Of course, under the assumption that we regard only those free falling bodies that are close to the surface of the Earth, so that they all have very nearly the same distance from the center of the Earth, *we can explain why g appears (mistakenly) to be a constant.*

The situation with Kepler's laws is closely similar.

For each system of *two* bodies of which one is very heavy and the other negligibly light we can derive the three laws of Kepler from Newton's theory and thereby explain them. But since Kepler formulated his laws for a many-body system consisting of the sun and several planets, it is, from the point of view of Newton's theory, invalid. It could therefore form neither a complete system nor a partial system of premises (inductive or deductive) of Newton's theory.

So much for an inductive or deductive derivation of Newton's theory from Kepler's theory or from Galileo's theory.

(4)   Of course, it was a decisively important success of Newton's theory that it could explain Galileo's and Kepler's theories: that is to say, that these theories could be deduced from Newton's under certain simplifying (and, strictly speaking, false) assumptions.

But since the alleged inductivist premises are, strictly speaking, inconsistent with the alleged inductivist conclusions, it is most misleading to speak in this case of an inductive inference, or of an inductive probabilistic relationship.

(5)   This kind of situation is typical of the history of science. The relationship between Newton's theory of gravitation and that of

Einstein is another very similar and important example of such a case.

(6) To my knowledge, no serious answer has been given to this argument, so far, especially not by the defenders of any of the current probabilistic theories of induction.

16. *Difficulties of Metaphysical Realism. By a Metaphysical Realist.*

It would be wrong to leave the topic of metaphysical realism without at least alluding to the difficulties of this position. These difficulties are grave. To me they seem to pose insoluble problems. And yet, they are of such a character that they do not in the least affect my faith in realism. They are on a different plane, as it were, from the problems and arguments by which I can support my faith in realism. It is a less rational plane, perhaps—one on which arguments become somewhat vague and less manageable.

Newton was led, by his theory of action at a distance, to the belief that space was the sensorium of God. The argument is somewhat fantastic, no doubt; but there is more to it than meets the eye. For the difficulty is very real. Distances in the universe are tremendous. Action at a distance would mean that gravitational effects were, like the Deity, omnipresent in the whole world. Newton, like Einstein, felt unable to accept action at a distance as a property of the mechanics of nature. He felt its mystery and attributed it to God.

Einstein solved this problem, or so it seems, by his theory which makes gravitational disturbances spread with the velocity of light. This solution is highly satisfactory, especially from the point of view of our discussion in the last section: it indicates a possible unification of the theories of light and gravity, and it does so by interpreting light, and gravitational disturbances, in terms of structural properties—field properties—of the universe, of our world.

And yet, we are still faced with Newton's problem. For what about these structural properties of our world themselves? They are, we believe, the same everywhere and at all times. How are we to understand this?

When we speak of structural properties of our world, we speak of the world, metaphorically at least, as if it were a thing, such as a crystal, or a balloon, or perhaps a machine. But according to present physical theory, the structural properties of a crystal, or a balloon or a machine, are due to interactions between its parts; and these interactions involve finite velocities, up to and including the velocity of light. Interactions keep the crystal in shape; they determine the gas pressure in the balloon; they keep the machine together.

But the structural properties of the world which we describe by laws of nature cannot be thus understood. They cannot, it seems, be explained as due to interaction, since they are the basis of all interaction. They are—this is their deepest characteristic—the same throughout the world, at any place and at any time: they are *omnipresent*. And so Newton's problem turns up again.

Moreover, the structures of the various things which we have mentioned can be understood, or explained, in terms of laws. But the structure of the world is different in that it is what the law describes rather than explains. This difficulty is, however, less serious. For we may explain the laws—and thus the structure of the world—by deeper laws: here it is a help to have given up the essentialist theory of ultimate explanation. Moreover, this is just the place to remind ourselves that we must not become victims of our own metaphor; that 'the structure of our world' was only a metaphor, designed to help us to envisage what laws describe; that the metaphor was bound to break down somewhere; and that it is quite satisfactory to have found the place where it does break down.

All this should be kept in mind. But it does not resolve Newton's difficulty. We must, I believe, accept the existence of laws of nature; but we must do so, I fear, as a mystery which has become perhaps even more impenetrable since Einstein; for the laws of nature themselves which postulate, according to Einstein, that no effect can spread with super-luminar velocity make it impossible to understand the omnipresent structural homogeneity of the world.

The theory of the expanding universe may be a help here; but not if it is supposed that, within a second (or a fraction of a second) from the zero moment, the radius of the world was of the order of many light years. (That one could not speak of 'thermal equilibrium' in this case seems, at least to me, quite obvious.)

It has often been said that—to use Wittgenstein's words—'Not *how* the world is, is the mystical, but *that* it is'.[1] Yet our discussion

---

[1] L. Wittgenstein, *Tractatus Logico-Philosophicus*, 6.44.

shows that *how* the world is—that it has a structure, or that its vastly distant regions are all subject to the same structural laws—seems to be inexplicable in principle and thus 'mystical', if we wish to use this term. This, at any rate, seems to be the predicament in which the realist finds himself. The idealist may have a way out of this—an explanation through which he may reduce this mystery to that of the sheer existence of the world. For he may say, with Kant, that our intellect imposes its laws upon nature; or in Wittgenstein's words, that 'only *law-like* connections are *thinkable*'.[2] Although the realist may perhaps agree, at least in part, with these views, they do not help him in the least to explain or understand why, if there was to be a world, it had to be a thinkable world, regulated by law—a world understandable to some intellect; a world inhabitable by life.

Some further illustration of the difficulty may perhaps be helpful. We may adopt the terminology of Comte and Mill who distinguish between two kinds of laws or regularities—regularities of succession and regularities of co-existence. Laws of succession are those 'causal' laws in which *time* plays an essential part, for example laws determining changes (such as accelerations) of the state of a system. Laws of co-existence are, for example, the laws which describe the anatomical or *structural* regularities of an animal, or of a molecule, or of an atom.

Now the structural laws of co-existence of animals, or molecules, or even of atoms *may* in principle be reduced to 'causal' laws—those causal laws in accordance with which these structures are produced, and keep for a time (relatively) stable. We seem to be able to understand, at least in principle, the conditions of the stability of a molecule, for example, with the help of the theory of resonance; that is, of a causal interaction of the constituent parts. But we do not understand as yet such structural laws, or laws of co-existence, as the absolute constancy of the electronic charge or mass; or more generally, the absolute qualitative and quantitative identity of the properties of the elementary particles. These cannot be understood, it seems, as due to interaction: according to general relativity there can be no equalizing interaction between simultaneous electronic charges whose distance may be measured in light years.

[2]*Loc. cit.*, 6.361.

Still, one day we might perhaps derive, and explain, the equilibrium conditions of electrons and of other elementary particles in a manner which apparently is not altogether different from the way we explain the equilibrium conditions of molecules; that is to say, in terms of causal 'laws of succession'—say, of field equations which determine a spectrum of discrete solutions. But would this mean that we have transcended the dualism between laws of succession and laws of co-existence by reducing the latter to the former? By no means; for instead of worrying about the inexplicability of the fact—the structural law—that all electrons in our vast world appear to have, *without causally influencing one another*, absolutely the same charge, we ought now to worry about the fact that all parts of our vast world are governed by the same laws determining the identity of electronic charges. And this structural fact—which is clearly the same as before, though differently expressed—seems quite obviously beyond any hope of being ever causally explained since any causal explanation would have to be in terms of laws just like those whose universal validity we should like to explain, and to understand.

Thus the assertion of the universality or constancy of our causal laws through space (and time) amounts to asserting a structural regularity—a regularity of co-existence which seems to be in principle causally inexplicable since it cannot be explained by any causal law, or law of succession. Indeed, if we speak of a law of succession, we intend to say that all successions, including all co-existing successions, exhibit the same regularity, the same structure. So we see once more that the structural homogeneity of the world seems to resist any 'deeper' explanation: it remains a mystery.

I do not think that this mystery can be solved by thinking that the world is what it is by a kind of logical necessity. The hope of reducing natural science to logic seems to me both absurd and repulsive. Nor do I think that the mystery can be solved by idealism. Idealism in its various forms, and especially conventionalism and instrumentalism, all offer solutions; but these solutions seem clearly not true, and it is better to face a mystery than to try to escape from it by false solutions, especially if they are cheap.

Take Kant's ingenious solution—that our intellect does not read the laws in nature's open book, but imposes its own laws upon nature. This is true, up to a point; our theories are of our own making, and we cannot ever describe empirical facts (or otherwise react to them) without interpreting these facts in terms of our theories (or of our, perhaps unconscious, expectations). But this does not mean, as Kant believed, that laws of nature, such as Newton's theory, are *a priori* valid, and irrefutable, even though it is true that we impose them on those very empirical facts to which we would have to appeal for a refutation. On the contrary, we have learned from Einstein that our intellect may form, at least tentatively, alternative theories; that it may re-interpret the facts alternatively in terms of each of these new theories; that, in the competition of these theories, we can decide freely, sounding their depth, and weighing the result of our criticism, including our tests; and that only in this way can we hope to get nearer to the truth.

Kant's epistemology is also refuted, I believe, by the very fact that it is, but *only within narrow limits*, highly successful. Kant believed that we are furnished with a mental apparatus, with a psychological and perhaps physiological digestive system, which allows us to digest the stimuli reaching our senses from the external world, and that, by digesting, assimilating, and absorbing them, we impose its structural characteristics upon them; that is, it is the imprint of our mind upon them which gives rise to *a priori* valid truths about the world. Now it is undeniable that there are such genetically *a priori* truths. But they are strangely unimportant. For we are *not* compelled to interpret the world of things in their terms; on the contrary, we easily become aware of their subjective character, and treat them for what they are. Excellent examples of this Kantian mechanism are provided by certain regular and inescapable optical illusions. Another is furnished by the order of the colours; that is to say by the fact that we experience red to be more similar to orange, yellow, purple, and blue, than to green; yellow more similar to green, and to red, than to blue; etc. . . . These are *a priori* truths: it may be an empirical fact that we do perceive colours—that we are not colour-blind; it may also be an empirical fact that our mechanism of colour perception is based upon a red-green and a yellow-blue component rather than on a red-blue and a yellow-green one (in this sentence I am using the colour names for wave lengths rather than for perceptions). But once we see the colours with the help of this mechanism, they are necessarily and intrinsically ordered by these relations of similarity and dissimilarity.

The explanation of all this is no doubt the Kantian psycho-physiological digestive mechanism with which we are endowed. Our physiology explains these similarities. We see with the help of a red-green and of a yellow-blue component; and if one of them is missing, we are colour-blind—either red-green or yellow-blue colour-blind. Red cannot, physiologically, blend with green, nor yellow with blue. Thus they become opposites. The other colours can blend, and one can by shades turn into another, because the two components can be stimulated independently at the same time.

Kant was right about this. But the point turns out to be comparatively trifling. Nobody is misled by these *a priori* valid relations between colours into imposing the corresponding laws upon nature, upon the world of coloured things. We neither believe that red *things* have, physically considered, a greater affinity to yellow or to blue things than they have to green things; nor do we think that there are laws of nature to be discovered here—except, of course, laws of our own psychology and physiology. But, according to Kant, these laws of our own digestive mental apparatus are imposed by us upon the world, in the sense that they are bound to become 'objective' laws of things which we perceive.

Optical illusions belong to the same category. And it might be conjectured that our belief in real *things* is similarly physiologically founded. But in this case, the physiological mechanisms, and the beliefs which spring from them (both are, we may conjecture, the results of a long evolution and adaptation) seem to withstand criticism, and to win in competition with alternative theories. And when they do mislead us, as in a cinema—especially in cartoons—they do not lead adults to assert seriously that we have before us a world of things. Thus we are not (as Kant and also Hume thought) the victims of our 'human nature' or of our mental digestive apparatus, of our psychology or physiology. We are not for ever the prisoners of our minds. We can learn to criticize ourselves, and so to transcend ourselves. We do have our limitations; but we are freer than Kant thought.

Similar considerations apply to other forms of human bondage, such as the tendency to accept the valuations, beliefs, and dogmas of our social group. This tendency is strong and may also have a physiological basis. But we can break away from it. To do so we may need at first, perhaps, the stimulus of culture clash or of conversion. But later we may develop the habit of searching, critical, rational discussion. Rational discussion and critical thinking are not like the more primitive systems of interpreting the world; they are not a framework to which we are bound and tied. On the contrary, they are the means of breaking out of the prison—of liberating ourselves.

Hume taught that fundamental human beliefs were no more than irrationally acquired habits which men cannot transcend but are bound to obey. Kant, in a sense, accepted this pessimistic idea although his 'Copernican revolution' gave it an optimistic turn. He taught that since the objective world of experience was a world formed by our experiencing intellect (which played its role even on the level of perception), our beliefs about this world could be objectively true and rational: though they might be described as habits, they were not only habits, since our reason played its part in their formation just as much as our senses and our associations. Any further development of this line of thought must take note of the fact that the *growth of knowledge* consists fundamentally in the critical revision of our beliefs; a fact that establishes that we are not bound to our fundamental beliefs to the extent envisaged by Hume and Kant. It also establishes that both Hume and Kant were partly right. Hume was right in being sceptical about the validity of our beliefs: this is shown by the fact that we have transcended some of them (which suggests the possibility that we may transcend others). But great critic as he was, he was wrong in overlooking, precisely, the significance of our ability to transcend our beliefs through criticizing them. And Kant was right in pointing out, against Hume, that reasoning was involved in the formation of our beliefs—even of our habitual beliefs. He was right, moreover, in pointing out the significance of the growth of our knowledge, and in teaching that any growth of knowledge needs a theoretical framework which must precede the growth. But he was wrong in believing that this framework could not possibly be transcended in its turn, and that it was, therefore, *a priori valid*. To continue our story we may perhaps say that Hegel was right in pointing out (if we may so interpret his obscure teaching) that the framework, too, was subject to growth, and could be transcended. But he was wrong in suggesting (if again we may so interpret his teaching) that truth is essentially relative to some framework, and that it is not *our* active criticism, *our* discovery of a contradiction, of a refutation, which forces us to change our ideas or our beliefs, but that these ideas transcend themselves, so that we are dependent upon the evolving ideas, rather than these upon us, upon our rational criticism. This makes our criticism dependent upon the historically inherited framework, and thus leads, again, to relativism—to historical relativism.

This philosophy of human bondage has exerted a strange fascination upon the post-Kantian theory of knowledge. (It has played a major part in the decline of rationalism and liberalism.[3]) Whatever Kant may have done to correct Hume was undone by the doctrine that our habits of belief—our valuations, our attitudes, our dogmas, and thus our world of experience—depend upon our historical period or our social group (Hegel, Marx). This doctrine which has become most fashionable in the hands of modern social science is, of course, true as long as we do not attribute to it any particularly deep epistemological significance. It is true that we are dependent upon our upbringing, our beliefs, our knowledge, our expectations. But it is also true that we are not completely dependent upon them. No doubt we can only slowly and partially liberate ourselves from this bondage. But there is no natural limit to this process of liberation, to the growth of knowledge. It is of course possible to deny that we can ever break through our intellectual fetters: it is possible to assert that we deceive ourselves if we believe that we are less bound by the framework of our prejudices than people two thousand years ago— that the framework, the fashion, has changed, but not its power over us. Although this fashionable view of the matter may be irrefutable, it is simply untrue. Since the Renaissance, there has been a most striking increase in the critical attitude.

A special form of this philosophy of human bondage is linguistic relativism, a view which has been most forcefully presented by Benjamin Lee Whorf.[4] In our present context it may be formulated as the view that our human languages may incorporate (or fail to incorporate) in their structures beliefs, theories, and expectations to

[3]This topic is developed, though mainly for pre-Kantian philosophy, in my lecture 'On the Sources of Knowledge and of Ignorance', *Proceedings of the British Academy*, **46**, 1960, pp. 39–71; reprinted in *Conjectures and Refutations*, pp. 3–30.
[4]B. L. Whorf: *Language, Thought and Reality*, 1956.

such an extent that we cannot break out of these ideological fetters by criticism, since criticism must always make use of language. In this formulation, the use of the plural, 'languages', may indicate how to break out of these bonds: they are not as strong as one might think, for it may be possible for men to free each other by criticizing one language, or system of beliefs, in another (culture clash). There is no reason whatever to think, as some people do, that Whorf, or anybody else, has shown the incommensurability of sets of beliefs (or that all assertions are relative to irreducibly different sets of fundamental beliefs). Others have been led by Whorf's fascinating analyses to think that, as all languages have something in common, they will have a common set of beliefs which must be undetectable by that method of mutual criticism which is essentially dependent upon linguistic divergence. Admittedly, this is true up to a point; and it can be more simply expressed by saying that we shall always harbour prejudices of which we are unaware. But this does not mean that we cannot detect, at times, by some method or other (or by no method at all), some of our prejudices, and get rid of them through criticism. Nor does it mean that this perhaps slow but unending process of intellectual liberation cannot be sped up by the practice of critical thinking and of rational discussion.

Rational discussion must not be practised, however, as a mere game to while away our time. It cannot exist without real problems, without the search for objective truth, without a task of discovery which we set ourselves: without a reality to be discovered—a reality to be explained by *structural universal laws*.

Thus we are back to our problem, and to Newton's problem. Idealism offers an easy way out, but even in Kant's form it hardly offers a convincing solution. At any rate, we realists have to live with the difficulty. But we should face it.[5]

---

[5]When I wrote this sentence, I felt convinced that 'Newton's problem', as I called it here, was insoluble—or that it had, at best, an unsatisfactory religious solution, somewhat on Newton's own lines. I personally had no hope of solving it, and no intention even of tackling it. It therefore came as a complete surprise to me when later, in the course of an attempt to re-interpret general relativity in the sense of indeterminism, I stumbled upon what looked like a solution of a part—a tiny part—of Newton's problem. As a consequence, I am no longer convinced of its insolubility. (See also *The Open Universe*, Vol. II of this *Postscript*, section 19; and *Quantum Theory and the Schism in Physics*, Vol. III of the *Postscript*, section 27.)

With this I conclude the metaphysical discussion which I began in section 7, encouraged by a passage of Einstein's which I selected there as my motto; and with it, I also conclude the discussion of the 'fourth phase'—the metaphysical phase—of the problem of induction.

CHAPTER II

DEMARCATION

17. *The Significance of the Problem of Demarcation.*
After presenting again my solution of the problem of induction, I have tried to follow its ramifications into its metaphysical stage, as I call it—far beyond its original scope. Yet my exploration of the ramifications of the problem of induction would be incomplete were I to neglect *the problem of the demarcation between science and metaphysics.* Indeed, there is a question which is almost always put to me as soon as people realize that I really do not believe in induction, and that I do not even believe induction to play a significant part in the sciences. It is this: if you abandon induction, *how can you distinguish the theories of the empirical sciences from pseudo-scientific or non-scientific or metaphysical speculations?*

This is the problem of demarcation. It is to be solved, I suggest, by accepting testability, or refutability, or falsifiability, as the distinguishing characteristic of scientific theories. From the formulation given, it is hardly possible to gauge its significance. At first sight, it may even look more like a pedant's question than like a problem of real interest. For what is in a name, or in a distinction, or in a classification, or in a demarcation? If we are anxious to know, if it is our aim to learn about the world, we do not care much for the compartments or departments to which our prospective knowledge may have to be assigned. As I said in the Introduction, subject matters and other divisions of learning are fictitious and badly misleading, convenient though they may be as administrative units. As far as science and metaphysics are concerned, I certainly do not believe in anything like a sharp demarcation. Science has at all times been profoundly influenced by metaphysical ideas; certain metaphysical ideas and problems (such as the problem of change, or the Cartesian programme of explaining all change by action at vanishing distances) have dominated the development of science for centuries, as regulative ideas; while others (such as atomism, another attempt to solve the problem of change[1]) have by degrees turned into scientific theories. Of course, there have been developments in the opposite direction too: as some positivists are fond of saying, a considerable number of metaphysical doctrines can be shown to be the echoes of obsolete doctrines of science.

We can illustrate this with the help of the history of positivism itself. Mach's own positivism and phenomenalism may be said to have been, originally, a respectable scientific theory, designed to explain the lack of success of atomism, and of other theories of the structure of matter, by the hypothesis that there simply was no physical entity such as matter or 'substance'. Mach could point to the success of phenomenalist physics—especially of phenomenalist thermodynamics—and to the fundamental logical difficulties in the way of Boltzmann's attempts to explain the second law in terms of an atomic or molecular structure. Mach's proposed solution implied that these problems, and all others pertaining to 'substance' or 'matter', were pseudo-problems; including, of course, all problems concerning the 'structure of matter'. But owing to Einstein's work of 1905, on Brownian movement, the full physical significance of Maxwell's and Boltzmann's theories was established. Brownian movement achieved, through Einstein's interpretation of it, the status of a crucial experiment. And, as Einstein himself pointed out, the existence of Brownian movement refuted the phenomenalist version of the second law of thermodynamics.[2] With this, the problem of the atomic structure of matter was shown to be a genuine physical problem. Thus after 1905, Machian positivism and phenomenalism became increasingly metaphysical, in one of the positivists' favourite senses: it became a piece of obsolete physics which scientists *qua* scientists had abandoned, but which continued

[1] See section 23 below (text before footnote 1).

[2] This is thus another case of a theory correcting its own observational basis; see section 15 above, especially the text after footnote 11, and also my paper, 'Irreversibility, or Entropy since 1905', *Brit. Journ. Philos. Science* 8, 1957. [See also Popper: 'The Arrow of Time', *Nature*, March 17, 1956, p. 538; 'Irreversibility and Mechanics', *Nature*, August 18, 1956, pp. 381–2; 'Irreversible Processes in Physical Theory', *Nature*, June 22, 1957, pp. 1296–7; 'Irreversible Processes in Physical Theory', *Nature*, February 8, 1958, pp. 402–3; 'Time's Arrow and Entropy', *Nature*, July 17, 1965, pp. 233–4; 'Time's Arrow and Feeding on Negentropy', *Nature*, January 21, 1967, p. 320; 'Structural Information and the Arrow of Time', *Nature*, April 15, 1967, p. 322; and *Unended Quest, op. cit.*, section 35. Ed.]

to linger on among philosophers, and among scientists when they turned into philosophers—or into apologists, as they sometimes do when their theories run into trouble. (See also section *113 below, i.e., section 21 of *Quantum Theory and the Schism in Physics*, Vol. III of the *Postscript.*)

As these examples show, there cannot be any sharp demarcation between science and metaphysics; and the significance of the demarcation, if any, should not be overrated. In spite of this, I contend that the problem of demarcation is highly significant. It is so, not because there is any intrinsic merit in classifying theories, but because a number of genuine and important problems are closely linked with it; in fact, all the main problems of the logic of science.

At the beginning of this section, I alluded to one of these links: to the view that the inductive method provides us with a criterion of demarcation. Another one mentioned before—the problem of the arguability, and thus of the rationality, of scientific hypotheses—is, of course, linked with the problem of their testability. We may look upon testability as a certain kind of arguability: arguability by means of *empirical* arguments, arguments appealing to observation and experiment. A third link with the problem of induction is shown by the way in which I distinguished the fourth or metaphysical stage of the problem of induction from its three logical and methodological stages. In its fourth stage—that is, as the problem of whether true natural laws exist—the problem assumed a character strikingly different from that of the previous stages, and this difference urgently needed elucidation. For this elucidation, the existential character of the problem provided us with the clue: purely existential statements are empirically irrefutable. If they are to be argued at all, we must always keep their empirical irrefutability in mind. The fact that metaphysical statements and problems may nevertheless be arguable (even though inconclusively), I have tried to establish by the simple device of arguing about them.

The problem of demarcation is also, of course, closely related, historically as well as logically, to what I called, at the beginning of section 2, the central problem of the philosophy of knowledge. For the problem of how to adjudicate or decide among competing theories or beliefs leads, as I said there, to the problem of deciding whether it is possible or impossible to justify a theory rationally; and this, in its turn, leads to the problem of distinguishing between, or of demarcating, rational theories and irrational beliefs; a problem that is often identified (perhaps a little rashly) with the problem of distinguishing between, or demarcating, empirical or 'scientific' theories from 'metaphysical' ones.

Thus the problem of demarcation is more than a question of classifying theories in order to be able to call them either 'scientific' or 'metaphysical'. Indeed, it provides an access to some of the most fundamental problems of the theory of knowledge, and thus of philosophy.

But the problem of demarcation is also of considerable practical importance. I stumbled upon this problem, and upon its solution, several years before I had become interested in the problem of induction, and before I had perceived these links between the problems of induction and demarcation to which I have just referred. This was in 1919, when I became suspicious of various psychological and political theories which claimed the status of empirical sciences, especially Freud's 'psychoanalysis', Adler's 'individual psychology', and Marx's 'materialist interpretation of history'.[3] All these theories were argued in an *uncritical* manner, it appeared to me. A great number of arguments were marshalled in their support. But criticism and counter arguments were regarded as hostile, as symptoms of a wilful refusal to admit the manifest truth; and they were therefore met with hostility rather than with arguments.

What I found so striking about these theories, and so dangerous, was the claim that they were 'verified' or 'confirmed' by an incessant stream of observational evidence. And indeed, once your eyes were opened, you could see verifying instances everywhere. A Marxist could not look at a newspaper without finding verifying evidence of the class struggle on every page, from the leaders to the advertisements; and he also would find it, especially, in what the paper failed to say. And a psychoanalyst, whether Freudian or Adlerian, assuredly would tell you that he finds his theories daily, even hourly, verified by his clinical observations.

But were these theories testable? Were these analyses really better tested than, say, the frequently 'verified' horoscopes of the astrologers? What conceivable event would falsify them in the eyes of their adherents? Was not every conceivable event a 'verification'? It was

[3]I have told this story from a somewhat different angle in my paper, 'Philosophy of Science: A Personal Report'. See also *Unended Quest*, section 8.

precisely this fact—that they always fitted, that they were always 'verified'—which impressed their adherents. It began to dawn on me that this apparent strength was in fact a weakness, and that all these 'verifications' were too cheap to count as arguments.

The *method of looking for verifications* seemed to me unsound—indeed, it seemed to me to be the typical method of a pseudo-science. I realized the need for distinguishing this method as clearly as possible from that other method—the method of testing a theory as severely as we can—that is, the method of criticism, the *method of looking for falsifying instances.*

The method of looking for verifications was not only uncritical: it also furthered an uncritical attitude in both expositor and reader. It thus threatened to destroy the attitude of rationality, of critical argument.

Freud was by far the most lucid and persuasive of the expositors of the theories to which I am referring. But what was his method of argument? He gave examples; he analyzed them, and showed that they fitted his theory, or that his theory might be described as a generalization of the cases analyzed. He sometimes appealed to his readers to postpone their criticism, and he indicated that he would answer all reasonable criticism on a later occasion. But when I looked a little more closely at a number of important cases, I found that the answers never came. Yet strangely enough, many readers were satisfied.

In order to show that these are not mere assertions or empty accusations I will substantiate them in some detail by an analysis of Freud's discussion of the fundamental thesis of his great book, *The Interpretation of Dreams*, rightly considered by him and others his most important work. *Was his approach critical?*

18. *A Case of Verificationism.*

> If an otherwise highly intelligent patient rejects a suggestion on not too intelligent grounds, then his imperfect logic is evidence for the existence of a . . . strong motive for his rejection.
>
> SIGMUND FREUD

The purpose of this section is to show, by analyzing a famous case, that the problem of demarcation is not merely one of classifying theories into scientific and non-scientific ones, but that its solution is urgently needed for a critical appraisal of scientific theories, or allegedly scientific theories. I have selected for this purpose Freud's great work, *The Interpretation of Dreams,* for two reasons. First, because my attempts to analyze its arguments played a considerable part in the development of my views on demarcation.[1] Secondly, because, in spite of severe shortcomings, some of which I shall try to expose here, it contains, beyond any reasonable doubt, a great discovery. I at least feel convinced that there is a world of the unconscious, and that Freud's analyses of dreams given in his book are fundamentally correct, though no doubt incomplete (as Freud himself makes clear) and, necessarily, somewhat lopsided. I say 'necessarily' because even 'pure' observation is never neutral—it is necessarily the result of interpretation. (Observations are always collected, ordered, deciphered, weighed, in the light of our theories. Partly for this reason, our observations tend to support our theories. This support is of little or no value unless we consciously adopt a critical attitude and look out for refutations of our theories rather than for 'verifications'.) What holds even for the most detached observations will also hold for the interpretation of dreams.

What I propose to do in this section is to analyze Freud's way of arguing in support of his central thesis in *The Interpretation of Dreams.*

Freud's main aim in this book is that of 'proving that, in their essential nature, dreams represent fulfilments of wishes'.[2] Freud is,

[1]Another theory which played a similar part (see 'Philosophy of Science: A Personal Report') was Marxism (see *Unended Quest,* section 8); but while I have discussed Marxism in great detail in my *Open Society,* and historicism in general in my *Poverty of Historicism,* I have not previously published any detailed analysis of Freud's method of dealing with falsifying instances and critical suggestions.

[2]*Cf.* Sigmund Freud, *The Interpretation of Dreams* (first published in 1899), translated and edited by James Strachey, 1954, p. 127. See also pp. 119, 121. In what follows I translate, in one or two places, directly from Freud's *Gesammelte Schriften,* ii and iii, 1925. I may say here that an analysis of the *Introductory Lectures* (1916–1917) would have led to the same results. (*Cf.* especially the fourteenth lecture.) For a psycho-analytic criticism and re-establishment of Freud's main thesis, see J. O. Wisdom, 'A Hypothesis to Explain Trauma-Re-Enactment Dreams', *Intern. J. of Psycho-Anal.* **30**, 1949, pp. 13 *ff.* The passages quoted there from Freud complement those quoted here. Compare especially Wisdom's reference on pp. 13 and 15 (notes 2 and 8) to Freud's *New Introductory Lectures,* 1937, pp. 43–4, where Freud introduced, in order to explain trauma-re-

of course, aware that there is a most obvious objection to this theory—the existence of nightmares and of *anxiety dreams;* yet he rejects this objection. 'It does in fact look', Freud writes in a passage in which he formulates what is to be our main problem here, 'as though anxiety-dreams make it impossible to assert the general proposition (based on the examples quoted in my last chapter) that dreams are wish-fulfilments; indeed they seem to stamp any such proposition as an absurdity. Nevertheless there is *no great difficulty* in meeting this objection.'[3] The method of meeting this objection, he explains,[4] consists in showing that, what in its *appearance* (in its *'manifest* content') seems to be an anxiety dream, is in *reality* (in its *'latent* content') a wish-fulfilment. This leads Freud to a very slight 'modification' of his main thesis concerning 'the essential nature of dreams', which he formulates as follows: *'a dream is a (disguised) fulfilment of a (suppressed or repressed) wish.'*[5]

Freud repeatedly re-affirms his programme of revealing the latent content of every anxiety dream as a wish-fulfilment. Thus the programme is re-affirmed, for example, on p. 550, and even more fully on p. 557 where we read: 'Thus there is *no difficulty* in seeing that unpleasurable dreams and anxiety-dreams are just as much wish-fulfilments in the sense of our theory as are straightforward dreams of satisfaction.'[6] *Yet Freud never carries out his programme; and in the end he gives it up altogether*—without, however, explicitly saying so. The evidence for this assertion is as follows.

Freud begins early in his book (on p. 157) to discuss 'the very

enactment dreams, the concept of *'attempted'* wish fulfilment; 'but he did not regard this', Wisdom writes on p. 15, 'as saying anything essentially new'. Compare also Wisdom's reference on p. 14 (notes 4 and 5) to Freud's explanation of 'painful dreams' by the wish-fulfilment of *punishment wishes,* as indicated in the *Introductory Lectures,* 1943, p. 185, and *Beyond the Pleasure Principle,* 1922, p. 38. See also note 21 below.

[3]*Op. cit.,* p. 135. (The italics are mine.)

[4]*Loc. cit.*

[5]*Op. cit.,* p. 160. Freud's main thesis is closely connected with another fundamental one (*cp.* pp. 123 and 233 *ff.*): that it is 'the function' of a dream, or at any rate, its 'normal' function, to be a *guardian* of sleep against disturbances; even though, at times, it may also have 'to appear in the role of a *disturber* of sleep' (p. 580).

[6]*Op. cit.,* p. 557. The italics are mine. Compare this quotation with the text to my footnote 3, above. Yet Freud had spoken of anxiety dreams in very different and less confident terms before this passage; for example, on pp. 161 and 236 (see below).

frequent dreams which appear to stand in contradiction to my theory'[7]; and very soon we get an inkling that the programme of reducing anxiety dreams to wish-fulfilment dreams may have to remain an unfulfilled wish dream; for on p. 161 we learn that in *anxiety dreams, the anxiety has to be separated from the dream* to whose content it is only 'superficially attached'. (See also p. 162, and the editor's footnote thereto.) On p. 236, we learn that anxiety 'may be psycho-neurotic. . . . Where this is so . . . *we come near to the limit at which the wish-fulfilling purpose of dreams breaks down.*' (Italics mine; see also the bottom of p. 487.) So there is a limit, after all. On p. 580, Freud himself becomes conscious that so far he has only evaded the issue of reducing anxiety dreams to wish-fulfilment dreams: 'I am alluding, of course', he writes, 'to the issue of the anxiety-dream; and in order not to confirm the impression that I am trying to evade the evidence of this chief witness against the theory of wish-fulfilment whenever I am confronted with it, I will now give at least some hints towards an explanation of the anxiety-dream.'[8] But the hints are unsatisfactory; at least, they do not satisfy Freud. For after two pages from which nothing more enlightening emerges concerning our problem than a repetition of the old assertion that 'there is no longer anything contradictory in the notion that a physical process which develops anxiety can nevertheless be the fulfilment of a wish', Freud gives up the attempt altogether. He finally tells us, on p. 582, that the whole topic of anxiety dreams falls definitely 'outside the psychological framework of dream formation. If it were not for the fact that our topic [the theory of dreams] is connected with anxiety by the single factor of the liberation of the unconscious during sleep, *I should be able to omit any discussion of anxiety-dreams and avoid the necessity for entering in these pages into all the obscurities surrounding them.*'[9] In 1911, but not in subsequent editions, Freud summarized his elabo-

[7] *Op. cit.*, p. 157. The discussion of 'unpleasurable dreams' is continued on pp. 556 *ff.*; see the quotation to my preceding footnote. Incidentally, I am quite ready to believe Freud's hypothesis (p. 157) that some of his patients fulfilled, in their dreams, their *wish* to refute Freud's theory. And yet, with this hypothesis we are getting dangerously close to a conventionalist stratagem (*cp. L.Sc.D.*, section 20), as I shall try to argue.

[8] *Ges. Schriften*, ii, 1925, p. 497; corresponding to p. 580 (the last five lines before the new paragraph) of Strachey's edition.

[9] Strachey's edition, p. 582; the italics are mine.

rate though only implicit and apparently unconscious repudiation of his programme in a single sentence: 'Anxiety in dreams, I should like to insist, is an anxiety problem and not a dream problem.'[10]

On the next four pages Freud discusses, and partly analyzes, three anxiety dreams. His purpose is no longer to prove that they are wish-fulfilments, but merely to support his assertion 'that neurotic anxiety arises from sexual sources' (p. 582). This, clearly, entails the view that anxiety is connected with certain *wishes*. But it does not justify the inference that all anxiety dreams must have the character of wish-*fulfilments*. (This mistaken inference seems to have been drawn by some of Freud's readers; but it should be noted that Freud himself merely suggests that the first of the three dreams may have been, *in part*, a wish-*fulfilment*, and that he suggests nothing of the kind in connection with the second and third of the dreams.)

The reason why Freud does not carry out his original programme of showing (by way of detailed analyses such as he is wont to give) that all anxiety dreams are wish-fulfilments is, clearly, that in the end he no longer believes in it. So the anxiety dream becomes an anxiety problem: it now 'forms part of the psychology of the neuroses' (p. 582) rather than of the theory of dreams; that is to say, of the theory of wish-fulfilment. I should be the last to criticize such a change of mind. But the change is not a conscious correction, or the admission of a mistake. On the contrary, nine years after these passages were written, Freud added to the page (135) on which he first introduced his programme of reducing anxiety dreams to wish-fulfilment dreams a sharp rebuke to the 'readers and critics of this book'. He accuses them of failing to agree with his thesis that all dreams, including anxiety dreams, are wish-fulfilments, and of failing to understand his programme (abandoned years ago, though only at the end of the book) according to which 'anxiety dreams, when they have been interpreted, may turn out to be fulfilments of

---

[10]*Loc. cit.*, footnote 2; the italics are mine. A quite unambiguous statement (which, however, does not contain the term 'anxiety' but the term 'traumatic neurosis' in its stead), to the effect that some anxiety dreams are not wish-fulfilments but 'are the only *genuine* exceptions', is to be found in the first sentence of section ix of Freud's paper of 1923 referred to in note 13, below; *cp.* also the last four pages of the first chapter of the *New Introductory Lectures*, 1933, especially the remark, 'I will not have recourse to the saying that the exception proves the rule . . .'

wishes' (p. 135). 'It is almost impossible to credit the obstinacy', Freud writes, 'with which readers and critics of this book shut their eyes to this consideration, and to the fundamental distinction between the manifest and latent content of dreams.'[11]

Now my point is not so much that, as a matter of fact, it was not the readers and critics who were obstinate; that readers and critics could hardly fail to see the problem of anxiety dreams; and that they were perfectly right if they were dissatisfied with being first told that the reduction of anxiety dreams to wish-fulfilment dreams presented 'no great difficulty' (pp. 135 and 557) and finding in the end (p. 582) that this reduction was not even attempted, but instead dismissed as being 'not a dream problem'. Rather I wish to criticize Freud's way of rejecting criticism.[12] Indeed I am convinced that Freud could have vastly improved his theory, had his attitude towards criticism been different—especially that towards *uninformed criticism*', as psycho-analysts like to call it. And yet, there can be no doubt that Freud was far less dogmatic than most of his followers, who were inclined to make a religion out of the new theory, complete with martyrs, heretics, and schisms, and who looked on any critic as a foe—or at least as 'uninformed' (that is, in need of being analyzed).

This self-defensive attitude is of a piece with the attitude of looking for verifications; of finding them everywhere in abundance; of refusing to admit that certain cases do not fit the theory (and, at the same time, dismissing them as 'not a dream problem but an anxiety problem'—indeed a typical 'conventionalist stratagem' as discussed in section 20 of my *L.Sc.D.*; an 'immunization', as Hans Albert calls it).

Once this attitude is adopted, *every conceivable case will become a verifying instance.* I illustrated this, in 1919, by the following example of two radically opposite cases of behaviour. A man pushes

---

[11]*Ges. Schriften* iii, p. 25; corresponding to footnote 2, p. 135 of Strachey's translation. See also the remarks on 'the laity' in the fourteenth of the *Introductory Lectures.*

[12]See, in addition to the just quoted footnote 2, *op. cit.*, p. 135, also the footnote added in 1925 to p. 160. Freud says there of his critics that they 'make little use of their moral conscience' (*Ges. Schriften* iii, p. 31), suggesting that they are moved by 'aggressive inclinations' when attributing the doctrine that 'all dreams have a sexual content' to 'psycho-analysis'. Yet did not Otto Rank—who, as Freud explains, asserted just this—belong to the ranks of 'psycho-analysis'?

a child into the water with the intention of drowning it; and another sacrifices his life in an attempt to save the child. Each of these radically different cases of behaviour can be explained with ease in Freudian terms—and, incidentally, in Adlerian terms as well. According to Freud, the first man suffered from repression (say, of some component of his Oedipus complex) while the second man had achieved sublimation. (And, as the psycho-analyst S. Bernfeld once wrote, psycho-analysis can predict that a man will either repress or sublimate, but it cannot say whether he will do the one or the other.) According to Adler, the first man suffered from feelings of inferiority (producing, perhaps, the need to prove to himself that he dared to commit a crime); and so did the second man (whose need was to prove to himself that he dared to risk his life). I cannot think of any conceivable instance of human behaviour which might not be interpreted in terms of either theory, and which might not be claimed, by either theory, as a 'verification'.

[* (Added 1980) The last sentence of the preceding paragraph is, I now believe, too strong. As Bartley has pointed out to me, there are certain kinds of possible behaviour which are incompatible with Freudian theory—that is, which are excluded by Freudian theory. Thus Freud's explanation of paranoia in terms of repressed homosexuality would seem to exclude the possibility of active homosexuality in a paranoid individual. But this is not part of the basic theory I was criticizing. Besides, Freud could say of any apparently paranoid active homosexual that he is not *really* paranoid, or not *fully* active.]

That such radically opposite cases have in fact been interpreted as verifications may be shown in detail by analyzing Freud's treatment of certain objections to his theory. In *The Interpretation of Dreams*, Freud mentions the 'very frequent dreams which appear to contradict my theory because their subject-matter is the frustration of a wish, or the occurrence of something clearly unwished-for' (p. 157). One group of these 'counter-wish dreams', as he calls them, can be explained, he says, as dreams fulfilling a patient's wish that Freud's theory may be wrong. (There is another group which does not concern us here.) Thus the apparent falsification turns into a 'verification'. But what about the radically opposite case, of a patient whose dreams are dreamt in order to oblige the analyst and to confirm him rather than to refute him? These 'obliging dreams' (as Freud sometimes calls them[13]) are, of course, verifications too; for they are wish-fulfilments in precisely the same sense as were the others.

A more critical attitude towards these 'obliging dreams' would be this. They are (as Freud himself says) due to suggestion by the analyst—to the fact that the analyst has imposed his ideas upon a suggestible patient. Should we not therefore seriously consider the possibility that some other 'clinical verifications', of which analysts like to speak, or indeed all of them, are due to a mechanism of this kind? And does not the mere possibility of such a mechanism invalidate these 'verifications'?

Freud himself sees this problem, and it is interesting to see how he deals with it.[14]

'The analyst will perhaps get a shock at first', Freud begins the discussion, 'when he is first reminded of this possibility'—that is, of the possibility of thus influencing the patient.[15] This is an interesting remark: 'The analyst', like Freud, gets a shock because he sees that his whole edifice of 'clinical verifications' is threatening to collapse. But the analyst's anxiety subsides as soon as he is told that it is merely 'the sceptic' who reminds him of this shocking possibility: 'The sceptic may say that these things appear in the dream because the dreamer knows that he ought to produce them—that they are expected by the analyst', Freud writes; and he adds: 'the analyst himself will, with justice, think differently.'[16]

No doubt he will. But why 'with justice'? No reason is given. On the contrary, when the sceptic reappears three pages later for the last time—he is then called 'somebody'—even Freud himself no longer 'thinks differently'; for he now writes: 'Should somebody maintain that most of the dreams which can be made use of in an analysis are in fact obliging dreams which have been produced upon [the analyst's] suggestion, *then nothing can be said against this opinion from the point of view of analytic theory.* In this case, I need only refer to

[13]'Obliging dreams' *('Gefälligkeitsträume')* or 'compliant dreams'—dreamt to please or oblige the analyst by confirming his theory—are described and discussed by Freud on pp. 312–314 of *Ges. Schriften* iii (corresponding to 'Remarks on the Theory and Practice of Dream Interpretation', 1923, sections vii*f., Collected Papers* 5, 1950, pp. 141–145). See also the twenty-seventh and twenty-eighth of the *Introductory Lectures.*

[14]*Op. cit., Ges. Schriften* iii, pp. 310–314.

[15]*Op. cit.,* p. 130.

[16]*Op. cit.,* p. 311. (*Coll. Papers* 5, p. 142.)

the considerations in my *Introductory Lectures* where . . . it is shown how little the trustworthiness of our results is impaired by an understanding of the effect of suggestion, in our sense.'[17] I am afraid that the reference to the *Introductory Lectures* will hardly help anybody to get over the contradiction between the last two quotations. If somebody can think critically, he must remain in a state of 'shock'; especially if he reads, between the lines of the fifteenth of these *Introductory Lectures* (*cf.* there the first six lines of point 4), that the shock originated with the discovery that Freud's, Adler's and Stekel's patients dreamt, respectively, 'mainly of sexual impulses, . . . of mastery . . . [and] of rebirth', adapting in this way, as Freud puts it, 'the contents of their dreams to the favourite theories of their physicians'.

But returning from the *Introductory Lectures* to the passage which led me to cite them, the only argument worthy of the name in these four pages of apologetics is the *jigsaw puzzle argument* (pp. 312–313, and 143 respectively). It asserts that, if the analyst succeeds in piecing together the whole intricate picture, 'so that the drawing becomes meaningful, and no gap is left anywhere . . . then he knows that the solution is found, and that there is no other solution'.

Nothing could be more dangerous than this argument if, as in the present context, it is used in order to dispel the analyst's doubts concerning the results of suggestion. For what the analyst was frightened of, to begin with, was just the possibility that the puzzle might be put together under pressure—by squeezing the little pieces (which turn out to be elastic or plastic rather than rigid) into place, or perhaps by his unconscious suggestion to an obliging patient that he might produce some new pieces, specially made to measure, so as to fit nicely into the various 'gaps'.

Even without this decisive objection, the jigsaw puzzle argument

[17]*Op. cit.*, p. 314. (The italics are mine. *Cp. Coll. Papers*, p. 145.) The reference is to point four of the fifteenth of the *Introductory Lectures* which provides at best a circular argument as reply, and to the last (twenty-eighth) which provides an argument that shows no more than that some of the more general tenets of analysis are supported by independent evidence, so that not all can be due to suggestion. (I readily grant this; what is perhaps the most striking evidence of this kind can be found in Plato's *Republic*, e.g., 571–575. These passages are not mentioned by Freud. They and some other passages are discussed in my *Open Society*, note 59 to Chapter 10.) The passage in the twenty-eighth lecture also makes use of what is essentially the jig-saw puzzle argument; see below.

is acceptable only if we have a theory before us which can be severely tested: other theories can *always* make their puzzles fit. Consider, for example, interpretations of history in terms of struggling races or struggling classes: how well they all 'solve' the puzzle of history, and of current policy. The same holds for the astrological interpretation of history, or for Homer's interpretation of it in terms of domestic squabbles on Mount Olympus, or for the Old Testament interpretation in terms of collective guilt, punishment, and atonement. Each of these succeeds in 'solving' their puzzle. But their belief—and Freud's—*'that there is no other solution'* is shown to be baseless: they all succeed. (And so did Adler and Stekel.)

I do not wish to be misunderstood. I think that Freud's *Interpretation of Dreams* is a great achievement. Yet it is more of the character of pre-Democritean atomism—or perhaps of Homer's collected stories from Olympus—than of a testable science. It certainly shows that even a metaphysical theory is infinitely better than no theory; and it is, I suppose, a programme for a psychological science, comparable to atomism or materialism, or the electromagnetic theory of matter, or Faraday's field theory, which were all programmes for physical science. *But it is a fundamental mistake to believe that, because it is constantly being 'verified', it must be a science, based on experience.*

A dangerous dogmatism always goes hand in hand with verificationism. I myself do not think that the question, 'What is the essential nature of dreams?', is a good question to ask; but if it is raised, then answers other than Freud's wish-fulfilment theory seem to be at least as appropriate. For example, all of Freud's material as well as his analyses would fit very well the following reply: 'All dreams are the result of conflicts—either of conflicting wishes, or of conflicts between wishes and obstacles threatening to frustrate them, and creating worries or problems.'[18] Now since, in a dream, wishes can hardly be expressed in any other way than by a

---

[18]Dreams representing work on a problem are discussed in *The Interpretation of Dreams*, footnote 2 (added 1919) on p. 181, where Freud refers to experiments carried out by Pötzl. (See also p. 569.) Some of Pötzl's results were anticipated by Samuel Butler in a marvellous passage in *Erewhon Revisited* (1901), chapters 27 and 28. 'I wish someone would write a book about dreams', Butler wrote in Chapter 28, unaware of Freud's book published two years before.

representation of what is wished for[19]—that is, of their fulfilment—
a representation of this fulfilment is to be found in most dreams. Yet
although some dreams may culminate in a fulfilment, conflict and
frustration are always as strongly represented (even in the simplest
dreams of childhood and in hunger dreams); and they become
dominant in an anxiety dream, which need *not* be a symptom of an
anxiety neurosis.[20]

Nothing is further from my intention than to offer this theory—
which in any case would owe everything to Freud—as an alternative
to his own theory.[21] What I wish to point out is that Freud nowhere
discusses an alternative theory (such as the one sketched here)
which takes notice of the simple fact, now admitted, that *anxiety
dreams constitute a refutation of the general formula of wish-
fulfilment*—as suggested long ago by 'obstinate' readers and 'uninformed' critics. He nowhere compares his theory with a promising
competitor to it, weighing the one against the other, in the light of
the evidence; and he never criticizes it: he has got his theory and
tries to verify it; and he makes it fit, as long as possible, and—as the
example of the anxiety dream has shown—even beyond what he
himself thought possible when he first published his great book on
The Interpretation of Dreams.

[19]This idea is taken from the last sentence of section iii of Freud's *On Dreams*,
1901; *Standard Ed.* **5**, pp. 629 *ff.*

[20]See Butler, *loc. cit.,* for an interesting description and analysis of an anxiety
dream. Hunger dreams—of Shackleton and Wilson—which do not fit Freud's
theory but which fit the one proposed here are reported by Captain Scott in *The
Voyage of the Discovery* (Journal entry of December 22nd, 1902): 'My companions
get very bad "food dreams"; in fact, they have become the regular breakfast
conversation. It appears to be a sort of nightmare; they are either sitting at a well-
spread table with their arms tied, or they grasp at a dish and it slips out of their
hand, or they are in the act of lifting a dainty morsel to their mouth when they fall
over a precipice. Whatever the details may be, something interferes at the last
moment and they wake.'

[21]As another alternative to Freud's wish-fulfilment formula, its elaboration and
re-interpretation were proposed by J. O. Wisdom in the article referred to in note
2, above. By attributing great weight to Freud's idea of punishment wishes (an idea
of which Freud himself had made only little use), Wisdom proposes to explain *all*
dreams—even anxiety dreams—in terms of wish-fulfilment. (Wisdom himself
prefers the term 'need-fulfilment', but I cannot see any reason, from a psycho-
analytic point of view, why there should not be an 'unconscious wish' corresponding to every 'need' in Wisdom's sense, including what he calls 'punishment-
needs'.) Wisdom's theory might be said to admit conflicts in explaining dreams,
but, as far as I can see, it admits only *one* kind of conflict—that due to feelings of
guilt.

Such were the reasons, more or less, which led me in 1919 to reject the claims of Freudians, Adlerians, and Marxists that their theories were 'based upon experience' in the same way as were those of other sciences—experimental neurology, say, or bio-chemistry. I rejected their claims because I found that their theories failed to satisfy the criterion of testability, or refutability, or falsifiability. Today, this criterion is becoming widely accepted as a criterion of demarcation; but the three theories mentioned are rarely discussed in terms of it. Instead, they continue to be discussed in terms of confirming evidence—of 'verifications'.

This is how I first came to see the problem of demarcation. In the present context, it hardly matters whether or not I am right concerning the irrefutability of any of these three theories: here they serve merely as examples, as illustrations. For my purpose is to show that my 'problem of demarcation' was from the beginning the practical problem of assessing theories, and of judging their claims. It certainly was not a problem of classifying or distinguishing some subject matters called 'science' and 'metaphysics'. It was, rather, an urgent practical problem: under what conditions is a *critical appeal to experience* possible—one that could bear some fruit?

19. *Testability but Not Meaning.*
I have told part of the story of how I first came to formulate the problem of demarcation because I want to show that neither the philosophical dogma of falsifiability nor the philosophical difficulties of verifiability led me to it. Rather, it was a highly practical and urgent problem—that of deciding whether a theory was acceptable: whether it was arguable by means of empirical arguments (that is, arguments appealing to observation and experiment), and whether these arguments should be considered as serious tests. My problem turned out to be, all at once, a logical problem, a methodological problem, and even a problem of science itself. For it is the scientist's task to judge theories; and one way of passing judgment on a theory is to say that it cannot be judged by ordinary scientific standards (that is, by assessing how it stands up to tests) because it is irrefutable, and therefore not testable.

Hence I suggested that testability or refutability or falsifiability should be accepted as a criterion of the scientific character of theoretical systems; that is to say, *as a criterion of demarcation between empirical science on the one hand and pure mathematics, logic, metaphysics, and pseudo-science on the other.*

It never occurred to me, either in those days or later, to propose testability or refutability or falsifiability as a criterion of *meaning* (as opposed to 'meaningless nonsense'); and when I first heard, in 1927 or thereabouts, that the Vienna Circle had accepted verifiability as a criterion of meaning,[1] I at once objected to this procedure on two entirely different grounds: first, because taking *meaningfulness* as a criterion of demarcation meant branding metaphysics as meaningless gibberish: a dogma which I felt unable to accept; and secondly, because *verifiability* was proposed as a criterion of meaning or sense or significance, and thereby indirectly as a criterion of demarcation: a solution which was entirely inadequate, and indeed the opposite of what was needed. For I could show that it was both *too narrow and too wide*: it (unintentionally) declared scientific theories to be meaningless, and it thereby placed them (again unintentionally) on the same level as metaphysics (*cp.* section 4 of my *L.Sc.D.* and Appendix *ii).

Moreover, it was 'verificationist' in the important sense that it overlooked the fact that scientific discussion (as a certain kind of rational discussion) was *critical* discussion, and that its fundamental attitude was to seek refutations rather than to seek verifications or confirmations.

The broad line of demarcation between empirical science on the one hand, and pseudo-science or metaphysics or logic or pure mathematics on the other, has to be drawn right through the very

---

[1]The formulation is due to F. Waismann (see note 2 to section 6 of my *L.Sc.D.*), although the idea may be said, I suppose, to be Wittgenstein's, more or less. It is now often forgotten that an expression is called 'meaningless' or 'nonsensical', in the somewhat technical sense (due to Russell and accepted by Wittgenstein and the Vienna Circle) which we are discussing here, only if it is sheer gibberish. A good example of a meaningless pseudo-sentence from Wittgenstein's *Tractatus* is: 'Socrates is identical'. As opposed to the customary sense of 'nonsense', manifestly silly assertions such as: 'Wee bees sneeze when three trees squeeze one cheese in a breeze', or '3 + 11 = 33', are not 'nonsensical' or 'meaningless' in this technical sense, but *false*. Moreover, *every* false proposition, however silly, is always meaningful (i.e., it has 'sense'); and its negation, however trite, is always true. (Concerning Russell's classification of expressions into true, false, and nonsensical, and its influence upon Wittgenstein, see also my papers 'The Nature of Philosophical Problems and their Roots in Science', and 'Self Reference and Meaning in Ordinary Language', now both in my *Conjectures and Refutations*.)

heart of the region of sense—with meaningful theories on both sides of the dividing line—rather than between the regions of sense and of nonsense. I reject, more especially, the dogma that metaphysics must be meaningless. For as we have seen, some theories, such as atomism, were for a long time non-testable and irrefutable (and, incidentally, non-verifiable also) and thus far 'metaphysical'. But later they became part of physical science.[2] And others suffered the opposite fate. It is clearly inadequate to describe them as nonsensical. I am ready to admit that some metaphysicians (I have in mind, especially, Hegel and the Hegelians) have indulged in talking non-sense and, what is worse, pretentious nonsense. Yet scientists are not quite free from this malady. At any rate, it seems better not to take nonsense too seriously. Surely, it is a little unhealthy and a little unwise to select the problem of nonsense, or lack of meaning, as the basic problem of one's philosophy, and its exposure as one's major, or perhaps one's only, task: meaning-analysis, like psycho-analysis, may easily turn into 'an affliction that mistakes itelf for its cure'.[3]

I could not have been more explicit in my rejection of the whole problem of sense or meaningfulness *versus* nonsense or meaning-lessness than I have been from the beginning, as any reader of my *L.Sc.D.* can see. I denounced it as a pseudo-problem, as a mistaken attempt to formulate the problem of demarcation, and as a mistaken solution of this problem. And I have consistently and repeatedly re-affirmed my position.[4] Yet in spite of all this, I am often labelled 'the positivist who proposed to modify the verifiability criterion of *meaning*, and to replace it by the falsifiability criterion of *meaning*'. By 1949 a rider had been added to this label: 'But I understand that

[2]The metaphysical character of the 'corpuscular theory' of matter (atomism) is briefly explained in section 23 below (text following footnote 1).

[3]I am alluding to an aphorism of the Viennese poet Karl Kraus, who wrote: '*Die Psychoanalyse ist jene Geisteskrankheit die sich für ihre Therapie hält.*' (*Nachts,* 1924, p. 80.) Or in free translation: 'Psycho-analysis is the affliction that mistakes itself for its cure.'

[4]See, apart from Appendix i, and sections 4 and 10 of my *L.Sc.D.*, my *Open Society* (1945 and later editions), especially notes 46, 51 and 52 to Chapter 11, and other places there mentioned; also note 2 to chapter I; see also my papers 'Indeter-minism in Quantum Physics and in Classical Physics', *The British Journal for the Philosophy of Science* 1, 1950–51; 'The Nature of Philosophical Problems, etc.', *ibid.,* 3, 1952; 'Three Views Concerning Human Knowledge', 1956; and 'Philoso-phy of Science: A Personal Report', 1957. [*Cf. Unended Quest,* sections 8 and 16, and *Conjectures and Refutations,* Chapters 1, 2, 3. Ed.]

he would not *now* call himself a positivist.'[5] I can only say that I have always severely criticised positivism, and that I have never changed my mind about these matters; and I ask the reader to ascertain this fact by looking up Appendix *i (1933) and sections 4, 10, 79 and 85 of my *L.Sc.D.* (1934).

I am resigned to the fact that, in spite of this, this label will stick to me to the end of my days. (And perhaps beyond, should the history of philosophy continue to notice us last laggards of the Enlightenment; which seems a remote contingency, in view of the already overwhelming and still increasing demand for an irrational and anti-rational philosophical messianism *à la* Heidegger from the one side, and for a 'mathematically exact' philosophical method from the other.[6]) Yet I intend to repeat, to the end of my days, that the confusion of the problem of demarcation with that of meaning is one of the major mistakes of the positivist school of thought.

20. *Non-Testable Statements.*

The mistaken assumption, corrected in the preceding section, that my criterion of demarcation is intended as a criterion of meaning, is liable to create much confusion. It not only misinterprets my intention, but is formally inconsistent with my theory. Thus if it is superimposed upon my theory, the result is self-contradictory. This may be shown as follows.

If we take any meaningful statement and form its *negation*, then the result will, clearly, again be a meaningful statement. The negation may always be formed by prefixing to the statement the words 'It is not the case that'. Now take some meaningless expression, and prefix these words: the result will, clearly, again be a meaningless expression. Thus we arrive at the following requirement which ought to be satisfied by any consistent and adequate criterion of meaningfulness and meaninglessness: *the negation of a meaningful statement must be meaningful, and that of a meaningless expression, of a meaningless sequence of words, must be meaningless.*

This requirement will be violated if my criterion of demarcation is interpreted as a criterion of the meaningfulness of an expression.

[5]*Cf.* note 3 on p. 76 of W. Kneale's *Probability and Induction.* [*Cf. Unended Quest*, section 17. Ed.]

[6]This 'exactness' is disappointing. See the Preliminary Remarks to Chapter III.

For the negation of a falsifiable universal statement is a non-falsifiable existential statement. For example, 'All ravens are black' is falsifiable since we may one day discover a white raven. But its negation is the existential statement, 'There exists (somewhere in the universe, either in the past, present or future) a raven which is not black'; and this existential statement is non-testable, since it cannot be falsified by any amount of observation reports. Thus if my criterion of demarcation were one of meaning, this existential statement would have to be regarded as meaningless, although it is the negation of a meaningful universal statement. Consequently the requirement would be violated. (It was, in fact, violated by the verifiability criterion, since this was a criterion of meaning.)

This does not show that my criterion of demarcation is at fault. It merely shows that absurd consequences must follow from any attempt to interpret my criterion of demarcation between science and metaphysics as a criterion of meaning. No wonder that some philosophers—those who were constitutionally unable to think of a demarcation between science and metaphysics which was not at the same time a demarcation between sense and nonsense—found my criterion of demarcation unsatisfactory.

Now the status of existential statements in my theory is quite simple and straightforward. As to their meaning there can be no question: they are as meaningful as their negations, the universal statements. Thus the requirement formulated above is satisfied as a matter of course (even though I am far from trying to produce a theory of meaning).

As to the question of the scientific or metaphysical character of existential statements, it is important to remember that my criterion applies to *theoretical systems* rather than to statements picked out from the context of a theoretical system (as I pointed out from the beginning: see *L.Sc.D.*, Appendix i). This is due to the fact that I am much more interested in *what* a theory says than in *how* it says it. But one and the same theory may be formulated in many different ways, with the help of a greater or a smaller number of hypotheses (premises). I therefore asserted in my *L.Sc.D.* (in section 15) that, according to my criterion, some existential statements are scientific, that is to say those which belong to a testable context. An example given there is the statement, 'There exists an element with the atomic number 72'. It is scientific as part of a highly testable theory, and a theory *which gives indications of how to find this element*. If, on the other hand, we took this existential statement in isolation, or as part of a theory which does not give us any hint as to how and where this element can be found, then we would have to describe it as metaphysical simply because it would not be testable: even a great number of failures to detect the element could not be interpreted as a failure of the statement to pass a test, for we could never know whether the next attempt might not produce the element, and verify the theory conclusively.

Thus some existential statements will be testable, and others will be non-testable. And sometimes this will depend upon the context, and perhaps change with a change of the context.[1]

21. *The Problem of 'Eliminating' Metaphysics*.

That my criterion of demarcation between science and metaphysics is intended to be applied only to *theoretical systems* is a point of considerable importance for its appraisal—more important than the question of existential statements.

This point, emphasized from the beginning, distinguished my criterion from that of the positivists; for they believed that their criterion of meaning could be applied to any *linguistic expression,* without reference to its *context.* (They thought that all that was needed was a knowledge of the rules of the language to which the expression belonged.) They believed that their criterion of meaning would enable them to *detect nonsense* wherever it might occur. Thus they sometimes described their aim as 'the elimination of metaphysics by way of language analysis'.[1] And they believed that they had a method, a technique, which would allow them to eliminate metaphysical elements—that is to say nonsense—also from scientific theories.

I do not believe that metaphysics is nonsense, and I do not think it possible to eliminate *all* 'metaphysical elements' from science: they are too closely interwoven with the rest. Nevertheless, I believe that whenever it is possible to find a metaphysical element in science which *can* be eliminated, the elimination will be all to the good. For

[1][*Cf.* 'Replies to my Critics', in P. A. Schilpp: *The Philosophy of Karl Popper,* Vol. II, section ii, pp. 976–1013, and pp. 1037–9. Ed.]
[1]This was Carnap's research programme.

the elimination of a non-testable element from science removes a means of avoiding refutations; and this will tend to increase the testability, or refutability, of the remaining theory. And indeed, in a number of cases a scientific theory has gained considerably by the discovery that there were metaphysical elements in it which could be eliminated, and by the attempt to eliminate them.

My criterion of demarcation is, however, not intended as an instrument for the detection of such elements. I do not intend to imply that it cannot be used as such. I have in fact found it quite useful in a number of cases. But I do not think that the detection of metaphysical elements and their elimination from science can ever become part of a routine or of a *technique*.

Positivists take a different view about the possibility of developing such a technique. To a positivist it would involve merely looking for grammatical and similar linguistic mistakes (which, after all, are made by most authors).

Yet where the content of theories becomes all-important, as when a theory is being improved by the elimination of *meaningful* metaphysical elements, the task will be part of rational criticism; and rational criticism is always an imaginative and creative process, rather than a mere technique. The 'elimination of metaphysical elements' never consists in the mere omission of a sentence or two, but always involves a reconstruction of the theory, inspired, as a rule, by a new idea concerning its interpretation.

We should also remember that one and the same theory can often be interpreted either in the conventionalist sense (as a set of irrefutable definitions), or empirically, as pointed out in my discussion of conventionalism (*L.Sc.D.*, sections 19 *ff.*). This shows how much depends on the *interpretation* of a theory, if we wish to judge its empirical character by applying our criterion of demarcation; and it shows, consequently, that the task of discovering 'metaphysical elements' cannot be solved just by scrutinizing the formalism. In fact, it is just part of the general task of improving a theory, by criticism, and by trial and error.

An interesting case, in this connection, is Berkeley's criticism of Newton. Berkeley succeeded not only in giving an excellent criticism of the 'occult' or metaphysical character of Newton's absolute space and absolute time, but he even anticipated, in his *De motu*, Mach's famous suggestion for a reform of the theory, later taken up by Einstein in his theory of general relativity.[2] Yet it should be remembered that in conjunction with the wave theory of light the theory of absolute space and time became testable, so that Michelson's experiment could be interpreted as a refutation. It was not, therefore, the intrinsically metaphysical character of these concepts which made their elimination desirable, but the fact that, given *only* the context of Newton's mechanics, they represented non-testable elements.

[*It may perhaps also be mentioned that my interpretation of Heisenberg's indeterminacy formulae as scatter relations was both an attempt to criticize Heisenberg's Machian positivism ('observables') and to eliminate what I regarded as his metaphysical dogmatism: his theory that the indeterminacy formulae indicated the limits of human knowledge.]

22. *The Asymmetry Between Falsification and Verification.*

There is, as pointed out in the *L.Sc.D.*, a fundamental logical asymmetry between empirical falsification and verification. Although some of my critics have denied the existence of this asymmetry, their arguments were anticipated and fully answered in my *L.Sc.D.*

This fundamental asymmetry cannot, I think, be seriously denied: a set of singular observation statements ('basic statements', as I called them) may at times falsify or refute a universal law; but it cannot possibly verify a law, in the sense of establishing it. Precisely the same fact may be expressed by saying that it can verify an existential statement (which means falsifying a universal law) but that it cannot falsify it. This is the fundamental logical situation; and it shows a striking asymmetry.

Of the various objections which have been raised against my claim that this asymmetry exists—and thereby against my criterion of demarcation—the one which at first sight looks the most striking is the following. Whenever we falsify a statement we thereby automatically verify its negation, for the falsification of a statement *a* can always be interpreted as the verification of its negation, non-*a*. Accordingly, we can always speak, if we like, of verification instead

[2]See my paper 'A Note on Berkeley as Precursor of Mach', *British Journal for the Philosophy of Science* 4, 1953, pp. 26 ff., in my *Conjectures and Refutations*.

of falsification, and *vice versa:* the difference between these two ways of putting things is merely verbal, and they are therefore, for logical reasons, completely symmetrical.

For example, if we can describe an *empirical test* as an attempted falsification or as a search for a negative instance (of the statement *a*), then we can also describe it as an attempted verification or as a search for a positive instance (of the statement non-*a*). Similarly, any obstacle to the verification of a statement *a* must, for logical reasons, also be an obstacle to the falsification of non-*a*, and *vice versa*. Thus it is said that it is difficult to verify an existential statement such as 'there exists a perpetual motion machine', that is to say, 'a machine which continues to emit energy without ever absorbing energy from its environment', since we might have to search the whole world for it (and incidentally to examine each candidate for an indefinite period of time); but obviously, it must be just as difficult to falsify the negation of this existential statement—that is, the universal statement 'all machines which continue to emit energy must, after a finite time, absorb energy from their environment'. For the verification of the one statement is nothing but the falsification of the other.

Now it follows from these obvious premises, my critic concludes, that it is, for purely logical reasons, pointless to distinguish between falsification and verification, or between falsifiability and verifiability, or to demarcate a class of 'falsifiable' or 'testable' statements as 'scientific' and to distinguish them from another class of non-falsifiable (though perhaps unilaterally verifiable) statements, which are called 'metaphysical'. Thus the falsificationist who asserts that the statement 'all swans are white' can be falsified will have to admit that every falsification or refutation of this universal statement will be equivalent to the verification, and acceptance, of the existential statement 'There exists a non-white swan'. It must be wrong, therefore, to call the universal statement 'scientific' and the existential statement 'metaphysical'. (The lesson usually drawn from this is that the distinction between scientific and metaphysical statements does not depend on such things as testability—that is, on a relation between statements—but rather on the *concepts*—observable or otherwise—which occur in the statements.[1])

[1]See especially sections 24 and 25; see also note 5 to section 11.

In answer to this criticism, I wish to say that I accept all my critic's premises as true—indeed, as trivially true; but I reject all his conclusions (which are stated in the last paragraph), none of which, incidentally, follow from his premises.

First I should like to get one point out of the way—a point with which I agree, since it belongs to my critic's premises, but which nevertheless betrays a misunderstanding. I mean the reference, in the concluding part of my critic's premises, to 'obstacles' or 'difficulties' which sometimes stand in the way of verifying a purely existential statement such as 'there exists a perpetual motion machine'. Of course, what my critic says is true—the empirical verification of this statement is for obvious reasons exactly as difficult—or exactly as easy—as the empirical falsification of the universal law which is its negation. But I have never worried about this 'difficulty', and I have never referred to it, or drawn any conclusion from it. I do not call an isolated purely existential statement 'metaphysical' because it is 'difficult' to verify, but because it is *logically impossible to falsify* it empirically, or to test it. And I have of course always stressed that the logical impossibility of falsifying an existential statement of this kind is exactly the same thing as the logical impossibility of verifying its universal negation. My critic's reference to obstacles or difficulties is therefore irrelevant. Moreover, it seems to betray a verificationist attitude: verificationists, it seems, cannot imagine any difficulty about purely existential statements, apart from the difficulty of verifying them.

The point concerning the 'obstacles' or 'difficulties' may thus be dismissed as irrelevant; and I can proceed to other and perhaps more relevant points.

No doubt one can say that problems of falsification and verification are in certain respects 'symmetrical'. The fact that there are certain symmetries here hardly precludes the existence of a fundamental asymmetry—any more than the existence of a far-reaching symmetry between positive and negative numbers precludes a fundamental asymmetry in the system of integers: for example, that a positive number has real square roots while a negative number has no real square root.

Thus one can certainly say that falsifiability and verifiability are 'symmetrical' in the sense that the negation of a falsifiable statement must be verifiable, and *vice versa*. But this fact, which I repeatedly stressed in my *L.Sc.D.* (where I even described universal statements as negative existential ones), and which is all that my critic establishes in his premises, is no argument against the fundamental asymmetry whose existence I have pointed out.

This asymmetry has a purely logical and also a methodological or heuristic aspect.

As to its logical aspect, there can be no doubt that a (unilaterally falsifiable) universal statement is logically much stronger than the corresponding (unilaterally verifiable) existential statement. For the following is a well-known logical rule. From a universal statement, pertaining to all things of a certain kind, or to all elements of a certain non-empty universe of discourse,

(1) All things have the property P,

we can derive, for any individual thing a belonging to this kind or universe,

(2) The thing a has the property P;

and from (2), in turn, we can derive

(3) There exists a thing that has the property P.

Thus (1) entails (2) and (3), and (2) entails (3). But (3) does not entail either (1) or (2), and (2) does not entail (1).

Or in other words, (1) is logically stronger than (2) and (3), and (2) is logically stronger than (3).

This is the source of the important asymmetry in the case of unilaterally falsifiable universal and unilaterally verifiable existential statements; and the situation is the same with more complex statements. (See section 24.)

Owing to their logical power, universal statements may be important as *explanatory hypotheses*: they may explain (especially in conjunction with singular initial conditions) singular events or statements. Purely existential statements, on the other hand, in isolation, or even in conjunction with singular statements, are usually too weak to explain anything.

This is why scientists are interested in universal hypotheses rather than in (isolated) existential hypotheses.

This leads us to the methodological or heuristic aspect of the asymmetry—to the difference between the critical or falsificationist attitude and the verificationist attitude.

The verificationist's view of science is somewhat like this: ideally, science consists of all true statements. Since we do not know all these, it must at least consist of all those which we have verified (or perhaps 'confirmed' or shown to be 'probable'). Thus verified , existential statements should, *for this reason,* belong to science.

The falsificationist's attitude is different. For him, science consists of daring explanatory hypotheses—'daring' in the sense that they assert so much that they may easily turn out to be false. And he tries his best to find fault with them, hoping to detect and to eliminate faulty candidates for the status of an explanatory theory, and also hoping thereby to gain other insights. As to purely existential statements, he is not interested in them because of their weakness, and because they cannot be falsified unless they form an integral part of a theoretical system. He is ready to admit them into science *if* they are entailed by an accepted basic statement; but even then their interest lies solely in the fact that their acceptance is equivalent to the rejection of their universal negations.

Another objection often raised against asymmetry is this: no falsification can be absolutely certain, owing to the fact that we can never be quite certain that the basic statements which we accept are true. I discussed this fully in the chapter of my *L.Sc.D.* on the 'empirical basis' (Chapter v), and I do not think that any other epistemology has taken as much account of it as has mine. (Theorists of induction, for example, never discuss this problem fully—in spite of the fact that their theories collapse if their empirical basis turns out not to be firm.) As to the claim that this fact refutes the asymmetry between falsification and verification, the situation is really very simple. Take a basic statement or a finite set of basic statements. It remains forever an open question whether or not the statements are true: if we accept them as true we may have made a mistake. But *no matter whether they are true or whether they are false,* a universal law may not be derived from them. Even if we knew for sure that they were true, a universal law could still not be derived from them.

However, if we assume that they are true, a universal law may be falsified by them.

Hence the asymmetry is that a finite set of basic statements, *if true,* may falsify a universal law; whereas, *under no condition* could it verify a universal law: there exists a condition wherein it could falsify a general law, but there exists no condition wherein it could verify a general law.

Thus, if we accept as true the statement, 'This swan here is black', then we are bound, by logic, to admit that we have refuted the universal theory 'All swans are white'; and if we accept as true the the statement, 'This planet is now more distant from the sun than it was a month ago', then we are bound, by logic, to admit that we have refuted the theory 'All planets move in circles with the sun as their common centre'. Now it is true—especially in the second case—that we may have made a mistake when we accepted the singular statement in question; and for this reason, the falsification of the theory *is not* 'absolutely certain'. But it *is* absolutely certain that, *if* we accept any singular statement ('basic statement') that contradicts a theory which we have accepted, we must have made a mistake somewhere—a mistake that must be corrected. And it is absolutely certain that if we accept a basic statement that contradicts the theory we are testing, then we are bound to reject this theory as falsified. And it is thus also absolutely certain (since every basic statement contradicts some theories) that, whenever we accept *any* basic statement, *some theories* are thereby implicitly declared to be falsified, *so that we are logically committed to rejecting them.* But *no* theory has been verified: there is none which we are bound to accept as true. Hence the asymmetry.

In fact, the asymmetry is even stronger than indicated so far. A traditional principle of empiricism which I accept is that theories are to be judged in the light of observational evidence. But this means that we have at least sometimes to make up our minds to accept some basic statement—if only tentatively, and after many tests and deliberations. And once we do accept it, we are, as we have seen, logically bound to *reject* some theory. There is nothing analogous to this as far as the *acceptance of a theory* is concerned, or as far as its *verification* is concerned.

Thus, the logical relation between basic statements and theories, and the uncertainty of basic statements, enforce rather than cancel each other: *both operate against verification; and neither operates unilaterally against falsification.*

The question of the uncertainty of the empirical basis is on an altogether different plane from the question of the logical relationship between basic statements and theories. Its character is like that of the truism that we can always err (even in a mathematical proof); while the character of the non-verifiability of theories is like that of

186

the more interesting remark that we can never have sufficient empirical premises—assuming them to be true—for establishing the truth of a universal law. Thus the two questions must be treated separately. To sum up, it is true—as emphasized in my *L.Sc.D.*—that even falsifications are never absolute, or quite certain; but the reasons for this uncertainty are utterly different from the reasons that render any verification of a theory *in principle impossible,* and not merely somewhat doubtful. The principle of empiricism itself implies both: the asymmetry, and the possibility of falsifying theories.

All this is so trivial that I should not have restated it were it not that the existence of a logical asymmetry between falsification and verification has been constantly denied by most of my positivist critics.

More serious is an objection closely connected with the problem of *context,* and the fact that my criterion of demarcation applies to *systems of theories* rather than to statements out of context. This objection may be put as follows. No single hypothesis, it may be said, is falsifiable, because every refutation of a conclusion may hit any single premise of the set of all premises used in deriving the refuted conclusion. The attribution of the falsity to some particular hypothesis that belongs to this set of premises is therefore risky, especially if we consider the great number of assumptions which enter into every experiment.

This objection belongs to those which were discussed, although briefly, in *L.Sc.D.* (in sections 16, last paragraph, especially note 2; and in 18, 19, 20, and elsewhere). The answer is that we can indeed falsify only *systems of theories* and that any attribution of falsity to any particular statement within such a system is always highly uncertain.

This does not, of course, affect the fundamental asymmetry which I have pointed out. But it leaves me with the task of explaining the undeniable fact that we are sometimes highly successful in attributing to a single hypothesis the responsibility for the falsification of a complex theory, or of a system of theories.

Many aspects of our actual methodological procedures are understandable as due to our efforts to make such attributions more successful. There is first the layered structure of our theories—the layers of depth, of universality, and of precision. This structure allows us to distinguish between more risky or exposed parts of our theory, and other parts which we may—*comparatively speaking*—take for granted in testing an exposed hypothesis. It explains the fact that we consciously test, as a rule, a certain chosen hypothesis, treating the rest of the theories involved in the test as more or less unproblematic—as a kind of 'background knowledge'.[2] This background knowledge is usually *varied* by us during the tests, which tends to neutralize mistakes that might be involved in it.

Perhaps the most important aspect of this procedure is that we always try to discover how we might arrange for *crucial tests* between the new hypothesis under investigation—the one we are trying to test—and some others. This is a consequence of the fact that our tests are *attempted refutations;* that they are designed—*designed in the light of some competing hypothesis*—with the aim of refuting, if possible, the theory which we wish to test. And we always try, in a crucial test, to make the background knowledge play exactly the same part—so far as this is possible—with respect to each of the hypotheses between which we try to force a decision by the crucial test. (Duhem criticized crucial experiments, showing that they cannot establish or prove one of the competing hypotheses, as they were supposed to do; but although he discussed refutation—pointing out that its attribution to *one* hypothesis rather than to another was always arbitrary—he never discussed the function which I hold to be that of crucial tests—that of *refuting* one of the competing theories.)

All this, clearly, cannot absolutely prevent a miscarriage of justice: it may happen that we condemn an innocent hypothesis. As I have shown (especially in sections 19 and 20, and also in 29 of *L.Sc.D.*), an element of free choice and of decision is always involved in accepting a refutation, or in attributing it to one hypothesis rather than to another.

Our scientific procedures are never based entirely on rules; guesses and hunches are always involved: we cannot remove from science the element of conjecture and of risk.

How can we reduce this risk? Only by trying to think out, as well as we can, the consequences of every decision—that is to say, of every adjustment to our theory—which looks promising. (We have

[2][For the idea of 'background knowledge', see *Conjectures and Refutations*, Chapter 10 and Addendum 3. Ed.]

to think out, as it were, all promising combinations.) In this respect, the situation is the same as in any other case in which we have to think out a new theory: the decision to ascribe the refutation of a theory to any particular part of it amounts, indeed, to the adoption of a hypothesis; and the risk involved is precisely the same. To meet it, we need ingenuity, daring—and some luck.[3]

Thus there is no routine procedure, no automatic mechanism, for solving the problem of attributing the falsification to any particular part of a system of theories—just as there is no routine procedure for designing new theories. The fact that not all is logic in our never-ending search for truth is, however, no reason why we should not use logic to throw as much light on this search as we can, by pointing out both where our arguments break down and how far they reach. The fundamental logical asymmetry which I have described can certainly throw some light on this question.

[*All this is about *empirical falsification* and its uncertainties. It must be distinguished from the purely *logical* criterion of *falsifiability*; that is, the *existence* (*not* the truth) of potential falsifiers of a theory. There are no similar difficulties connected with *falsifiability*. Falsifiability is untouched by the problems that may affect empirical falsifications.]

23. Why Even Pseudo-Sciences May Well Be Meaningful. Metaphysical Programmes for Science.

My criterion of demarcation—that is, testability—is needed by the scientist as well as the philosopher in certain concrete difficulties. It singles out those theories which can be seriously discussed in terms of experience. It warns the scientist that there are other theories which cannot be so discussed; and it draws his attention to the fact that these other theories, since they are not testable, must be examined by methods other than testing. If he finds no other way of examining them critically, he may regard himself as well justified in dismissing them. ('Irrefutability is not a virtue but a vice.' Cp.

---

[3]John W. N. Watkins has pointed out 'that the decision *where* the error lies may well be a matter of prolonged critical controversy', and that this may be all to the good, since this controversy is likely to help in clarifying the consequences of the various possible changes to the theory; see his paper, 'Epistemology and Politics', *Proc. Aristot. Society*, 1957, especially the text to the last footnote (referring to Duhem, *The Aim and Structure of Physical Theory*, Chapter vi).

section 7 above.) Yet in doing so, he will always be running a risk; for it is possible at times to learn something of real interest even from a pseudo-scientific or from a metaphysical theory.

As a classical example of a pseudo-science we may consider astrology. Its history may be traced, together with that of astronomy, to the religious belief that the planets are gods (as even Plato asserted). This polytheistic belief was given up in both astrology and astronomy, so that both agreed in the view that the planets were merely *named* after the gods. But astrology, while abandoning polytheism, continued not only to attach a magic significance to the old divine names, but also to attribute to the planets typically divine powers which it treated as calculable 'influences'. No wonder that it was rejected by Aristotelians and other rationalists. Yet they rejected it partly for the wrong reasons; and they took their rejection too far. The lunar theory of tides, for example, was historically an offspring of astrological lore. Prior to its acceptance by Newton, it was rejected by most rationalists as an example of astrological superstition. Yet Newton's theory of universal gravitation showed not only that the moon could influence 'sub-lunar events' but, in addition, that some of the super-lunar heavenly bodies did exert an influence, a gravitational pull, upon the earth, and thus upon sub-lunar events, in contradiction to the Aristotelian doctrine. Thus Newton accepted, reluctantly but consciously, a doctrine which had been rejected by some of the best brains, including Galileo, as part of a discredited pseudo-science.

This shows how easily we can miss a most important idea by rejecting out of hand a pseudo-scientific theory.

A good example of how complicated all this may be is furnished by Kepler, whose theories were a curious mixture of science and astrology. Unlike Newton, who accepted astrological ideas only reluctantly, Kepler belonged to the astrological tradition. Like Copernicus, Kepler belonged to the Platonic-Pythagorean tradition, and believed in astral 'influences', especially that of the sun upon the planets. Nonetheless, Kepler was a highly sophisticated and self-critical astronomer: he never tired of submitting his hypotheses to ingenious and highly critical tests, examining their consequences in the light of the best available astronomical evidence. His wonderfully self-critical attitude ('What a fool I was', he wrote) enabled him to make his great contributions to science—in spite of the fantastic character of some of his beautiful hypotheses.[1]

Astrologers have, incidentally, always boasted that their theories were based on an enormous number of verifications—upon overwhelming inductive evidence. This claim has never been seriously investigated and exploded, and I do not see why it should not be true. Yet it is hardly interesting to know how often astrology has been verified; the question is whether it has ever been seriously tested, by sincere attempts to falsify it.

We may now turn from the pseudo-science of astrology to a highly important metaphysical theory—to atomism. The metaphysical character of the 'corpuscular theory', before Avogadro at any rate, is clear. There was no possibility of refuting it. Failure to detect the corpuscles, or any evidence of them, could always be explained by pointing out that they were too small to be detected. Only with a theory that led to an estimate of the size of the molecules was this line of escape more or less blocked, so that refutation became in principle possible. ('Verifications' were, in principle, possible before this: the invention of the microscope, say, might conceivably have led to the discovery of microscopically visible molecules.) Thus atomism became testable as soon as it was committed to an estimate of the size of a molecule. This example shows that a non-testable theory—a metaphysical theory—may be developed and strengthened until it becomes testable. But if this is so, it seems grossly misleading to describe it as meaningless; and very risky to reject it out of hand as did Mach.

I have repeatedly referred to the example of atomism because it is so highly characteristic and so highly important.[2] Early atomism was a metaphysical system not only in the sense that it was not testable, but also in the sense that it conceived the world in terms of a vast generalization, on the grandest scale: *'There is nothing but*

[1]We see here, as so often (for example in the case of Copernicus), that important hypotheses may originate from truly fantastic ideas: the origin never matters, as long as the hypothesis is testable. Galileo's coolness towards Kepler and his theories is understandable: Galileo belonged to the rationalist camp (like the Roman Church) and was opposed to astrology. This is, of course, also why he was so obsessed with his anti-astrological yet mistaken theory of the tides, and with his oversimplified version of Copernicanism.

[2]*Cf.* sections 4 and 85 of my *L.Sc.D.*, and sections 17 and 19 above.

atoms and the void' (Leucippus, Democritus).[3] Both of its fundamental concepts, atoms and the void, were unobservable and therefore unknown—as Democritus pointed out with devastating logic.[4]
Thus atomism *explained the known by the unknown*: it constructed
an unknown and invisible world behind our known world.[5] And it
was, for precisely this reason, consistently attacked by positivists
(even after 1905), by all inductivists from Bacon to Mach, and by all
instrumentalists from Berkeley to Duhem. Since 1905, positivists
have understandably become more reticent on this point. Yet they
have never explained how it can happen that meaningless gibberish
can be transubstantiated into sense. In fact, the example of atomism
establishes the inadequacy of the doctrine that metaphysics is mere
meaningless gibberish. And it establishes the inadequacy of the
policy of making little surreptitious changes here and there to the
doctrine of meaninglessness, in the vain hope of rescuing it.[6]

However this may be, atomism is an excellent example of a non-
testable metaphysical theory whose influence upon science has
exceeded that of many testable scientific theories. Another grand
theory of this kind was Descartes's clockwork theory of the world
(as I like to call it because it was based on the doctrine that all
physical causation was by push), or, as it may be called, the programme of Hobbes, Descartes, and Boyle, of interpreting the physical world in terms of extended matter in motion. But the latest, and
so far the greatest, was the programme of Faraday, Maxwell, Einstein, de Broglie, and Schrödinger, of conceiving the world—the
atoms as well as the void—in terms of continuous fields.[7] I say 'was'
because this wonderful programme has been destroyed by some
other great physicists. (See Volume III of the *Postscript*.)

Each of these metaphysical theories served, before it became

---

[3]Diels, *Vorsokratiker* ii, 6th ed., 1952, pp. 79 (15); 84 (10); 168 (6).

[4]*Cp.* Democritus, Fragm. 125. (Diels, *op. cit.*, p. 168.)

[5]'But in fact, nothing do we know from having seen it; for the truth is hidden in
the deep.' Democritus, Fragm. 117. (Diels, *op. cit.*, p. 166; *cf.* the motto to this
volume.)

[6]*Cf.* section 26, below.

[7]I asked J. Agassi to study this topic, and am indebted to an as yet unpublished
work by him for many new and interesting details about Boyle's and Faraday's
metaphysical programmes. [*Cf.* Joseph Agassi: 'The Function of Interpretations in
Physics', University of London Library, 1956; and Agassi's book: *Faraday as a
Natural Philosopher*, 1971. Ed.]

testable, as a research programme for science. It indicated the direction of our search, and the kind of explanation that might satisfy us; and it made possible something like an appraisal of the depth of a theory. In biology, the theory of evolution, the theory of the cell, and the theory of bacterial infection, have all played similar parts, at least for a time. In psychology, sensationalism, which may take the form of a kind of psychological atomism (that is the theory that all experiences are composed of unanalyzable ultimate elements, such as, for example, sense data), and psycho-analysis, should also be mentioned as metaphysical research programmes.

Important as these metaphysical programmes have been for science, they have to be distinguished from testable theories which the scientist uses in a different way. From these programmes he derives his aim—what he would consider a satisfactory explanation, a real discovery of what is 'hidden in the deep'. Although empirically irrefutable, these metaphysical research programmes are open to discussion; they may be changed in the light of hopes they inspire or of the disappointments for which they may be held responsible.

Of course there are also metaphysical ideas that are valueless (or so I believe) for science. Some of the 'purely existential statements' discussed in section 15 (see also 27) of the *L.Sc.D.* belong here. The assertion 'there exists a sea-serpent' is not particularly interesting to the scientist as long as no indication whatever is added to it which gives a clue to possible tests. Nevertheless, such purely existential assertions have sometimes proved suggestive and even fruitful in the history of science even if they never became part of it. Indeed, few metaphysical theories exerted a greater influence upon the development of science than the purely metaphysical one, 'there exists a philosopher's stone (that is, a substance which can turn base metals into gold)', although it is non-falsifiable, was never verified, and is now believed by nobody. The example of the philosopher's stone shows, incidentally, that in this case at least, it is the existential character of the statement and not the occurrence of vague or meaningless *terms* which is responsible for the lack of testability: 'base metal' and 'gold' are good enough as far as *concepts* go.

CHAPTER III

METAPHYSICS: SENSE OR NONSENSE?

> The correct method of philosophy should really be like this. To
> say nothing . . . : and then, whenever someone else tries to say
> something metaphysical, to show him that he has given no mean-
> ing to some of the signs in his sentences.
>
> LUDWIG WITTGENSTEIN

Preliminary Remark (1982). This quotation from one of the last
paragraphs of the *Tractatus* became the programme of a world-wide
movement in philosophy. Its aim was to show its own senselessness.
Metaphysics is nonsense! Philosophy is gibberish!

Now I too often get impatient when reading philosophical writings. I am quite ready to admit that much of it is hardly better than
gibberish: philosophizing without genuine problems. So I am not
entirely out of sympathy with the tendency of Wittgenstein's *Tractatus.*

But as we shall see, metaphysical utterances may well be meaningful and interesting. I mean here by 'metaphysical' something like
'not empirically testable'. Wittgenstein meant 'not completely verifiable'.[1] These and related ideas will be discussed in the present

---

[1] According to some fashionable legends, Wittgenstein never upheld the verifiability criterion of meaning: he was misunderstood and misinterpreted by Schlick
and by Waismann (who stated the verifiability criterion in *Erkenntnis* 1, 1930, pp.
228*ff.*). That the legend is untrue can be best seen from Schlick's paper in *Die
Naturwissenschaften* 19, 1931, pp. 145*ff.*; see especially p. 156 where Schlick says
that a natural law 'may be modified in the light of further experience' and so can
never be conclusively or absolutely verified. It therefore 'does not have the logical
character of a statement': it is not a statement but (if I may use Ryle's formulation of
a theory due to J. S. Mill) an inference ticket. Schlick explicitly attributes this
theory to a personal communication of Wittgenstein's; and he undoubtedly obtained Wittgenstein's approval before publishing the relevant passages of his paper.
This shows that, at any rate in 1931, Wittgenstein demanded of 'genuine statements' (*'echte Sätze'*) that they can be 'conclusively' or 'absolutely' verified. All
this can be checked by those publications of Schlick's that were written under the
direct influence of his discussions with Wittgenstein and approved by him. (I may
add that this was confirmed to me personally by Schlick in a conversation in which

chapter. It turns out that two programmes suggested by Wittgenstein are mistaken: (1) that non-verifiable utterances are meaningless because they violate the grammatical rules of language: *Tractatus* 5.473 to 5.47321; (2) that they are meaningless because they employ words or expressions although a meaning has not been given to them: *Tractatus* 5.473 to 5.4733, and 6.53.[2]

24. *Logical Remarks on Testability and Metaphysics.*

In this section I intend to explain, with the help of a number of examples, that testable and non-testable (or 'metaphysical') statements may have the same logical form. Moreover, the same expressions—words, or symbols—may occur in them (perhaps in a slightly different order) and these expressions may have precisely the same meaning. Thus neither the logical form of a statement nor the kind of expression occurring in it suffices to determine whether a statement is testable or non-testable. This distinction is therefore not one between well-formed statements and ill-formed pseudo-statements (as the logical positivist programme assumed). Yet it is not beyond the scope of logical analysis.

Apart from straightforward universal and existential statements we often meet, inside science as well as outside, statements of a slightly more complex structure. The statement asserting the existence of the philosopher's stone is a case in point. For it may be written in the form:

'There exists an x such that x is a substance and for every y the following holds: if y is a base metal and a small amount of x is added to, or mixed with, y, then y turns into gold.'

We may call a statement of this kind an *existential-universal*

he passionately defended Wittgenstein's and his own theories against my criticism. It was also personally confirmed by Waismann.)

[2]The two programmes became for a considerable time the main occupation of the Vienna Circle and especially of its leading logician, Carnap. See his *Pseudoproblems in Philosophy* (*Scheinprobleme in der Philosophie*); 'The Overthrow (*Überwindung*) of Metaphysics through the Logical Analysis of Language'; *The Logical Syntax of Language* (see especially p. 278 where it is asserted that 'the logical analysis shows' that the spurious statements of metaphysics 'are pseudo sentences'). The interest continues in *Testability and Meaning* (where my testability criterion of demarcation is misinterpreted as a meaning criterion). So far as I know, Carnap discussed some examples of bad metaphysics, but he never showed by 'logical analysis' as he intended that non-verifiability or non-testability are always due to either (1) or (2). As I show in this chapter, they are not.

statement, because it asserts the *existence* of some *x* which is characterized by its relation to *every y* of a certain kind.

As a different type of example, we may consider the equally non-testable statement: 'Every event has a cause.' This might be written:

'For every event *x* there exist a *y* and a *z* such that *y* is a regularity describable by some (true) universal law *u*, and *z* is an event (a set of initial conditions) preceding *x*, and *x* is predictable (deducible) from *z* in the presence of *y* (or of *u*).'

We may describe this as a *universal-existential statement*. It is non-testable because there is nothing here to suggest to us how the existence of the regularity *y* (or the universal law *u*) and the preceding event *z* could be ascertained; and if we fail to find an appropriate *y* and *z*, there is nothing here to tell us whether the failure was due to lack of competence or due to the fact that the *y* or the *z* were well hidden, or due to the fact that no such *y* and *z* existed (that is to say, that our universal-existential statement 'Every event has a cause' was false).

I drew attention to statements of these forms in section 66 of *L.Sc.D.*, when discussing the so-called 'Axiom of Convergence' or 'Limit Axiom' of von Mises.[1]

Since my discussion has given rise to misunderstandings (*cp.* footnote *2 to section 66 of *L.Sc.D.*), a new and more thorough discussion seems appropriate here.

I wish to draw attention both to the method of the discussion and to its results. Both seem to me important, and to entail serious criticisms of the fashionable methods: of the method of constructing model languages for the sake of 'precision', and of the method of linguistic or conceptual analysis.

My assertion is that all the expressions or symbols used in two statements, one of which is testable and one of which is metaphysical, may be not only the same, but may have *precisely the same*

[1] von Mises's 'axiom' (which postulates the existence of a limit of the relative frequency of the occurrence of a property *P* in any probabilistic sequence of events or 'collective') may be written as a universal-existential-universal-existential-universal statement, of the following form: '*For every* probabilistic sequence, *there exists* a real number *x* between 0 and 1, called the limit of relative frequency, such that *for every* given fraction *y*, however small, for which *y* > 0 holds, *there exists* a natural number n_y such that *for every* natural number *n* (for which $n > n_y$ holds) the relative frequency, *m/n*, of *m* occurrences of the property *P* up to the *n*th event of the sequence does not deviate from *x* by more than *y*; that is to say, $-y \leqslant x - (m/n) \leqslant +y$.'

meaning. This is of considerable importance. So is the fact that my method makes it possible to assert that the meaning is precisely the same—even though we may not know what 'meaning' is and can ignore the problem (if there is any problem). The method which I am going to adopt is the following: I shall analyze certain simple *conjectures of pure mathematics*, certain statements of number theory. Since they belong to number theory, these conjectures *may* be different in character from conjectures in the natural sciences, but for our purposes these differences do not matter. (I still believe that there are differences, although this belief has been challenged by my colleague, Dr. I. Lakatos.)[2]

We can consider the natural numbers 1, 2, 3, . . . as constituting our infinite universe of discourse: they are the individual things in which we are interested, and about whose properties we are proposing to theorize. For the present purposes, we shall consider a statement to be testable if, and only if, it can be refuted (falsified) by inspecting and determining the properties of a finite number of these individuals, and non-testable (or metaphysical) if we cannot refute it, however many of the natural numbers we may inspect. (It is assumed that we cannot inspect infinitely many of them.)

The only important properties of these numbers which I shall consider here are the properties 'prime' and 'composite': a natural number is called 'prime' if and only if it is greater than 1 and not divisible by any natural number except itself and 1; otherwise it is called 'composite'. (Thus the number 1 is the only non-divisible natural number which is composite, and 2 is the only even number which is prime.)

I choose for my analysis the following two very simple and interesting conjectures about prime numbers:

(G) Goldbach's conjecture

(H) the twin-prime conjecture.

Goldbach conjectured that *every* even number greater than 2 (that is, *every* even number which is non-prime) is the sum of two primes. This conjecture may also be expressed as follows, since all even numbers greater than 2 can be written in the form $2 + 2x$:

For every natural number x, the following holds: $2 + 2x$ is the sum of two primes.

[2][Cf. his *Proofs and Refutations*, 1976. Ed.]

Still another way of expressing the same conjecture is this:

(G) For every natural number x there exists at least one natural number y such that the two numbers,

$$x + y, \text{ and } (2 + x) - y,$$

are both prime.[3]

Goldbach's conjecture may one day be proved. (An important corollary has been proved by Vinogradov.) Today we cannot prove it; but so far we have failed to find a counter-example—a natural number x, that has not the 'Goldbach property', that is, the property ascribed to it by (G).

We now turn to our second example (H), the so-called twin-prime conjecture. Two primes which, like 3 and 5, or 5 and 7, or 11 and 13, or 17 and 19, or 29 and 31, are separated by just *one* (even) natural number, are called 'twin primes'. The twin-prime conjecture states that there are infinitely many twin primes, or, in other words, that there is no greatest pair of twin primes. It can be expressed in a form similar to Goldbach's conjecture, as follows:

(H) For every natural number x there exists at least one natural number y such that the two numbers,

[3]Note that the sum of these two numbers equals $2 + 2x$. Writing '$P(x)$' for 'x is a prime', and using the symbol '(x)' to mean 'for every natural number x', '(Ey)' to mean 'there exists a natural number y such that', and '&' to mean 'and', we can write Goldbach's conjecture as follows (note that the sum of $(x + y)$ and $(2 + x) - y$ is $2 + 2x$, that is, an even number greater than 2):

(G) $\qquad\qquad (x)(Ey)(P(x+y) \ \& \ P((2+x)-y)).$

This is about the simplest way of writing Goldbach's conjecture in this formalism. We can also define (writing '$\leftrightarrow$' for 'if and only if') a property Gdb (to be read 'Goldbachian') of natural numbers such that a number x has the property Gdb if, and only if, $2 + 2x$ is the sum of two primes:

Def 1 $\qquad\qquad Gdb(x) \leftrightarrow (Ey)(P(x+y) \ \& \ P((2+x)-y)).$

With the help of this predicate 'Gdb' we can now express Goldbach's conjecture by

(G′) $\qquad\qquad\qquad (x)\,Gdb(x)\,,$

or in words: 'Every natural number is Goldbachian'.

198

$$x + y, \text{ and } (2 + x) + y,$$

are both prime.[4]

Our two conjectures, (G) and (H), are in their wording (though not in their content) strikingly similar. The only difference is the occurrence of a minus-sign in (G) in a place where (H) has a plus-sign. Apart from this, they are exactly alike. Moreover, (H) uses *only* expressions which occur also in (G); even the plus-sign occurs in (G), although only once. It is therefore clear that if (G) is 'meaningful' (in any more or less acceptable sense), (H) must also be 'meaningful' in precisely the same sense.

Yet if we are interested in *testability*, this small difference makes *all* the difference: for while (G) is testable (falsifiable), (H) is not. We can decide of any given number x whether x has the Goldbach property; but not whether x is exceeded by twin primes.

To test (G) for any given number x it is sufficient to consider those numbers y which are smaller than x. (Only in case $x = 1$ do we consider $y = x$; in all other cases we check the y's between 1 and $x - 1$, these bounds included.) Each of the numbers y for which $x + y$

---

[4]Using the same symbols as in footnote 3, the twin-prime conjecture may be written, very much like (G):

(H) $\qquad\qquad (x)(Ey)(P(x+y) \ \& \ P((2+x)+y)).$

This is, again, about the simplest way of writing the twin-prime conjecture in this formalism. We can also define a property *Etp* (to be read 'exceeded by at least one pair of twin primes') of natural numbers, such that a number x has the property *Etp* if, and only if, it is smaller than some pair of twin primes:

Def 2 $\qquad\qquad Etp(x) \leftrightarrow (Ey)(P(x+y) \ \& \ P((2+x)+y)).$

With the help of the new predicate '*Etp*' we can now express the twin-prime conjecture by

(H') $\qquad\qquad (x) \ Etp(x),$

or in words: 'Every natural number is exceeded by some pair of twin primes'.

Incidentally, we may attribute the property 'twinprimal', or '*Twp*', to one—say the lower—of a pair of twin primes; that is to say, we may define

Def 3 $\qquad\qquad Twp(x) \leftrightarrow P(x) \ \& \ P(2+x)$

which would allow us to replace Def 2 by

Def 2' $\qquad\qquad Etp(x) \leftrightarrow (Ey)Twp(x+y).$

is prime is then checked to see whether or not $2 + x - y$ is also prime. If one of the numbers y makes both $x + y$ and $2 + x - y$ prime, then x has the Goldbach property; otherwise it has not this this property. Thus we can decide in at most $x - 1$ steps of this checking process whether or not some given x is Goldbachian.

As against this, to test (H) for some given number x involves checking all the numbers which exceed x by y, for $y = 1$, $y = 2$, . . . , in order to see whether we find a y such that $x + y$ and $2 + x + y$ are both prime. Of course, if we find, for some given x, a y of this kind, then we have corroborated (H) for this particular given number x. But should we not find a satisfactory y up to, say, $x + y = 2x$, or $x + y = x^2$, then we just have to go on checking larger and larger y's. And even if we never find a satisfactory y, this does not mean that we have falsified (H) for the given number x; for it is always possible that a still greater y may do the trick, and may satisfy (H).

Thus (G) can be refuted by inspection, that is to say, by finding a refuting instance—an x that does not satisfy it. But (H) cannot be refuted in this way, since we can never make sure, by inspection, that a given x does not satisfy (H). This means that (G) is falsifiable, and that (H) is non-falsifiable. Of course, neither (G) nor (H) is verifiable.

It is interesting to note, however, that the status of (H) would change radically should we succeed one day in proving a theorem asserting, say, that if x is exceeded by at least one pair of twin primes, then the smallest of these pairs must lie between x and some calculable number beyond x. For example, we may succeed one day in proving the following falsifiable conjecture:

(H^f) For every natural number x, the following holds: provided $x > 1$, and provided x is exceeded by at least one pair of twin primes, there must be a pair of twin primes between x and $2 + 2x$; or in other words, there must be a y which is the lower of a pair of twin primes such that

$$x < y < 2x.$$

The conjecture (H^f) implies, among other statements, the following: for every pair of twin primes, x and $x + 2$—provided it is exceeded by some pair of twin primes and provided $x \geq 5$—there exists a larger pair of twin primes such that the lower twin lies between $x + 6$ and $2x - 1$, these bounds included.

If we succeed in proving the conjecture (H^f) (or perhaps a weaker one, say with $x<y<x^2$, or with $x<y<2^x$, instead of $x<y<2x$) then the twin-prime conjecture (H) would become testable (in the quasi-empirical sense discussed here). For a proof of (H^f) would make it in principle possible to establish, by inspecting of a finite number of different numbers, of any given number x that there are no twin primes greater than x; or in other words, that x falsifies the twin-prime conjecture.

It is interesting to note that (H^f) in its turn is a *falsifiable* conjecture: it could be falsified, for example, by finding two *consecutive* pairs of twin primes, a, $a + 2$, and b, $b + 2$, such that $b>2a$.

Yet (H^f), though falsifiable, is not systematically testable in the same way as, for example, (G). For if we find, after some number a, no further twin prime, although we have tried out all numbers up to, say, $3a$, or $4a$, then we do not know whether a is a refuting instance of (H^f) or not: $4a + 1$ may be the first of a pair of twin primes (in which case (H^f) is false), or there may be no further twin primes (in which case (H^f) may be true).

We have seen that (H^f) is falsifiable; and in the presence of (H^f), (H) becomes falsifiable also. We can therefore say that the metaphysical character of (H) is *relative*—relative with respect to the actual state of our knowledge (*not* with respect to the 'language system' used). For although (H) is at present non-testable and metaphysical, a new mathematical discovery (the discovery of a proof of (H^f)) would make it testable. Moreover, a system of conjectures consisting of both (H^f) and (H) constitutes a testable theoretical system.

The question arises whether we could replace, in this theoretical system, (H^f) by a non-testable or metaphysical conjecture (which we may call (H^m), while retaining the testability of the *conjunction*—that is, of (H) and (H^m). The answer is: yes, and very easily. (H^m) may be, for example, the following conjecture:

(H^m) If every number is exceeded by some pair of twin primes then, for every number x, if x and $x + 2$ are twin primes, there occurs at least one (larger) pair of twin primes whose lower twin lies between x and $2x$.

Although (H^f) and (H^m) differ in their logical form, the difference in the wording is very small, and they certainly both use the same expressions or 'concepts'; this will be seen clearly if we write (H^f) as follows.

(H^f) For every number x the following holds: if x is exceeded by some pair of twin primes, then, if x and $x + 2$ are twin primes, there occurs at least one (larger) pair of twin primes whose lower twin lies between x and $2x$.

(H^m) is obtained, very simply, by weakening (H^f) sufficiently to make it non-testable. Yet although (H^f) is testable, and (H^m) is metaphysical, the conjunction of (H) and (H^f) has precisely the same logical force or content as the conjunction of (H) and (H^m), as can easily be seen.[5]

Our examples (G), (H), (H^f), (H^m) may be used to show a number of important points.

(1) The difference between testable and metaphysical statements does not necessarily depend upon their logical form[6], but rather

---

[5]To make all this quite clear we write

(H^f) $(x)(Etp(x) \rightarrow (Twp(x) \rightarrow (Ey)(Twp(y) \ \& \ x < y < 2x)))$.

(H^m) $(x) (Etp (x)) \rightarrow (x) (Twp (x) \rightarrow (Ey) (Twp (y) \ \& \ x < y < 2x))$.

It will be clear that (H^m) follows from (H^f) in accordance with the well-known logical principle $((x) (A(x) \rightarrow B(x))) \rightarrow ((x)A(x) \rightarrow (x)B(x))$. It is also clear that from (H), that is, $(x)Etp(x)$, together with (H^m), we obtain (H^f); thus the conjunction of (H) and (H^f) has precisely the same force as the conjunction of (H) and (H^m).

[6]We can, of course, construct formalisms into which (G) and (H) can be translated and with respect to which their difference may be said to be a difference of logical form. For example, we may stipulate that the symbol '(Ex)' may be followed by a subscript stating an upper bound for the variable x. In this case, (G) and (H) might become, say,

(G′) $(x)(Ey)_{x-1}(x > 1 \rightarrow P(x+y) \ \& \ P((2+x)-y))$.

(H′) $(x)(Ey) (x > 1 \rightarrow P(x+y) \ \& \ P((2+x)+y))$.

202

upon the information they convey—something that can be expressed in the most different ways and that is (if it is adequately expressible in some language) not dependent on the language chosen: if not testable in language L_1, it may be so in language L_2.

(2) The difference certainly does not depend upon the occurrence, in the metaphysical statement, of meaningless or ill-defined symbols or expressions. Only such symbols occur in (H) or (H^m) as occur also in (G) or in (H^f) respectively. In (G) there occurs one symbol, ' $-$ ', which does *not* occur in (H); but it is obvious that this fact—which might conceivably make (G) meaningless—cannot possibly make (H) meaningless if (G) is meaningful.

(3) The metaphysical character of both (H) and (H^m) is relative to our other conjectures: in the presence of (H^f), (H) becomes testable. Also, in the presence of (H), (H^m) becomes testable (and *vice versa*).

(4) The metaphysical character of a statement may be the result of its logical weakness—its lack of content: (H^m) is obtained from (H^f) by weakening it (decreasing its content).

Similarly, we can obtain a metaphysical statement by weakening (G). For example we may get, by weakening (G), the non-falsifiable conjecture that *almost all* natural numbers (say, all except a finite set) are Goldbachian (a conjecture closely related to one which has been proved by Vinogradov). Or we may get the still weaker conjecture that, for every number x, there exists a block of at least x consecutive Goldbachian numbers. Both these conjectures are, obviously, non-testable consequences of (G). (A still weaker consequence—that there exist infinitely many Goldbachian numbers—is demonstrable, since it is an immediate consequence of the theorem, proved by Euclid, that there exist infinitely many prime numbers.)

(5) The metaphysical character of a statement is not *merely* a question of its logical weakness, or lack of logical content. Though a (non-tautological) statement whose *logical* content is too small

---

The difference between them would now be exhibited by the fact that the operator '(Ex)' is bounded in (G) but not in (H), which, it could be said, makes a difference in their logical form; and we can, of course, exclude (H) as meaningless by stipulating that all existential operators must be bound (by some finite constant or some term other than the one occurring in the operator itself). But the fact remains that we may express the two conjectures perfectly adequately in languages in which their logical form is identical.

cannot have any empirical content and must therefore be metaphysical, the converse does not hold: a statement may be metaphysical (which means that its *testable content* is zero) even though it may have a high *logical content*. This fact will be established in the next point, (6).[6a]

(6) We can illustrate all this with the help of statements employing not only purely mathematical but also empirical terms. The simplest example, perhaps, is obtained by replacing some of our mathematical terms by empirical ones.

Thus taking as our universe of discourse a sequence of tosses with a penny, we can replace 'x is prime' by 'the toss number x results in heads turning up'. We thus obtain from our two conjectures, (G) and (H), the following *conjectures about our sequence of penny tosses*:

(GP) For every natural number x, there exists at least one natural number y such that the two numbers

$$x + y \text{ and } (2 + x) - y$$

are both numbers of tosses resulting in heads turning up.

(HP) For every natural number x there exists at least one natural number y such that the two numbers

$$x + y \text{ and } (2 + x) + y$$

are both numbers of tosses resulting in heads turning up.

[6a][*(Added 1982) It is possible to make (G) and (H) even more nearly similar. We can use the symbol $|x|$ for the 'absolute value' of x; that is $|x| = x$ if and only if $x \geqslant 0$; otherwise $|x| = -x$ (and therefore positive). So we may write:

(F) $\qquad\qquad (x)\,(Ey)\,(P(|x| + |y|) \,\&\, P(2 + |x| + y)).$

This formula states (G) when we let the variables x and y range over the negative integers, $x,\ y < 0$; and the same formula (F) states (H) when we let the variables x and y range over the positive integers, $x,\ y > 0$; and it states

(I) $\qquad\qquad (x)\,(Gdb(x) \lor Etp(x))$

when we let x and y range over the positive and negative integers except 0: $x,\ y,\ \neq 0$.

It is interesting that even though this range includes the others (I) is by far the weakest of the three conjectures (G), (H), and (I). It is far weaker even than the disjunction of (G) and of (H).]

Now (G) is a very dubious conjecture about all x, since its holding good for a small x is a sheer matter of luck; though with increasing x it becomes more and more probable that any particular instance x will conform to (GP).

But (HP) is 'almost certainly true' for every x and every series of penny tosses, because it merely asserts that we shall find from time to time, if we continue tossing, two tosses of heads separated by just *one* toss (no matter whether heads or tails). This is a very weak conjecture: given the information that the probability of heads turning up equals $1/2$, and that the tosses are independent, the logical content of (HP) turns out to be zero.[7]

It follows that the logical content of the negation of (HP), given the same information, equals 1. This is so because the negation of (HP) is 'almost certainly false', on the information given: it 'almost contradicts' this information.[8] Nevertheless, the negation of (HP) has no *empirical* content: it is not testable, precisely because it asserts *the bare existence* of a toss-number after which there will be no more twin tosses of heads (separated by precisely one toss). Thus it will not be refuted, even if we find such twin tosses again and again, after any number we examine.

The example is important. It shows the existence of 'completely metaphysical' statements (that is, non-testable statements whose negations are also non-testable, or in other words, statements which are neither falsifiable nor verifiable). Moreover it shows that we may have reasons—perhaps even empirical reasons—for accepting even a 'completely metaphysical' statement, and thus tentatively rejecting its negation. (This, of course, is an ordinary occurrence with unilaterally verifiable metaphysical statements such as 'there exists at least one non-white swan'; though it is a matter of choice whether to call this statement 'metaphysical' or 'empirical' in so far as it may be regarded not as 'isolated' but as forming part of an

---

[7] I assume here that the logical content of a given b is complementary to the probability of a given b; that is to say, $Ct(a,b) = 1 - p(a,b)$.

[8] In the theory of von Mises, the negation of (HP) actually contradicts the information; for in this theory, the fact that an occurrence of twin tosses of heads, separated by precisely *one* toss, has the probability of $1/4$, entails that occurrences of this kind must turn up again and again. Thus, in this theory, (HP) actually follows from the 'given' information, while in classical and in measure theoretical probability theories it does not follow (though it 'almost follows'). (See below, sections 22ff.)

empirical system. Of course, if a non-testable statement *adds* to the logical content of a testable system—as in the case of the conjunction of (H) and (H^m), or their empirical counterparts—we no longer call it 'metaphysical' when regarding it as part of that system.)

There is no need to formulate explicitly the penny-tossing conjectures which are counterparts of (H^f) and (H^m). Instead, it may be worth mentioning that we can construct other very close (though not very interesting) counterparts of our mathematical conjectures in empirical terms. We may choose, for example, conjectures about sun-spot activity, interpreting our natural numbers x as numbers of years (counted from some zero moment) and replacing our predicate 'prime' by 'a year of low sun-spot activity'.

(7) Both the testable statement (G) and the metaphysical statement (H) give rise, as we have seen, to the definition of a characteristic predicate—the predicates 'Goldbachian' and 'exceeded by some pair of twin primes'. The first of these predicates may be called 'completely decidable' or 'normal', or 'ordinary', or 'recursive' because we can decide in a finite number of steps of inspection whether or not any given number n is Goldbachian. It is in this respect very similar to many quite ordinary empirical predicates (see (8) below). The second predicate, 'exceeded by some twin primes', may be 'unilaterally decidable' or 'semi-normal', or 'semi-ordinary', or 'recursively enumerable' because it is verifiable but not falsifiable: given some number n, we *may* find a pair of twin primes greater than n, and thereby verify the statement 'n is exceeded by some twin primes'; but failure to find a pair of twin primes does not falsify the statement.

We see here the skeleton of a classification into categories (a classification which could be much refined, should we so wish):

(a) 'Ordinary' predicates which, if applied to single individuals (for example, individual numbers), yield *completely decidable* singular statements.

(b) 'Semi-ordinary' predicates which, if so applied, yield *unilaterally verifiable* (that is, non-falsifiable) singular statements. (These predicates may be called 'metaphysical predicates'.)

(c) 'Semi-ordinary' predicates yielding *unilaterally falsifiable* singular statements.

(d) 'Extraordinary' predicates yielding singular statements which are *neither verifiable nor falsifiable*. (These predicates may also be called 'completely metaphysical predicates'.)

It will be seen that the complement or negation of a predicate of category (a) again belongs to category (a), and that the negation of a predicate of category (d) again belongs to (d); but the negation of a predicate of category (b) belongs to (c), and *vice versa*.

(8) Some examples and construction methods for predicates (both mathematical and empirical) of the four categories are implicit in our discussion.

(a) As examples of 'ordinary' or completely decidable predicates, we may take the predicates 'prime' or 'Goldbachian' or 'red' or 'toss resulting in heads turning up', and, of course, the complements or negations of these predicates.

While the *singular statements* obtained from these predicates are completely decidable, the *universal statements* are, in an infinite universe, unilaterally falsifiable, and the *existential statements* unilaterally verifiable.

(b) We may now construct some unilaterally verifiable predicates from those of category (a); for example the predicate 'exceeded by some Goldbachian number' and the predicate 'exceeded by some non-Goldbachian number' are both unilaterally verifiable (and so is the predicate 'exceeded by some twin primes').

As to 'red', we may assume a universe of motor cars, and we may call a car *c* 'succeeded by another car' if *c* is of an earlier date or has a lower chassis number. The predicate 'succeeded by a red-painted car' will then be a verifiable predicate, and non-falsifiable as long as the production of motor cars continues. Analogous remarks hold for 'succeeded by a year of low sun-spot activity' or 'succeeded by five tosses of heads in a row', etc.

These predicates yield unilaterally verifiable singular statements and unilaterally verifiable existential statements; but the universal statements are neither verifiable nor falsifiable: they are completely metaphysical, like (H).

(c) Unilaterally falsifiable predicates may be constructed in the same way. For example, 'exceeded only by Goldbachian numbers' or 'succeeded only by red-painted cars' yields, clearly, singular statements which are unilaterally falsifiable. (Also, the complements of predicates of category (b) may be referred to here, and *vice versa*.) The universal statements are also unilaterally falsifiable, but the existential statements are completely undecidable.

(d) Finally, we may construct completely undecidable predicates from those in categories (b) or (c), by a kind of universalization of those in (b) and by a kind of particularization (existentialization) of those in (c). We thus obtain two sub-categories, (d_b) and (d_c), but there are many others belonging to (d). For example, the (demonstrably universal) predicate 'exceeded only by numbers which in turn are each exceeded by some Goldbachian number' is a completely metaphysical predicate of sub-category (d_b), while 'exceeded by some number which in turn is exceeded only by Goldbachian numbers' belongs to (d_c). The other predicates in (b) and (c) yield analogous predicates such as 'succeeded only by cars, each of which is in turn succeeded by a red-painted car' (d_b), or 'succeeded by some car which in turn is succeeded only by red-painted cars' (d_c).

The predicates of (d_b) and (d_c) constructed according to this scheme yield singular, universal, and existential statements which are all completely untestable or 'completely metaphysical'.

But none of these predicates and none of these statements should be described as 'meaningless' in any sense whatever. On the contrary, the predicates are well-defined, in terms of obviously meaningful concepts; and the statements are grammatically well-formed.[9] Moreover, some of the statements are demonstrably true and others are demonstrably false.

For example, 'exceeded by some Goldbachian number', which belongs to (b), is a demonstrably universal predicate. That is to say, it demonstrably holds for every number n (simply because there is no greatest prime, and therefore no greatest sum of two primes). The same holds for 'exceeded only by numbers which are exceeded by some Goldbachian numbers', which belongs to (d_b) and is therefore 'completely metaphysical'. In fact, all the singular, universal, existential statements obtained from predicates of category (d) formed in the manner here described are logically weak consequences of statements of category (a). For example, 'exceeded by

---

[9]In order to show this in some standard symbolic language, we can start with the completely decidable predicate 'A' ('red' or 'Goldbachian' or 'non-Goldbachian', etc.), and the relation '$<$' ('is exceeded by'). We define

$$B(x) \leftrightarrow (Ey)\,(x < y \,\&\, A(y))$$
$$C(x) \leftrightarrow (y)(x < y \rightarrow A(y))$$
$$D_B(x) \leftrightarrow (y)(x < y \rightarrow B(y))$$
$$D_C(x) \leftrightarrow (Ey)\,(x < y \,\&\, C(y))$$

These seem to be about the simplest definitions yielding the desired results; but there are, of course, many other ways of achieving similar results.

some number which is exceeded only by Goldbachian numbers' (d$_c$) is a logically slightly weaker form of 'exceeded only by Goldbachian numbers' (c).[10]

It would be difficult for anyone to describe these examples as meaningless (or 'cognitively meaningless'); especially for those philosophers who argue, rightly, in favour of the meaningfulness of purely existential statements and who believe, wrongly, that they can thereby show the inadequacy of my criterion of demarcation (which indeed brands isolated existential statements as non-testable and 'metaphysical'—though not of course as meaningless).

The predicate 'exceeded by some pair of twin-primes' which is based on (H) is a non-testable predicate; yet in the presence of (H^i) it becomes testable.

Thus the *status of a predicate* with respect to testability may change with the progress of our knowledge; precisely as the status of a statement like (H) with respect to testability may change with the progress of our knowledge. But the *meaning* (whatever this may be[11]) of the predicate need not change at the same time. In our example it certainly does not change, since its meaning is fully defined throughout in terms of the predicates 'natural number', 'prime', 'greater', and some logical operators, etc. Analogous remarks hold for the predicate 'exceeded by two tosses of heads which are separated by precisely one toss'.

This result is of considerable interest; at least for myself it solves a problem about which I have been uneasy for many years. The problem is whether or not the meaning of our theoretical terms *always* changes when, due to the progress of science, their status with respect to testability changes. That the meaning of our theoretical terms changes very often in the course of scientific progress, is clear: one has only to think of such terms as 'atom' or 'planet' or 'light' or 'movement' or 'space' or 'force'. Influenced by these examples, I became doubtful whether there could be such a thing as 'meaning-invariance' (the term is due to Paul Feyerabend),[12] in a

[10][The text is defective here. Several lines containing further examples are missing, and could not be reconstructed. Ed.]

[11]I do not know *what* the 'meaning' of an expression *is* (and I do not believe in 'what-is? questions') but I assume that a (mathematical) definition may at least help to determine it.

[12][*Cf.* his 'Problems of Empiricism', in Robert G. Colodny, ed.: *Beyond the Edge of Certainty*, 1965, p. 164. See Popper: *L.Sc.D.*, section 38, note *3. Ed.]

growing science; and more especially, whether the meaning of a theoretical term was not bound to change whenever the progress of science affected its status with respect to testability.

Yet these doubts are unfounded, as our discussion shows. Though the meaning of our terms may change as our theories change, such change is not bound to follow even a change of status with respect to testability. Testability and meaning are, of course, related: whether a theory is testable will depend, partly, upon the meaning of the terms used (and partly upon the state of our knowledge). But there is no dependence in the other direction: even spectacular changes in testability need not, as such, affect meaning.

These results present a solution of what was for me an open problem of long standing. When I understood the solution I also thought that my doubts might perhaps be a symptom that I was not entirely free from the influence of positivist thinking even though I had argued long ago against the theory that the meaning of a term was determined by the method of verifying (or even of testing) the statement in which the term in question occurred. I had never been really interested in the problem of the meaning of terms, and I had uncritically, and indeed thoughtlessly, accepted the somewhat broader view that the meaning of a term was determined by its usage. Since the scientific usage of a term was often influenced by its testability status, it appeared that changes in this status might influence the meaning of the term in question.

It was only as the result of the analysis reported in this section that I realized that all this was mistaken; that it is not the 'usage' of a term which determines its meaning but, much more narrowly, *the relationships which, we assume,* hold good *between the entities* denoted by the term, and other entities. Here, as always, linguistic analysis is mistaken. Our usage (or its rules) may be symptomatic of our theoretical assumptions—it may reveal these assumptions—but it is not identical with them. And although our assumptions are often only implicit and not easy to formulate explicitly, they can be formulated explicitly as *conjectures about things.* Usage, on the other hand, or rules of usage, can at best be formulated as rules describing our own behaviour.

(For example, the meaning of the word 'between' may be said to be given by Hilbert's axioms. But these axioms are assumptions about things—and about special relationships between things. They are not rules laying down habits of speech.)

The fact that the meaning of our terms is not, in general, dependent on their testability status destroys, I believe, another very popular theory of the way in which theoretical terms acquire empirical meaning. I am alluding to the inductivist theory according to which it is through their (indirect) contact with observation and experiment that empirical meaning is bestowed upon theoretical terms. (The operationalist theory of meaning is a variant of this inductivist approach.) The intuitive idea underlying this view may be expressed as follows: non-theoretical terms, or empirical terms of a basic character, denote 'observables' and are therefore 'meaningful'; theoretical terms ought, if possible, to be defined by (or 'constituted' by, or 'reduced' to) these basic empirical terms; if this is not possible, they can still acquire *some* meaning through being employed in statements from which other statements that use basic empirical terms can be derived. In this way the undefined theoretical terms will absorb, as it were, some empirical meaning: empirical meaning will percolate from basic statements and basic terms up to the theoretical statements and the theoretical terms.

If this view were correct, then nothing could be of greater influence upon the meaning of a theoretical term than a change in its testability status—or more precisely, in the testability status of some of the statements in which it occurs—from 'non-testable' to 'testable', and on to 'tested'. However, this change does not, in general, affect the meaning, as we have seen; and so the inductivist theory of the absorption of empirical meaning by theoretical terms collapses.

This need not surprise us, considering that the distinction between basic empirical terms and theoretical terms is altogether mistaken. As I have said before, all terms are theoretical, though some are more theoretical than others.[13]

25. *Metaphysical Terms Can Be Defined by Empirical Terms.*

It may be worth while briefly to sketch a general procedure by which we can define, in empirical terms, metaphysical predicates; that is to say, predicates which are essentially non-testable (though perhaps verifiable).

Let x be a person (or an animal, or a plant, or another thing). We

[13]*L.Sc.D.*, end of section 25; see also *Conjectures and Refutations*, pp. 118, 388.

wish to define a metaphysical disposition of x—an ability of x which is not only 'occult' in Berkeley's sense but in principle not testable.

Take any empirical thing y. It may be a most ordinary thing. Let it be a glass of water.

Now what will x do with a glass of water? The simplest thing is to drink it. So we come to the schema, 'x has the disposition D if and only if *there exists* a y such that y is a glass of water and if x drinks y then . . . ' . Now insert for the dots something emotionally exciting, such as being rejuvenated, or being turned into a lion. This finishes the story: 'x is able to rejuvenate himself (or to turn himself into a lion) if and only if there exists a y such that y is a (very special) glass of water (or a golden beaker filled with water) and if x drinks y, x will look, and feel, as he did when he was young (or will look, and feel, like a lion), and x knows about the secret of y.'

If x is not a person, or an animal, but a thing, we can still continue to operate with our glass of water: x may then be convertible into gold, or into a pearl, if wetted by the water, or dropped into it. Moreover, the metaphysical disposition may be such that it is possessed by all men (or animals or plants, or other things) or merely by very special ones (people born on a Sunday, or under the sign of Gemini; or a herb planted during a new moon).

For example, let x be a certain disease. Then y may be the herb which cures it, if y is taken three times a day. Thus we arrive at a metaphysical property 'curable by *some* herb *existing somewhere*'.

It may be said that what I have described are not metaphysical predicates but rather magical ones. Yet primitive metaphysics is closely related to magic. They are magical formulae insured against refutation *by their existential character*.

The metaphysical predicates whose construction is described here are all meaningful; for they can all be *defined* in meaningful terms; and a definition is generally supposed to give meaning to the defined terms provided the defining terms have meaning. And since we may choose the most trivial empirical terms as defining terms, there is clearly no problem here.

More complex definitions may be designed, making use of more than one existential operator: two herbs may be said to have the dispositional property of being convertible into a cure-all if, and only if, there exists a constellation, *and* there exists a secret Latin formula, such that if under that constellation, the two herbs are brewed together while the formula is pronounced, the result will be a cure-all.

The main attack of the positivists is of course directed against religious metaphysics, especially against the possibility of a 'rational theology'. They assert that its terms must be meaningless. In order to refute this assertion, I attempted some time ago to formulate what I called 'the arch-metaphysical assertion' in the symbolism of Carnap's 'Testability and Meaning'[1]. By 'the arch-metaphysical assertion' I mean the assertion of the existence of a personal God, that is to say, of an omnipotent, omnipresent, omniscient, and personal spirit. 'x is omnipotent' can be defined by generalizing the idea 'x can put the thing y into the place z': this idea belongs to ordinary life and is therefore sufficiently empirical. And although its generalization, 'x can put anything anywhere' is, perhaps, no longer empirical, it is definable by the application of a very ordinary logical operation to an empirical term, and therefore clearly meaningful. Omnipresence is even easier to define by generalizing the empirical predicate 'x is located at the place y'. The result is 'x is located at all places'. The terms 'omniscience' and 'person' can both be defined together, with the help of the predicate 'x knows y'. We can say that x is a person if and only if there is a y and x knows y; and we can say that x is omniscient if and only if x is located somewhere, and if x knows every y. As to spirit, an omnipresent person may be said to be a spirit. In this way, we may give 'empirical meaning' to rational theology. Yet it remains metaphysical: not because it is meaningless, but because its assertions cannot be tested—because they are irrefutable.

Some positivists have accepted the meaningfulness of my reconstruction of the statement which I have called 'the arch-metaphysical assertion'; but have replied that my statement, since it is meaningful, is clearly not metaphysical but simply a *false empirical statement*; to which they add the rider that this is an old problem which was settled long ago. The rider is a necessary part of the reply, and it is false. For it can easily be shown, from the literature of the Vienna Circle, that the main attack of positivism was upon 'traditional metaphysics' of which the so-called 'rational theology',

[1]This attempt is made in my paper 'The Demarcation between Science and Metaphysics', in *Conjectures and Refutations*, Chapter 11. See notes 1 to 3 to the next section.

for instance of Spinoza, was the outstanding example. The positivist thesis was that all this was *irreparably meaningless*. This thesis is clearly discarded by those who now say that, by showing that it is possible to interpret some metaphysical theories as meaningful, I have merely shown that they are non-metaphysical.

Indeed, we may now well ask: '*What was this traditional philosophy which you once attacked so violently?* What was all the excitement about? And in any case, did you not connect testability and meaning? And do not our examples, and others, show that there are non-testable but meaningful statements? And why should the tautology 'A meaningless combination of words is a meaningless combination of words' be of the slightest interest to anybody under the sun? Yet does not your assertion that metaphysics is meaningless turn into this tautology if my examples are countered by saying that, as they are meaningful, they are not metaphysical?'

But leaving controversy aside, I may sum up my views as follows. While most *scientific concepts* cannot be defined in terms of what may be called 'phenomena', or even of what may be called the empirical terms of ordinary language, some *metaphysical concepts* can be so defined. Accordingly, an attempt to characterize science (as opposed to metaphysics) by a criterion like the empirical definability of its terms leads to a demarcation that is at once too narrow and too wide: it will exclude nearly all it is intended to include, and it will include much of what it is intended to exclude.

26. *The Changing Philosophy of Sense and Nonsense.*

Wittgenstein, in his *Tractatus*, like Berkeley, had a philosophy of meaning and meaninglessness, or of sense and nonsense, which was both vigorous and clear. On the one side were informative empirical statements, and on the other mere gibberish, mere verbiage; though mere verbiage that could look like an empirical statement. The signs constituting the empirical statements were words which had been given an empirical meaning: each was associated, by usage, with certain observable things, or events. The gibberish, on the other hand, was either ungrammatical (as in 'Socrates is identical') or it contained words to which we 'had given no meaning'. And the sole task of philosophy was—according to Wittgenstein—to 'demonstrate' to people who talked metaphysics that they were talking nonsense. This is how he summed up his message in the *Tractatus* (6.53): 'The right method of philosophy would be this. To say nothing except what can be said, i.e. the propositions of natural science, i.e. something that has nothing to do with philosophy: and then always, when someone else wished to say something metaphysical, to demonstrate to him that he had given no meaning to certain signs in his propositions.'

Here we had a challenging theory. Clearly, it was incomplete as long as it had not been shown that the words, or signs, or concepts which are used by the natural sciences can all be defined by empirical definitions—definitions based upon concepts expressing immediate experiences. This was the task which Carnap set himself in *Der logische Aufbau der Welt* (1928). In this book, he tried to develop the outlines of a system of definitions or 'constitutions' of scientific terms: a concept was 'constituted' if there existed a *chain of definitions* reducing it to terms of immediate experience.

This, of course, was a theory of induction from 'data', applied to concepts instead of statements. But it remained a programme, for the thing could not be done. The 'constitution of concepts' was no more possible than the 'induction of theories'.[1]

As a consequence, Wittgenstein's simple and straightforward theory of meaninglessness or nonsense had to be scrapped. It was scrapped; and the philosophy of meaninglessness or of nonsense, and with it, the positivist school, began to disintegrate.

Many little changes and adjustments have been tried over the years. The straightforward idea of meaninglessness or gibberish has been replaced successively by various more sophisticated ideas.[2] It has been said, for example, that an expression is 'metaphysical' if it is 'not a significant sentence of the language of science'. (As if any metaphysician would care whether or not he was using 'the language of science', or what somebody may claim to be this language.) Thus the thesis of meaninglessness has been watered down until it can no longer be recognized, except by experts. A similar thing has

[1]This was pointed out in my *L.Sc.D.* (end of section 25); and I had pointed it out even earlier in discussions with members of the Vienna Circle. See for example note 27, and the text between notes 25 and 27 of my paper, 'The Demarcation between Science and Metaphysics', *op. cit.*

[2]The story is told more fully in my paper, 'The Demarcation between Science and Metaphysics'; see the preceding note. [*Cf. Unended Quest*, sections 16 and 17. Ed.]

happened to the idea of metaphysics. Expressions which during the nineteen thirties would have been considered models of metaphysical nonsense (since they were neither verifiable nor falsifiable) are now considered perfectly respectable. And all these surreptitious changes have been introduced, *ad hoc*, in order to avoid the unintended consequences of a criterion of meaning which, however modified, has always remained inadequate. For it always was *too narrow and too wide* at the same time, as may easily be shown: it has always placed desirable scientific theories and undesirable metaphysical theories on the wrong sides of the demarcation line. For even rational theology, for example, can be 'formalized' in some of the proposed languages of science.[3]

Why not simply abandon the dogma that the criterion of meaning (or—now—of 'significance') must be equivalent to the criterion of demarcation? Attempts to rescue this dogma have caused endless trouble. I cannot imagine why the trouble was thought to be worth taking. Perhaps in order to save the venerable philosophy of positivism from being refuted? But it has been refuted; and it has been given up, in one way or another, by everybody who understands the situation. So why bother any longer?

With this I conclude my comments on the problem of demarcation. For most of its logical aspects, especially its connection with the idea of empirical *content*, have already been treated fairly satisfactorily in *L.Sc.D.*, and are not in urgent need of further elucidation.

---

[3]See note 1 to section 25, and my paper, 'The Demarcation', etc. Carnap has since written a paper, 'The Methodological Character of Theoretical Concepts' (see my references in notes 5 and 6 to section 11, note 12 to section 13), in which he revises some of the views which I criticized in my paper; more especially, he tries to give a new criterion of meaning (now called 'criterion of significance') and he offers two proofs, one establishing that his criterion is *not too wide* (pp. 54–57), and the second that it is *not too narrow* (pp. 58–9). Yet in the presence of certain assumptions made in the second proof, it can be shown that the first proof breaks down, and that the criterion turns out to be *too wide*; and similarly, in the presence of certain assumptions made in the first proof, the second proof breaks down, and the criterion turns out to be *too narrow*.

CHAPTER IV

CORROBORATION

27. *Corroboration: Certainty, Uncertainty, Probability.*
I have, in the preceding two chapters, explored the logical ramifications of the problem of induction. There is another ramification, however, which I have not yet touched on, because it is not *logically* connected to the problem of induction. But it is connected with it by ties that may prove even stronger than logic: by the inductive prejudice, and by a *mistaken solution of the problem of induction* which, unfortunately, is still widely accepted as valid. I am alluding, of course, to the view that although induction is unable to establish an induced hypothesis with *certainty,* it is able to do the next best thing: it can attribute to the induced hypothesis some degree of *probability.* (And a probability of 1 would be certainty.)

This view is radically mistaken; yet it can be supported by a highly persuasive argument. This argument may be presented as follows.

The whole problem of induction, the argument runs, clearly arises from the fact that inductive inferences are *not valid*: which is the same as saying that inductive conclusions do not follow deductively from the inductive premises. But there is no need to get alarmed about this somewhat trite fact; especially as there exists a large and important class of inferences in which the conclusion does not strictly follow from the premises. In fact, every deductive inference may be modified so as to yield an inference which is not valid, but only 'more or less valid', or 'valid to a degree'. Take the following example:

VALID	VALID TO A DEGREE
All men smoke	x percent of all men smoke
Jack is a man	Jack is a man
Jack smokes	Jack smokes

As the example suggests, we may say that, while in the *valid* inference the truth of the premises makes the truth of the conclusion *certain*, in the *more or less valid* inference, the truth of the premises leaves the truth of the conclusion *uncertain*. Yet the conclusion is made more or less *probable*—to be precise, probable to the degree of *x* per cent.

Inferences which establish a conclusion as probable—or probable inferences, as we may say—do exist. As the example shows, they are inferences in which the conclusion, somehow or other, goes beyond what is stated in the premises: it is not fully entailed by the premises. Clearly, this is so with all inductive conclusions. This seems to show that the problem of induction will be solved as soon as we have developed a theory of probability which allows us to assess the probability of an inductive conclusion—that is, of a hypothesis—given some inductive, or evidential, premises. If we write

$$p(a,b) = r,$$

to mean 'the probability of *a* given *b* equals *r*' (where *r* is some fraction between 0 and 1), then we can put the persuasive argument finally as follows.

Let *h* be a hypothesis (an 'inductive conclusion'), and *e* our 'inductive evidence'. Then the problem of induction—or so it seems—consists in determining the value of *r* in

$$p(h,e) = r:$$

that is to say, the value of the probability of the induced hypothesis *h* given the evidence *e*. Thus the problem of induction is to be solved by constructing a generalized logic—a logic of probability. *For according to this persuasive argument, inductive logic is nothing but probability logic.* It is the logic of uncertain inference, of uncertain knowledge; and $p(h,e)$ is the degree to which our certain knowledge of the evidence *e* rationally justifies our belief in the hypothesis *h*.[1]

[1] In *L.Sc.D.* (Appendix *iv) I gave a number of axiom systems for the formal calculus of probability (one of whose interpretations is the logical interpretation). Although I do not assume any acquaintance with these technicalities, I will, for the sake of completeness, state here one of the simplest axiom systems. It consists of three axioms only, the first introducing the idea $p(a,b)$, the second the idea of the product *ab* (to be read '*a*-and-*b*'), and the third the idea of the complement *ā* (to be read 'non-*a*').

A   There are at least two elements, *a* and *b*, such that

$$p(a,a) \neq p(a,b).$$

As I have said before, I believe that this argument is completely mistaken. The appeal to probability does not affect the problem of induction at all. (*Cf.* sections 1, 80 and 83 of my *L.Sc.D.*, and the passages from Hume in Appendix *vii, text to footnotes 4 to 6.) Formally, this may be supported by the remark that every universal hypothesis *h* goes so far beyond any empirical evidence *e* that its probability *p(h,e)* will always remain zero, because the universal hypothesis makes assertions about an infinite number of cases, while the number of observed cases can only be finite.

There are at least two ways of criticizing the alleged logical connection between the problems of induction and of probability. The full discussion of one of these I postpone to the second part of

B
$$p(ab,c) = p(a,d)p(b,c) \leqslant p(a,c),$$
provided $p(a,a) = p(bc,d)$ and $p(bc,c) = p(d,c)$.

C
$$p(\bar{a},b) = p(c,c) - p(a,b),$$
provided $p(d,d) \neq p(d,b)$ for some element *d*.

To this we may add a postulate introducing the (substitutional) identity, $a = b$, of two elements *a* and *b*:

(*) $a = b$ if, and only if, $p(a,c) = p(b,c)$ for every element *c*. (In this case, *a* and *b* may be substituted everywhere for each other.)

Further definitions may be added by adopting a rule of definition based upon $a = b$; for example, we may define, '*a* v *b*' (to be read '*a*-or-*b*'), as usual, by

$$\text{'}a \text{ v } b = \overline{\bar{a}\bar{b}}\text{'}.$$

This simple axiom system allows us to deduce, besides all the laws of Boolean algebra, the well-known laws of the calculus of probability, such as the *law of reflexivity*

(1)
$$p(a,a) = p(b,b) = 1.$$

Immediate consequence of (1) and B are the very important *law of monotony*

(2)
$$p(ab,c) \leqslant p(a,c)$$

and the *general multiplication law*,

(3)
$$p(ab,c) = p(a,bc)p(b,c).$$

A more remote consequence of the system is the *general addition law*,

(4)
$$p(a \text{ v } b,c) + p(ab,c) = p(a,c) + p(b,c).$$

We obtain, furthermore, a number of theorems which are not obtainable in the usual calculus, for example the following:

(5)
$$p(a,\overline{b\bar{b}}) = 1$$

(6)
$$ab = b \text{ if, and only if, } p(a,b\bar{a}) \neq 0.$$

Within the logical interpretation, the equality, '$a = b$' may be read 'is logically equivalent to' (or 'is interdeducible with'); and '$ab = b$' may then be read '*a* follows from *b*' which may of course also be written '$a \geqslant b$' or '$b \leqslant a$'.

this volume, where I intend to criticize the *subjective interpretation of probability*—that is to say, the interpretation of probability as degree of incomplete or uncertain knowledge. There I shall attempt to show that the idea of an inductive probability does not work, and why it cannot work.

The other way is by pointing out that *the intuitive belief in an inductive probability of a hypothesis in the light of evidence* is a mixture of at least two intuitive ideas. One of these ideas—to be discussed in this and the following sections—is defensible but of only limited importance, and logically unconnected with the problem of induction; while the other—the idea of an *inductive probability logic* which I have just expounded—is indefensible. That it is indefensible for logical and mathematical reasons will be shown more fully in the second part of this volume; here I intend to point out another aspect of this idea, one that appears to me in a way even more important. I mean the mistaken attitude towards science from which this idea seems to spring.

Let us turn first to that idea which, I believe, is defensible. It is the idea that hypotheses may be distinguished according to the results of their tests: the idea that some hypotheses are well tested by experience, and others are not so well tested; that there are, further, hypotheses which so far have not been tested at all, and hypotheses which have not stood up to tests, and which therefore may be regarded as falsified. If we look upon a number of hypotheses from this point of view, there can be no objection to grading them according to the degree to which they have passed their tests— exactly as we may grade students who have undergone a number of tests, some of them easy, some of them difficult.

The wish to grade hypotheses according to the tests passed by them is legitimate: I do not know of any serious objection. For reasons to be discussed in the next section, I propose to call the grade of a hypothesis, or the degree to which it has stood up to tests, its *'degree of corroboration'* (rather than its 'probability'). One might think that it is very important, since the *acceptability of a hypothesis*, obviously, will depend upon its degree of corroboration. But I do not believe that any uneasiness or difference of opinion regarding the acceptability of a hypothesis will ever be removed by an 'exact' determination of its degree of corroboration. Although I shall later give a definition of degree of corroboration—

one that permits us to compare rival theories such as Newton's and Einstein's—I doubt whether a numerical evaluation of this degree will ever be of practical significance. (Perhaps I am biased, for I am in general not a believer in examination marks.) However this may be, the idea of degree of corroboration is significant, in our own present context, mainly because of the help it can give in clearing up the great muddle created by mistaken views on inductive probabilities.

This brings me to my main and final point in this section—the general philosophical background of the belief in inductive probabilities.

The view of science from which this belief springs is fundamentally the old view of Science with a capital 'S'. It is the view of science as *scientia* or *epistēmē*—as certain, demonstrable, knowledge. No doubt, this view is now somewhat modified: everybody now realizes that full certainty is unattainable in the sciences which are called 'inductive'. But as induction is considered a kind of (weakened) generalization of deduction, the old ideal is only slightly modified.

This old ideal must, however be completely scrapped. It should be scrapped even if we consider purely deductive systems. We no longer look upon a deductive system as one that establishes the truth of its theorems by deducing them from 'axioms' whose truth is quite certain (or self-evident, or beyond doubt); rather, we consider a deductive system as *one that allows us to argue its various assumptions rationally and critically,* by systematically working out their consequences. Deduction is not used merely for purposes of *proving* conclusions; rather, it is used as an instrument of rational criticism. Within a purely mathematical theory, we deduce conclusions in order to investigate the power and the fruitfulness of our axioms. And within a physical theory, we deduce conclusions in order to criticize and, especially, to test the deduced conclusions, and thereby our hypotheses: as a rule, we have no intention of *establishing* our conclusions.

By looking upon inductive probability as a measure of the reasonableness of our beliefs or the reliability of our knowledge, the devotee of probable induction makes it clear that he still clings, like Bacon, to a weakened ideal of *epistēmē*. He conceives his evidential statements *e* as playing a part analogous to that of the self-evident axioms supposed to 'prove' our theorems. And he conceives his hypothesis h as playing a part analogous to that of theorems whose truth is made certain by deduction from the axioms; only that, induction being weaker than deduction, we now get merely an *Ersatz* certainty: probability comes in as the substitute, or surrogate, of certainty—not quite the thing, but at least the next best thing, and at any rate approaching it.

All this is unacceptable. The evidential statements e are themselves far from certain. (I have always stressed this point; nevertheless it was used by inductivists as an argument against my doctrine of the asymmetry between falsification and verification; *cf.* section 22. But no inductivist has ever explained how to interpret '$p(h,e)$' when e itself is uncertain and, presumably, 'only probable'.)[2] Nor are these evidential statements 'given' to us—by God, or by nature, or by our senses. Every observation and, to an even higher degree, every observation statement, is itself already an *interpretation in the light of our theories*.[3] Yet even though this fact is most important, it raises a minor issue compared with what I wish to criticize here: a general attitude; a general philosophy of science; a philosophy which makes its main problem that of explaining whence science derives its 'certainty', its rational reliability, its validity, or its authority. For I hold that science has no certainty, no rational reliability, no validity, no authority. The best we can say about it is that although it consists of our own guesses, of our own conjectures, we are doing our very best to test them; that is to say, to criticize them and refute them.

But the inductivist philosophy not only attributes authority to science, it also (perhaps quite unwittingly) attributes to science a cautious, and indeed timid approach which is entirely foreign to our real procedure. This philosophy, in regarding it as the aim of science to attain high probabilities for its theories, implies that science proceeds according to the rule: 'Go as little as possible beyond your evidence e!' For the content of our hypothesis h cannot go far beyond the evidence e without reducing $p(h,e)$ to a value very close to zero. For example, let e be the conjunction of many descriptions of an event of a certain kind: h does not need to assert many further

---

[2][See R. C. Jeffrey: *The Logic of Decision*, 1965, Chapter 11. Ed.]

[3][*Cf.* Donald T. Campbell: 'Evolutionary Epistemology', in P. A. Schilpp, ed.: *The Philosophy of Karl Popper*, pp. 413-63. Ed.]

events, not covered by e, in order to make $p(h,e)$ very small, according to the calculus of probability. This shows that a high probability is the dubious reward for saying very little, or nothing. In other words, the rule 'Obtain high probabilities!' puts a premium on *ad hoc* hypotheses.

All this presents a most uninspiring picture of science—a picture, moreover, that does not in the least resemble the original. Indeed, what makes the original so inspiring is its boldness; its boldly conceived hypotheses, boldly submitted to every kind of criticism—to every refutation we can think of, including the most severe tests which our imagination may help us to design. It is this boldness which helps us to transcend the limits, narrow at first, of our imagination, and of ordinary language.

28. *'Corroboration' or 'Probability'?*

In the foregoing section I introduced the term 'degree of corroboration' to characterize the degree to which a hypothesis has stood up to tests. In the present section I intend to discuss merely a terminological issue—my reasons for proposing to speak of 'degree of corroboration' rather than of the 'probability of a hypothesis in the light of tests'. My main reason is, of course, that the latter phrase—although in itself perfectly legitimate—is liable to lead to confusion.

I am ready to admit that the words 'probable' and 'probability' are often used in a sense similar to the one in which I propose to use the somewhat awkward expression 'degree of corroboration'. An expression like 'a probable guess' belongs to common parlance; and expressions like 'an improbable conjecture' or perhaps even 'an improbable hypothesis' are perfectly straightforward expressions of ordinary, though perhaps slightly 'learned', usage. Moreover, the ordinary meaning of a phrase such as, 'You must try to think of a more probable hypothesis' or, 'Of the various hypotheses so far offered, yours appears, in the light of tests, the one that is most probable' seems to me clear enough: the intended meaning is no doubt closely related to the one for which I propose to use the more artificial term 'degree of corroboration'; a proposal which would make us replace the last phrase by the following more awkward one: 'Of the various hypotheses so far offered, yours appears to be the one that has, in the light of tests, the highest degree of corroboration.' Nevertheless, I discarded in *L.Sc.D.* (sections 79 to 84) the well-established usage of the words 'probable' and 'probability', replacing these words by a more artificial terminology. This replacement is in need of an explicit defence. For I do believe that we should speak simply and clearly, and should not deviate from ordinary usage without compelling reasons.

There are compelling reasons, however, to do so here; and since they are only implicitly given in *L.Sc.D.* (in sections 81 to 83) I shall try to state them here in full.

(1) There is a second and even better established usage of 'probable' and 'probability' which intuitively is hardly distinguishable from the first. (By the first usage, I mean that 'probability' of hypotheses which I propose to re-name 'degree of corroboration'.) It occurs in phrases like: 'Ann will probably marry Arthur' or 'It is probable that Bob will just scrape through in the finals.' This probability may be described as the *probability of an event*. It has a characteristic feature: the probability that Ann will marry Arthur *and* that Bob will just scrape through in the finals cannot exceed, and is ordinarily less than, the probability of the less probable of these two events. Or more generally, *the probability of a complex event consisting of the concurrence of several single events will in general be less than, and at most equal to, the probability of any of the component events.*

For statements (hypotheses) describing events, the following holds, accordingly: the probability of a statement describing an event *decreases with increasing logical content of the statement.*

This 'rule of content', as I may call it (it corresponds to the 'axiom of monotony' of the calculus of probability; see previous section, footnote 1, and *L.Sc.D.*, Appendices *iv and *v), agrees with the well-established linguistic usage according to which a throw of 12 with two dice is said to be 'less probable' than a throw of 6 with one die.

No corresponding rule of content holds for the first sense of 'probability', which I call 'degree of corroboration'. On the contrary: most physicists will say that Maxwell's theory of light is 'more probable', in the first sense—that is to say, 'better corroborated' or 'better tested'—than Fresnel's theory of light. The reason is that Maxwell's theory has been *more widely and more severely tested*—even in fields in which Fresnel's theory cannot be tested. At the same time, Maxwell's theory has a much greater logical content than Fresnel's: Maxwell's is a wave theory of light *and* a theory of electromagnetism, while Fresnel's is merely a wave theory of light. Thus Maxwell's theory, although 'more probable' in the sense of being 'better corroborated' or better tested, is at the same time 'less probable' in the sense of the second usage of the word which is also very well established, especially if we are thinking not so much of the *tests successfully passed by a hypothesis,* but rather of the *chances that an event will occur.*

(2) The two usages—the probability of a hypothesis with respect to its tests, and the probability of an event (or a hypothesis) with respect to its chances—have rarely been distinguished, and are mostly treated on a par. This may be due to the fact that intuitively—that is to say, at least 'upon first appearances'—they are hard to distinguish. Yet it can be shown that *while the second usage agrees with the rules of the mathematical calculus of probability (especially with its 'axiom of monotony'), the first does not.*

An early and particularly instructive example of this failure to distinguish the two senses can be found in a famous letter of Leibniz's.[1] In this letter, Leibniz speaks very clearly of hypotheses which I should describe as having a great *content*; he speaks of their 'simplicity', their 'virtue', and their (explanatory) 'power'; and he says that a hypothesis is the more *'probable'* the greater its simplicity and power, and the greater the number of phenomena which it can 'solve' (explain) with the fewer (additional or *ad hoc*) 'assumptions'. Clearly, he has in mind our first sense of 'probability'—what I propose to call 'degree of corroboration'. Yet he seems to think that this 'probability' behaves like one that satisfies the probability calculus; for he suggests that this probability is a kind of surrogate or *Ersatz* of certainty or truth, since it may approach 'physical certainty' (or 'moral certainty', as Couturat says in his comments). There is no suggestion here that he realizes that simplicity, and explanatory power, are equivalent to logical *improbability,* in the sense of the probability calculus (*cf. L.Sc.D.,* sections 34 to 46, 83, 30 and 32).

(3) Now if I am right in all this, it will be important to avoid a

---

[1] Letter to Conring, March 19th, 1678; see vol. i, pp. 195 *f.* of Gerhart's edition of *Die philosophischen Schriften von G. W. Leibniz,* 1875–1890, and Louis Couturat, *La Logique de Leibniz,* 1901, p. 268.

confusion between these two senses: especially as nearly every writer on the probability of hypotheses has assumed *without further discussion* that the probability of a hypothesis with respect to the *tests* passed by it (or its 'degree of corroboration') can be treated in terms of the probability calculus.

If, on the other hand, I am mistaken in my assertion that degree of corroboration does not satisfy the probability calculus, then it cannot do any harm to distinguish provisionally the two usages as sharply as possible—perhaps in order to prove later that both may be treated in terms of the probability calculus, thereby establishing that I was mistaken. (I believe, however, that I can *prove* that I am right; see below, section 31.)

Thus everything speaks for a clear distinction—at least provisionally, and against a premature and uncritical assumption that degree of corroboration must satisfy the general rules of the calculus of probability.

(4) An additional reason arises from the fact that a third sense of 'probability'—the probability of an inference, briefly explained in the preceding section—can also be considered (like the second sense) as satisfying the probability calculus. It is this third sense which has been so often unconsciously and uncritically confused with degree of corroboration; for example by Locke and by Hume who distinguish a 'probable inference' or a 'probable argument' from a 'demonstrative argument' or a 'proof'.[2]

Now if I am right in my contention that degree of corroboration does not satisfy the probability calculus, then Locke's and Hume's probabilities have to be sharply distinguished from 'probability' in our first sense, that is to say, from degree of corroboration.

(5) If some new artificial terms are to be introduced in order to replace the words 'probable' and 'probability' in one or the other of the senses here discussed, it will be best to choose a new name for the first sense and to speak of, say, a 'corroborated hypothesis', for the following reason.

The probability of an event is the more popular concept: it is established in gambling as well as in the physical and social sciences and, of course, in mathematics. The 'probability' of a universal law in the light of tests is, in comparison, a logician's or a philosopher's

---

[2] In the *Enquiry* (*cp.* section vi, first footnote), though less sharply in the *Treatise*, Hume divides arguments into '*demonstrations, proofs,* and *probabilities*'.

concept. But it is less objectionable (if at all) to interfere with philosophical usage than with common usage. (Also, it is perhaps *slightly more probable* that the attempt may succeed, in the event.)

(6) All the above usages except the first satisfy the rules of the probability calculus. My suggestion, therefore, has the advantage that it leads to the following easily remembered terminological rule.

We retain the terms 'probable' and 'probability' in those cases in which there is no reason to doubt that the mathematical calculus of probability is satisfied (and especially the 'rule of content' or 'axiom of monotony'). For other cases—such as the 'probability of hypotheses' in our first sense—we choose *other terms*, such as 'corroboration', at least provisionally; that is to say, until the doubt that the rules of the calculus may not be applicable to them has been removed.

This concludes my defence of the terminological decisions tacitly adopted in sections 79 to 83 of *L.Sc.D.* A reading of these sections will show that the reasons given here are not only already implicit there, but that some of them take the form of a detailed criticism of authors such as Keynes who uncritically identify degree of corroboration and probability.

In the remaining sections of the present chapter, I shall be concerned only with the degree of corroboration of hypotheses—as I shall say from now on, avoiding the term 'probability' in accordance with my terminological rule.

The other probabilities, especially the probability of events, and the probability of inferences (probability logic) will be discussed in Part II of the present volume. (See also *L.Sc.D.*, Appendix *iv.)

29. *Corroboration or Confirmation?*

In the foregoing section, I briefly explained my reasons for suspecting that *degree of corroboration does not satisfy the calculus of probability.*

This suspicion creates our whole problem; and in order not to obscure this problem by a confusing terminology, I introduced the concept 'degree of corroboration', and strongly urged the use of 'probability' only for concepts that satisfy the probability calculus (and especially the 'rule of content', corresponding to what I have called the 'axiom of monotony' of the calculus).

There is, unfortunately, the danger of another terminological confusion. Until recently (in fact, until *L.Sc.D.* was in galley proof) I did not use the term 'degree of corroboration', but, in its place, the term 'degree of confirmation'. And I made use of this term for precisely the same reason: because of the need to avoid the term 'probability'. Therefore I must now make clear why I have decided to change my terminology, after using it in at least half a dozen publications.

The story is quite simple. Until recently I used the label 'degree of confirmation' because this was Carnap's translation, in his 'Testability and Meaning', of my term *'Grad der Bewährung'* or *'Bewährungsgrad'*—the term I introduced in *L.d.F.* in order to avoid 'probability', for the reasons just explained.[1] Since labels do not matter, I saw no reason why I should not accept Carnap's translation, even though I did not particularly like it. The term soon established itself. No doubt it was originally used by Carnap in order to denote what I had intended to denote: the degree to which a theory has stood up to *tests*. (Even in a recent paper, Carnap still speaks of 'the requirement of confirmability or testability', with a reference to 'Testability and Meaning', where both these terms are translations of terms used in *L.d.F.*—in the title of my section 83, for example.) However, the term was soon used with a new and different meaning; for Carnap assumed without further ado in the first sentence of his book *Logical Foundations of Probability* that the 'degree of confirmation' of a hypothesis satisfied the rules of the calculus of probability.

None of the participants in this development ever mentioned that I had argued at length against this assumption, and none of my arguments were ever answered. Apparently they had been forgotten.

I found this situation a little embarrassing, and when I published a paper (entitled 'Degree of Confirmation') in which I gave a definition of what I *now* propose to call *'degree of corroboration'*, I referred to this development in a footnote. In this footnote I mildly

---

[1]Carnap's translation will be found, with references to *L.d.F.*, in his 'Testability and Meaning', *Philosophy of Science* 3, 1936, p. 427. A year or two later I suggested to Carnap that 'degree of corroboration' might be a better translation, but since he did not think so, I accepted his translation. (The term 'degree of corroboration' was suggested to me by my friend, Professor Hugh N. Parton.)

protested against a remark of Carnap's to the effect that it was 'generally accepted' that degree of confirmation satisfied the rules of the calculus of probability.

My paper received a reply from J. Kemeny (who also uses 'degree of confirmation' in the sense of 'probability') in the *Journal of Symbolic Logic*. He said, in the course of his reply: 'It should be pointed out that Popper used the term 'degree of confirmation' first—twenty years ago—and hence it is unfortunate that in recent years it has been widely used in a sense not intended by Popper. But Popper does not seem to realize that the recent usage *has* been in a different sense.'[2]

It was this remark of Kemeny's which led me to drop the term 'degree of confirmation'. For I certainly did not regard this label as my personal property.

But it was a little surprising to hear that I did 'not seem to realize that recent usage *has* been in a different sense' when it was just this obvious fact that provoked my footnote. And recent users of the term do not seem to be aware of the possibility that *the degree to which a hypothesis has been tested may not satisfy the probability calculus;* nor the fact that I said so long ago; nor that my arguments have never been answered.

But as I have now chosen a brand new label, I think it best to ignore the history of the old one which I have discarded. What is important now is only that things should not be confused again.

As far as 'degree of confirmation' is concerned—in its more recent sense which makes it a probability—I cannot help feeling that the term is redundant. Why not stick to 'probability'? In addition, it has led to much confusion. For probability and 'degree of confirmation' (in its recent usage) do not deliver the goods. They give all universal laws, whether refuted or well corroborated, the same zero degree of probability or confirmation (*cp. L.Sc.D.*, section 80). And they lead to other insuperable difficulties of this kind, as Carnap later discovered.

Incidentally, I do prefer the label 'degree of corroboration' to 'degree of confirmation'. For the term 'confirmation' may easily

---

[2]*Journal of Symbolic Logic* **20**, 1955, p. 304 (italics in the original). The following correction should be made in Kemeny's reply, p. 304, line 16 from bottom: for 'measure of the support given by *y* to *x*' read 'measure of the explanatory power of *x* with respect to *y*'.

suggest a wrong idea. It contains the root 'firm', and it suggests either a process of making a hypothesis by degrees more certain, or even a process of making it finally secure.[3] In other words, *the term 'confirmation' has strong verificationist associations.* I therefore gladly surrender it to verificationists and believers in induction.

Should 'degree of corroboration' also become a probability in the course of the next twenty years or so,[4] I would suggest to my pupils (if any) to change over, for another decade or so, to some simple label like 'grade of a hypothesis', in place of 'degree of corroboration', and 'graded by tests' in place of 'corroborated'.

30. *The Problem of Degree of Corroboration.*

If we take it *as a purely practical problem,* the problem of induction may be reformulated thus: 'When do we—tentatively—accept a theory?' Our answer is, of course: 'When it has stood up to criticism, including the most severe tests we can design; and more especially when it has done this better than any competing theory.'

This is all: there is no need to carry this problem any further, although once we have solved it, we may raise another one—one which need not be raised, and which has little bearing on research practice. It is the problem: 'Can we assess the degree to which a theory has stood up to tests? In particular, can we compare two theories—say, Newton's and Einstein's theories of gravitation—and say precisely why Einstein's is better tested or better corroborated and therefore (tentatively) more acceptable?'

In its simplest form, this problem may be put as: How well is the theory *t* tested? And can we ascribe to it a number—a measure or degree—summing up the severity and number of tests, and the manner in which the theory has stood up to them, awarding to it something like an examination mark, to be called its 'degree of corroboration'?

I discussed the characteristic properties of degree of corroboration in Chapter X of *L.Sc.D.* The main points of my discussion were these:

(1) Degree of corroboration is closely related to the *testability* of a

[3]*Cf. Philosophy of Science* **12**, 1945, note 2 on p. 99.
[4][See Jaakko Hintikka: 'Induction by Enumeration and Induction by Elimination', in I. Lakatos, ed.: *The Problem of Inductive Logic,* 1968, pp. 191–216. Ed.]

theory. The fact itself is fairly obvious: a more testable theory can be better tested; and what we are looking for is a mark, or degree, expressing how severely the theory was tested, and how well it has stood up to its tests.

(2) Since testability in its turn can be measured by the *content* of the theory, and since content, in its turn, can be measured by the absolute logical *improbability* of the theory, content and improbability stand in the same close relation to degree of corroboration as does testability itself. (I may mention here in passing that the idea of the empirical content of a theory, as a measure of the class of its falsifiers, was perhaps the most important *logical* idea of *L.Sc.D.* It plays a decisive role in the theory of degrees of testability; of simplicity; of logical probability and improbability; and of corroboration.)

(3) Corroboration cannot possibly be a probability, since it is more closely related to the improbability of a theory than to its probability: a strong theory (such as Maxwell's electromagnetic wave theory of light) can be tested more widely and more severely than a weaker theory entailed by it (such as Fresnel's wave theory of light). Every test of the latter theory is also a test of the former, but not *vice versa*. The situation may be similar even in the case of theories which are logically related in a somewhat different way, such as Einstein's and Newton's: a test supporting Newton's theory also supports Einstein's, though some tests are, as a matter of fact, crucial tests between the two theories; but in addition, there are tests of Einstein's theory which simply are not tests of Newton's. (For example, red-shift in a strong gravitational field.) In all these cases it is the logically more improbable theory which not only has greater testability or content, but which will soon turn out, in fact, to be the better tested one—provided always that it is not refuted. (It is also the one with greater *explanatory power*.)

(4) The degree of corroboration of the weaker theory (such as Fresnel's or Newton's) might not only be exceeded by that of a stronger theory (Maxwell's or Einstein's), even without the falsification of the weaker theory. Indeed, the degree of corroboration of the weaker theory may actually decrease upon the emergence of a stronger theory.[1]

[1]This is stated briefly in the last sentence before the penultimate paragraph of section 82 of *L.Sc.D.* To justify this remark, degree of corroboration must be

(5) Everybody had taken it uncritically for granted that a hypothesis high in probability is something good, something we ought to aim at. But the highest probability will be that of a hypothesis which says nothing (like a tautology) or next to nothing (like certain purely existential statements), or which goes as little as possible beyond the facts it is expected to explain (that is to say, a hypothesis which is *ad hoc*). Not only has the alleged aim of obtaining high probabilities never been critically examined, but the intuitive principle that high probabilities are something good can be shown to clash with another intuitive principle: the principle that *ad hoc* hypotheses are something bad. And it is the latter principle that is adopted in actual critical discussions of scientific theories as well as in scientific practice, not the former.

There are some further points, but the five points summed up here will suffice, I hope, to show the following. My real problem in connection with degree of corroboration was *not* to give an 'adequate definition' of an intuitive idea of degree of corroboration. The real problem I had in mind was, rather, this. There is an intuitive idea which may at first sight look like a probability, and which many logicians have without examination taken to be a probability; at the same time, if we look more closely at our actual assessment of theories, this idea exhibits properties incompatible with the rules of the probability calculus. Thus we arrive at the following two questions.

The first question is this. Who is right? Those philosophers who say that by testing a hypothesis we establish its probability, in the sense of the probability calculus? Or I, who say that, whatever we establish by severely testing a hypothesis, it cannot, in general, be a probability in the sense of this calculus?

The second question may be put thus. Is my idea of degree of corroboration consistent? Does there exist a measure function which has the properties which I ascribe to degree of corroboration (and which accordingly does not satisfy the calculus of probability)?

Our method of solving these two questions will be to give a definition of degree of corroboration, *in terms of content*—the content of the theory and that of the test statements. Since content is

defined so that it is relative not only to evidence in the form of tests, but also to other theories. This is achieved by what I call its 'relativized form'; see below, section 32.

in its turn definable in terms of improbability and thus of probability, our definition of degree of corroboration will be put in the more familiar terms of probability. From this definition, we can read off immediately the solution to our two questions. Since the definiens will be a function of probabilities, we can at once decide the question whether 'degree of corroboration' is itself a probability. The answer is negative. And since we know that the calculus of probability is consistent, we can also answer the second question in the affirmative.

The very fact that I give a definition of degree of corroboration makes it necessary to emphasize here that my real problem—the one I wanted to solve—was *not* to define degree of corroboration. Rather, the task of finding an adequate definition of this idea (and of degree of acceptability) arose not from the wish to define, but from some real problems. In the last analysis, it arose from the problems of induction and demarcation. And genuine problems cannot be solved by definitions; even though definitions may sometimes be of help in clarifying certain issues.

31. *Corroboration.*

The trouble about people—uncritical people—who hold a theory is that they are inclined to take everything as supporting or 'verifying' it, and nothing as refuting it. Many empiricists have seen this danger, for example Bacon; and they have tried to counter it by counselling the scientist to abstain from theorising, and to rid his mind of all 'preconceived' theories—until, as the result of pure and unprejudiced observation, a theory forces itself on his mind. As we have seen, this counsel is impracticable, and can only lead to self-deception and to the habit of holding one's theories unconsciously (and therefore uncritically). The proper counsel to the scientist is that he will always hold, consciously or unconsciously, a host of theories and that he is well advised to adopt a critical attitude towards them—even though he cannot, as a rule, be actively critical of more than one theory at a time.

It amounts to adopting the uncritical attitude if one considers an event, or an observation (*e* say) as supporting or confirming a theory or a hypothesis (*h* say) whenever *e* 'agrees' with *h*, or is an instance of *h*.

Thus the occurrence of a white swan is, uncritically, believed to 'support' or 'confirm' the hypothesis h_1 that all swans are white. But what of the hypothesis h_2 that 90% of all swans are white? Clearly h_2 is inconsistent with the hypothesis h_1 that all are white; yet if the latter hypothesis is regarded as being 'supported' by the occurrence of a white swan, the former should also be so regarded.[1] It may perhaps be replied that, for the support of the statistical theory h_2, a sample of at least 10 swans is needed—nine white and one non-white. But if this were so, should we not also say that for the support of h_1, at least the same number of swans is needed, all of them white, of course? (But if so, then one instance could never support any theory.)

We shall, however, postpone the question of statistical theories like h_2 until later: at the moment we shall consider only universal theories.

With respect to these, we have two main attitudes to consider:

(a) The uncritical or verificationist attitude: one looks out for 'verification' or 'confirmation' or 'instantiation', and one finds it, as a rule. Every observed 'instance' of the theory is thought to 'confirm' the theory.

(b) The critical attitude, or falsificationist attitude: one looks for falsification, or for counter-instances.[2] Only if the most conscientious search for counter-instances does not succeed may we speak of a corroboration of the theory.

It may be asked whether it is really so uncritical to look upon all instances of a theory as confirmations of it. But it can be shown that, for logical reasons, to do so amounts to the belief that *everything is a confirmation*—with the sole exception of a counter-instance. Thus not only white swans but also black ravens and red shoes 'confirm' the theory that all swans are white. This may look strange; but if it is true then it must be admitted that those who take (almost) everything as confirming their theories are indeed acting quite properly from the verificationist point of view, and *vice versa*: that the

---

[1] C. G. Hempel, who holds an instantiationist theory of confirmation, also holds, strangely enough, the theory that no evidence e can confirm two hypotheses h_1 and h_2 if h_1 and h_2 are inconsistent. See note 13 to Appendix *vii of *L.Sc.D.*

[2] It is historically not quite without interest that until about 1600 'instance' itself meant 'counter-example' or 'counter-instance' (from *instantia*, that is an obstacle: that which stands in the way—in this case, of acceptance).

verificationist is indeed a man who tends to take everything as supporting his views.

It can easily be shown that everything is indeed an instance of 'All swans are white'—unless it is a counter-instance, that is, a non-white swan. For 'All swans are white' may be written: 'Anything that is a swan is white', or 'If something is a swan then it is white' or 'Everything is white unless it is not a swan'. Whenever a thing satisfies this, it is an *'instance'*. It is well known that (in extensional logic), anything that is not a swan satisfies these formulae, and also anything that is white; so that *only* a thing that is both a swan and not white (and therefore a counter-instance) does not satisfy it. This shows why verification, or instantiation, is too cheap to be of any significance, and why only such cases are interesting as might be expected to be counter-instances unless, indeed, the theory is true. More precisely, the interesting cases will be *crucial* cases—cases for which the theory to be tested predicts results which differ from results predicted by other significant theories, especially by those theories that have been so far accepted.

The fundamental difference between the verificationist and my own view of 'support' is therefore this. While the verificationist view leads to the claim that *every 'instance'* of h supports h, I assert that only the results of *genuine tests* can support h.

Thus an observed white swan will, for the verificationist, support the theory that all swans are white; and if he is consistent (like Hempel), then he will say that an observed black cormorant also supports the theory that all swans are white. In my view, on the other hand, neither of these observations necessarily supports this theory, although either *may* support it under special circumstances. Thus if, for example, we have *good* reason to think (in the light of previously accepted theories) that the thing in this pond is a black swan, then either the discovery that it is a white swan or that it is a black cormorant *might* indeed support the theory that all swans are white.

In the verificationist view of the matter, there is thus a simple formal-logical relationship—instantiation—whose presence or absence decides whether or not e supports h: if e is an instance of h, then e supports h. In my own view the situation is less simple: only if e is the result of genuine or sincere attempts to refute h can e be regarded as supporting h.

It seems to me unlikely that this relationship between e and h is capable of complete logical analysis: sincerity is not the kind of thing that lends itself to logical analysis. But there is no reason to be disheartened. First of all, we can do quite well without a logical analysis of support and of corroboration. And secondly, we can go pretty far towards analyzing it—further than one might expect.

I shall now try to analyze more fully the idea 'e supports h'; preparatory to analyzing the degree to which e corroborates h. To this end we shall use, besides e (the empirical evidence) and h (the hypothesis), a further variable: our background knowledge b. And we may then say that e supports h in the light of b.[3]

By our background knowledge b we mean any knowledge (relevant to the situation) which we accept—perhaps only tentatively—while we are testing h. Thus b may, for example, include initial conditions. It is important to realize that b must be consistent with h; thus, should we, before considering and testing h, accept some theory h' which, together with the rest of our background knowledge, is inconsistent with h, then we should have to exclude h' from the background knowledge b. (It may not always be easy to decide what we have to exclude from b, but this is a problem which I do not intend to discuss here.)

Let us suppose that there are certain experimentally testable consequences of hb, that is, the conjunction of h and b. Among these consequences there may be some events which we should certainly expect not to occur if h is false. Let e be such a consequence; that is to say, let e describe an observable event which can be predicted to occur (in the presence of b) if h is true, and which we should expect not to occur if h is false. We shall then be inclined to say that if e actually occurs, then this supports h.

This, however, is not a very satisfactory formulation, although it may serve as a starting point. The trouble is connected with the remark (made twice) that e is an event 'which we should expect not to occur if h is false'.

In view of the fact that we demanded that 'e follows from h (in the presence of b)' one might be inclined to interpret the remark in question by the demand 'non-e follows from non-h (in the presence of b)'. But this would amount to another form of verificationism: it

---

[3][The idea of 'background knowledge', introduced here by Popper, was first published by him in 1954 and 1957. See *L.Sc.D.*, pp. 401, 404. Ed.]

would make *e* and *h* equivalent (in the presence of *b*), and would thus allow us to verify *h* by observation, that is, by observing that *e* is true.

But quite apart from any hostility to verificationism, it is altogether implausible to demand that non-*e* should follow from non-*h* (and *b*). For let us assume that *e* is an event which supports *h*—something predicted by *h*, and something nobody would ever have considered without *h*. For example, let *e* be the first observation of a new planet (Neptune) by J. G. Galle, in a position predicted by Adams and Leverrier, and let *h* be Newton's theory upon which their prediction was based. Then *e* certainly supports *h*—and very strongly so. Yet in spite of this fact *e* also follows from theories which, like Einstein's, entail non-*h* (in the presence of *b*). It would therefore be a major mistake to demand that non-*e* should follow from non-*h* (and *b*). The mistake amounts to the belief that *e* (which, we have said, must be crucial between *h* and *some* other significant theory) must be crucial between *h* and *all* other theories that contradict *h*.

For this reason, our second demand 'non-*e* follows from non-*h* (and *b*)' is much too strong. But we cannot do entirely without something of its kind. For our first demand—that *e* follows from *h* (and *b*) is quite insufficient, considering that *e* follows from *h*, in the presence of *b*, if it follows from *b* alone.

But to demand that *e* does not follow from *b* alone is not enough either. For if *e* should be *probable*, in the presence of *b* alone ('probable' in the sense of the probability calculus), then its occurrence can hardly be considered as significant support of *h*.

Thus James Challis, to whom Adams had given the result of his calculations, actually observed Neptune close to the calculated orbit before Galle. But the star he saw did not seem to move, and he did not think his observation sufficiently significant to compare it with later observations of the same region which would have disclosed its motion. The presence of *some* unknown star of eighth magnitude, close to the calculated place, was in itself quite probable on his background knowledge and therefore did not appear significant to him. Only that of a *moving* star, a planet, would have been significant, because unexpected—though not on Adams's calculations.[4]

---

[4]For the facts see B. A. Gould, *Report to the Smithsonian Institute on the History of the Discovery of Neptune*, 1850.

So we shall assume that the unexpectedness of an event can be identified with a *low probability,* in the sense of the calculus of probability, on the background knowledge.

It is therefore necessary to raise a second demand; we may exclude by it, for example, those cases in which *e* would be probable, given the background knowledge *b* alone; 'probable' in the sense of the calculus of probability.

We therefore need the calculus of probability; and we shall write

$$p(a,b) = r,$$

which is to be read: 'The probability of *a*, given *b*, equals *r*.' Similarly, we shall write '*ab*' for the conjunction '*a* and *b*' and *ā* for 'not *a*'. Accordingly,

$$p(ab,c) = r$$

should be read: 'The probability of *a* and *b*, given *c*, equals *r*'; and, for example,

$$p(\bar{a},bc) = r$$

should be read: 'the probability of non-*a*, given *b* and *c*, equals *r*.'

Now we can express our demand that the empirical evidence *e*, if it is to support *h*, should not be probable (or expected) on the background knowledge *b* alone by

$$p(e,b) \ll 1/2.$$

This leads us at once to realize that *the smaller p(e,b), the stronger* will be the support which *e* renders to *h*—provided our first demand is satisfied, that is, provided *e* follows from *h* and *b*, or from *h* in the presence of *b*.

This result is interesting in two ways. First, it is intuitively highly satisfactory that only improbable evidence *e*—improbable on our background knowledge—will be accepted as significant, as our Neptune example shows. A soothsayer's prediction to a young lady 'you will soon meet an interesting young gentleman' is much too probable in itself to be accepted, if it comes true, as supporting the theories on which the soothsayer's art may be based. Secondly, we demand intuitively that only *severe* tests should count, and that the more severe they are, the more they should count. But this is the same as to demand that *e* should be improbable on our background knowledge. This clearly expresses in probabilistic language the demand that *e* should be the result of a *severe* test. And this, in its turn, involves the demand that *h* should be highly testable—that is, testable by *severe* tests—and therefore highly improbable, or of great empirical content.

All these considerations suggest that the following definition of '*e* supports *h*' would be reasonably adequate:

e supports *h* in the presence of the background knowledge *b* if, and only if,

(a) *e* follows from *hb*

(b) $p(e,b) \ll 1/2$.

It is clear that we can write (a), like (b), in probabilistic terms; that is to say, instead of '*e* follows from *hb*' we can here use the slightly weaker[5] formula 'the probability of *e*, given *hb* (that is to say, the conjunction of *h* and *b*) equals one. We write this as follows:

(a) $$p(e,hb) = 1.$$

In view of (a) and (b), we can further modify (and weaken) our definition by writing:

e supports *h* in the presence of *b* if, and only if,

(c) $$p(e,hb) - p(e,b) > 1/2.$$

It may be mentioned in parenthesis that, written in this form, 'supports' turns out to be a stronger version of 'is positively dependent upon'; for we define quite generally in probability theory:

a is positively dependent upon *b* in the presence of *c* if, and only if

$$p(a,bc) - p(a,c) > 0.$$

And indeed, we could use this as a definition of 'supports', or more precisely, we could replace the definiens (c) by the weaker definiens

(d) $$p(e,hb) - p(e,b) > 0.$$

In this way we come to the (relativised form of) the definition of 'supports' given in *L.Sc.D.*, Appendix *ix in the first of the three Notes (p. 396).

---

[5]'*e* follows from *hb*' can be represented, in probabilistic language, for example by the formula

$$p(e,hb\bar{e}) = 1.$$

From this formula the weaker formula (a) follows at once. See *L.Sc.D.*, Appendix *v, formula (⁺) on p. 356, and the discussion following this formula.

It seems to me that the definiens (d) is quite adequate, as long as we add to it the following rider which, as it were, replaces (c):

The support given by e to h becomes *significant* only when

(e) $$p(e,hb) - p(e,b) \geqslant \tfrac{1}{2}.$$

With this rider we have entered into the discussion of the problem of *degree* of support or *degree* of corroboration. For by accepting the definiens (d) and adding this rider, we say that it is not enough to define 'supports', but that we have, beyond this, to determine a *significant degree* of support; or in other words, that we have to introduce a *measure of degree of support* or *of corroboration* in order to distinguish significant degrees.

The first and obvious suggestion appears to be this: accepting the rider, we take the difference used in (c) and (d), that is to say

(f) $$p(e,hb) - p(e,b)$$

as a measure of the degree of support given by e to h, in the presence of b, or of the degree of corroboration (C) of h by e in the presence of b; that is to say, the suggestion is that we define

$$C(h,e,b),$$

that is, the degree of corroboration of h by e in the presence of the background knowledge b, by the difference expressed in definition (f). The suggested definition has a few blemishes which can be repaired by 'normalizing' (e), that is to say, by dividing (e) by the 'normalization factor'

(g) $$p(e,hb) - p(eh,b) + p(e,b).$$

Thus we arrive at the definition

D $$C(h,e,b) = \frac{p(e,hb) - p(e,b)}{p(e,hb) - p(eh,b) + p(e,b).}$$

While in this definition D, the numerator (e) of the fraction has a clear and simple intuitive significance, the denominator (f) has no such significance: it is chosen merely because it leads to satisfactory results—it removes the blemishes mentioned—and because it seems to be the simplest normalization factor to lead to these results.

Before pointing out one of the blemishes which the adoption of D removes, I will mention some of the general results of adopting either (f) or D.

Whether D is adopted as definition of degree of corroboration or (e) is chosen as definiens, we obtain the following highly intuitive results:

(i) if e supports h (given the background knowledge b) then $C(h,e,b)$ is positive. If e undermines h (so that non-e supports h) then $C(h,e,b)$ is negative. If e does neither, so that it is *independent* of h in the presence of b, then $C(h,e,b)$ equals zero.

For example, if e is a tautology, or a logical consequence of b, then e will not corroborate or undermine any theory h and $C(h,e,b)$ will be zero. Similarly, if h is a tautology or a logical consequence of b, it will be neither corroborated nor undermined by any evidence e, and again $C(h,e,b) = 0$.

(ii) The maximum value which $C(h,e,b)$ can reach is 1. This is reached if and only if

(aa) $\qquad\qquad\qquad p(e,hb) = 1$
(bb) $\qquad\qquad\qquad p(e,b) = 0.$

This result is highly satisfactory: only an e which is extremely improbable on the background knowledge b can give h maximum support, or a maximum degree of corroboration. Of course, to do this, it must also in the presence of b follow, (or 'almost follow') from h; that is $p(e,hb)$ must be equal to one. Ultimately, in order that h should be maximally corroborable, its content or degree of testability must be maximal; that is, the following formula is a further condition which must be satisfied if corroboration is to reach its maximal value 1:

(cc) $\qquad\qquad\qquad p(h,b) = 0.$

Indeed, (cc) is an immediate consequence of (aa) and (bb).[6]

(iii) More generally even, if $p(h,b) \neq 0$, the maximum value which $C(h,e,b)$ can attain is equal to $1 - p(h,b)$ and therefore equal to the content of h relative to b, or to its degree of testability. This makes the degree of testability equal to the maximal degree of corroboration of h, or to its 'degree of corroborability'.

[6](cc) follows from (aa) and (bb) in the presence of the formulae B1 and B2 of *L.Sc.D.*, Appendices *iv and *v:

$$p(e,b) \geq p(eh,b) = p(e,hb)\,p(h,b).$$

So far so good. But now to a blemish: what about an empirical evidence e which *falsifies* h in the presence of b ? Such an e will make p(e,hb) equal to zero. But if this e reports the result of a severe test—say, of a very precise measurement—then it may well be very improbable relative to b; that is to say, we may have not only p(e,hb) = 0 but, approximately p(e,b) ≈ 0. In this case, the degree of support or of corroboration will become zero (or only very slightly less) if we adopt (f) as definiens. But if we adopt D, then the degree of support or of corroboration will always be equal to − 1 when e falsifies h (in the presence of b); that is to say, we obtain from D:

If e is incompatible with hb, then

$$C(h,e,b) = -1.$$

This is clearly satisfactory, and it removes a blemish inherent in the choice of the definiens (f). It makes, for every h (provided it is consistent with b) minimal and maximal degrees of corroboration equal to − 1 and to the content or degree of testability of h (whose maximum is + 1).

(Other satisfactory aspects are mentioned at some length in the three notes reprinted in Appendix *ix of *L.Sc.D.*; see especially the nine 'desiderata' stated there in section 9 of the first note.)

There are definitions other than D, and some of them may be preferable to D from some points of view. In fact, my first definition[7] was slightly different from D. I prefer D because it seems to be the simplest and most lucid of the various formulae satisfying my desiderata. But certain logarithmic formulae may do just as well—or better for certain purposes.

What is important about all those formulae which are satisfactory is this: they all seem to be 'topologically equivalent'; that is to say, if by adopting D we have, for some h, e, b, and h', e', b',

$$C (h, e, b) \geqslant C (h', e', b'),$$

then the same relationship holds if we adopt any of the other satisfactory definitions. But this means that philosophically there is nothing to choose between them—except from the point of view of simplicity and perspicuity.

[7]See formulae 9.2 and 10.1 in *L.Sc.D.*, Appendix *ix, first note (pp. 400–401). The present definition D is to be found there as (10.1*) in note *2.

But for all 'satisfactory' definitions—that is to say for all those which are topologically equivalent to D—the following theorem holds:

Degree of corroboration is not a probability; that is to say, it does not satisfy the rules of the calculus of probability.[8]

This has by now been widely accepted, though some people still insist that 'acceptability', as opposed to 'corroboration', is a probability. Of course, we can so define 'acceptability' (or 'confirmation'). But if we mean by the degree of acceptability of a theory h the degree to which h is satisfactory from the point of view of *empirical science*—that is, from the point of view of *the aims of empirical science*—then acceptability will have to become topologically equivalent to corroboration. Tautologies have the maximum degree of probability but they are unacceptable to empirical science: establishing that an alleged empirical theory is tautological (or almost so) always amounts to a decisive criticism of it.

Thus I do not deny that the logical interpretation of probability, or the probability of statements, may be said to give the degree of probability, or likelihood, or chance, of a statement to be true; and insofar as it does so, it gives the degree of 'acceptability', *provided* that by calling a statement 'acceptable' we mean to say that we believe it is true. But this is not all we mean by accepting a statement in science. This can easily be seen from the fact that interesting scientific theories have always a negligible (if not zero) probability—including those which are at present generally accepted.

I conclude this section by giving a summary of my views concerning corroboration, in the form of seven points, the first of which contains the fundamental idea.

(1) The degree of corroboration of a theory is an evaluation of the results of the empirical tests it has undergone.

(2) There are two attitudes, two ways of looking at the relations between a theory and experience: one may look for confirmation, or for refutation. (These two attitudes are, obviously, variants of the apologetic [or dogmatic] and of the critical attitude.) *Scientific tests are always attempted refutations.*

[8]The same holds even for $l(h,e)$, the 'likelihood of h' in Fisher's sense, defined by
$$l(h,e) = p(e,h);$$
for even though it is a probability, it is not one of h.

(3) The difference between attempted confirmations and attempted refutations or tests is largely, though not completely, amenable to logical analysis.

(4) A theory will be said to be the better corroborated the more severe the tests it has passed (and the better it has passed them).

(5) A test will be said to be the more severe the greater the probability of failing it (the absolute or prior probability as well as the probability in the light of what I call our 'background knowledge', that is to say, knowledge which, by common agreement, is not questioned while testing the theory under investigation).

(6) Thus every genuine test may be described, intuitively, as an attempt to 'catch' the theory: it is not only a severe examination but, as an examination, it is an unfair one—it is undertaken with the aim of failing the examinee, rather than the aim of giving him a chance to show what he knows. The latter attitude would be that of the man who wants to confirm, or to 'verify' his theory.

(7) Assuming always that we are guided in our tests by a genuinely critical attitude and that *we exert ourselves in testing the theory* (an assumption which cannot be formalized)[9], we can say that the degree of corroboration of a theory will increase with the improbability (in the light of background knowledge) of the predicted test statements, provided the predictions derived with the help of the theory are successful.[10]

32. *Some further Comments on the Definition of Degree of Corroboration.*

We shall formulate in this section some further conditions or demands or requirements which the degree of corroboration, $C(h,e,b)$, should satisfy. (See also the list of desiderata in *L.Sc.D.*, pp. 400f.) These conditions, and especially the problem whether they are consistent, led me originally to the definition of $C(a,b)$, and later to $C(a,b,c)$ or, as I wrote in section 31 above, to $C(h,e,b)$.

---

[9]Although logic may . . . set up criteria for deciding whether a statement is testable, it certainly is not concerned with the question of whether anyone exerts himself to test it.' (See *L.Sc.D.*, section 11, p. 54.)

[10][Some lines appear to be missing from the text here, and could not be reconstructed. Ed.]

(1) The main requirement is that *our concept should give us an estimate of the degree to which a hypothesis or theory has been tested*. Almost all other demands follow from this central demand (apart from demands of a conventional character).

(2) The better, the more severely, a theory can be tested, the better it can be corroborated. We therefore demand that *testability and corroborability increase and decrease together*—for example that they be *proportional*. This would make corroborability inversely proportional to (absolute) logical probability; see *L.Sc.D.*, sections 34, 35, 83. (By 'corroborability' I mean the highest degree of corroboration theoretically attainable by the hypothesis in question.)

(3) The simplest convention will be to assume that the factor of proportionality equals 1; or in other words, that *corroborability equals testability and empirical content*.

(4) A tautology—an analytic statement—cannot be empirically tested. This may be expressed by the demand (which follows from the preceding one) that its degree of corroboration shall be zero, whatever the evidence may be, and consequently also its degree of corroborability.

(5) A self-contradictory hypothesis has, of course, maximal content; and it has, therefore, maximal testability—but only in the sense that every test may be considered as a refutation. We can deal with it either by excluding self-contradictory hypotheses or by the demand that a self-contradictory hypothesis should always have, whatever the evidence may be, that minimum degree of corroboration which is indicative of a *refutation*.

(6) In view of (3) and (4), this cannot be zero, and we thus have to take negative numbers in order to indicate unfavourable results. A simple convention is to use the number -1 for characterizing the degree of corroboration of a hypothesis with respect to tests that refute it.

[1]See Appendix *ix of *L.Sc.D.*, where my three papers on 'Degree of Confirmation' are reprinted. (*Brit. Journ. Phil. Science* 5, 1954, pp. 143 *ff*. See also corrections in 5, pp. 334 and 359; 6, 1955, pp. 157 *ff.*; 7, 1956, pp. 244 *f.*, 249 *ff.*, and 350 *ff.*) Note that in these papers I used the term 'confirmation' instead of 'corroboration', and that I used the symbol '$C(x)$' for the content of x, while I use here the symbol '$Ct(x)$'. In the first of these papers, several *desiderata* are stated which are not mentioned here. See also Appendices *ii and *iv. [See also the Addenda to *Conjectures and Refutations*. Ed.]

(7) This is advisable since in view of the convention of making the factor of proportionality equal to 1, formulated in point (3), the maximal degree of corroboration which even the most testable hypothesis can obtain is $+1$. Thus we obtain symmetrical limits, -1 and $+1$, for degree of corroboration, while 0 means no corroboration, as in the case of the tautology.

(8) If we now introduce the symbol '$C(a,b)$' for degree of corroboration of a by b, and '$Ct(a)$' for the empirical content or degree of testability of a, so that

$$Ct(a) = 1 - p(a),$$

we can express our demands, as so far formulated, by

(i) $$-1 \le C(a,b) \le Ct(a) \le +1.$$

Note that 'b' is *not* used here for 'background knowledge', which is, later in the section, denoted by 'e'.

(9) Writing '$\bar{a}$' for the negation of a, we can say that $\bar{a}$ refutes a. Thus $C(a,\bar{a}) = -1$. Similarly, if we write 'ab' for the conjunction of a and b, $C(a\bar{a},b) = -1$, at least when b is consistent, since $a\bar{a}$ is a contradictory statement. Moreover, a statement a—say, a singular hypothesis—will be completely corroborated if the evidence itself entails a. Thus $C(a,a) = C(a,ab)$ will be the maximal degree of corroboration which a consistent a is capable of, that is to say, $C(a,a)$ will be equal to what we have called the corroborability of a, and thus to its content $Ct(a)$. If we incorporate all this into our formula, admitting only consistent evidence b, we obtain:

(ii) $-1 = C(a\bar{a},b) = C(a,\bar{a}) \le C(a,b) \le C(a,a) \le Ct(a) \le +1.$

This determines the limits of $C(a,b)$. What is conventional here is merely the assumption (a) that $C(a,a)$, the maximal degree of corroboration which a is capable of (and thus its corroborability) equals $Ct(a)$, its content, instead of being merely proportional to it; and (b) that the minimum degree of corroboration (which is the degree of a refuted hypothesis) is the negative of the maximum degree which can be attained by the best testable statements; which means that it equals -1.

(10) Considering that $a\bar{\bar{a}}$ ('not $(a$-and-non-$a)$') is a tautology, we also should have, for every b,

(iii) $$C(a\overline{a},b) = 0.$$

(11) Having fixed the limits, and zero, we now demand that degree of corroboration of a hypothesis shall increase with every new genuine test it passes, and that it shall increase the more, the more severe the test.

Provided we know that b has been designed as a genuine attempt to refute a—that is to say, as a genuine test of a—we can measure the severity of the test b by the (absolute) *improbability* of b. For example, Adams's and Leverrier's predictions, which led to the discovery of Neptune, were such a wonderful corroboration of Newton's theory because of the exceeding *improbability* that an as yet unobserved planet would, by sheer accident, be found in that small region of the sky where their calculations had placed it. Another striking example is Einstein's eclipse prediction which concerns the angular distance between fixed stars appearing on opposite sides of the sun's disc. (He predicted that this distance would be greater than that obtained from measurements of the night sky.) Here an effect is predicted of which nobody[2] had thought before, and which, *in the light of earlier theories*, was exceedingly *improbable*. (Einstein's general relativity predicted twice the deviation from the night sky distance than what could be obtained from Newton's theory.)

These examples illustrate how we ascribe to bold and highly improbable predictions that turn out to be successful the power of increasing greatly the degree of corroboration. For in these cases, the success of the prediction could hardly be due to coincidence or chance: the improbability of the prediction measures the *severity of the test*. The argument can be used in the opposite direction also: the success of vague and cautious predictions such as astrologers or soothsayers are wont to produce has little corroborative power, since lack of precision goes with high absolute probability (*cf.* section 37 of *L.Sc.D.*). This kind of success can just be due to coincidence (as explained in section 83 of *L.Sc.D.*). The argument also applies, of course, to the result of a *multiplicity* of tests, since a statement describing many tests (especially if they are independent of one another) will be less probable than a statement describing

[2][*When writing this I was unaware that J. von Soldner had already in 1801 calculated the bending of light particles grazing the sun. This is revealed in a remarkable paper by Stanley L. Jaki: *Foundation of Physics* **8**, 1978, pp. 927*ff*.]

only some of these tests. (Thus a multiple test is more improbable—and accordingly also *more severe*—than its component tests.) We can sum up all this in the following rule:

(iv)   Provided *b* is a genuine test of *a*, *C(a,b)* increases with *Ct(b)*, that is to say, with the content, or absolute logical improbability, of *b*; or in other words, with the severity of the test *b*.

(12) The demand just formulated is in several respects of great importance in our theory. One reason for its importance is that *it holds not only for degree of corroboration but also for probability.* The logical probability of a hypothesis *a*, relative to the evidence *b*, also increases with the absolute improbability of *b*, provided the absolute logical probability of *a* was not zero, and that *b* was derivable or predictable by *a*. We may denote the relative probability of *a* given by *b* by

$$p(a,b)$$

and the absolute probability of *a* by

$$p(a).$$

(Here *p(a)* may be defined as $p(a,a\bar{a})$, that is, as the relative probability of *a*, given nothing—or else the tautology—as evidence.) We can now put the rule of the probability calculus (a form of Bayes's theorem) that corresponds to (iv) as follows:

(iv)$_p$ Provided $p(a) \neq 0$ and *b* is predictable from *a*, *p(a,b)* increases with $p(\bar{b})$, that is to say, with the absolute probability of non-*b*, which is the same as the logical improbability of *b*, or *Ct(b)*, the content of *b*.

It will be seen that this rule (iv)$_p$ corresponds exactly to (iv); and this may suggest that degree of corroboration may, after all, be a probability—that is, it may satisfy the probability calculus or perhaps a very similar calculus.[3] In other words, the analogy between (iv) and (iv)$_p$ shows the pertinence of *the first of our two problems—whether or not corroboration is a probability. (cf.* section 30.)

(13) But this analogy also raises *the second problem—whether the requirements explained here are consistent.* For (iv) may be taken as

[3]*Cf.* note 1 to section 81 of *L.Sc.D.*, with references to Janina Hosiasson-Lindenbaum.

indicating that corroboration, even if not a probability, must be very similar to a probability. But our demand (2), and even more clearly (4), which is the same as (iii), are *incompatible* with an interpretation of corroboration as a probability. For (4) and (iii) give zero degree of corroboration to all tautologies—which have, of course, the highest degree of probability, that is, 1.

Thus it is here that the importance of our two questions (*cf.* section 30) becomes really apparent.

(14) It is easy to show in more detail that our demand (2) contradicts the view that corroboration is a probability; but demand (2) follows immediately from the fundamental idea that degree of corroboration is simply a report about the severity of tests. (See also the introduction preceding section 79 of *L.Sc.D.*) Thus (2), and its consequences now to be shown, cannot be given up; which makes it still more urgent to prove that our demands are consistent, in spite of the analogy between (iv) and (iv)$_p$.

Let us assume that the statement b follows from the statement a but not *vice versa*, and that $0 \neq p(a) < p(b) \neq 1$. In this case, there will be a statement c such that a is equivalent to the conjunction bc, and $0 \neq p(c) \neq 1$.

Since the content of b is only a part of the content of a, we have $Ct(a) > Ct(b)$, so that the corroborability of a will be greater than that of b. In view of (8) above, this can also be written $C(a,a) > C(b,b) = C(b,bc)$, or since $a = bc$, we have

$$C(a,bc) > C(b,bc).$$

This result may be formulated as follows:

(v)   If b follows from a but not *vice versa*, and $0 \neq p(a) < p(b) \neq 1$, then there exists a statement d (in our case d may be, for example, the conjunction bc, that is, a itself) such that

$$C(a,d) > C(b,d).$$

This rule holds for degree of corroboration. *But the corresponding rule for probabilities is false.* Not only this, but the precise negation of this rule holds. For the following rule is a theorem of the probability calculus.

(v)$_p$ If b follows from a but not *vice versa*, and $0 \neq p(a) < p(b) \neq 1$, then for every statement d,

$$p(a,d) \leqslant p(b,d).$$

The clash between degree of corroborability and probability could not be more pronounced than by the conflict between (v) and (v)$_p$. But (v) follows directly from our fundamental idea that degree of corroboration is a report about the severity of tests, and that testability is equal to corroborability. (Nothing would change here if we reverted to saying 'proportional to' or even merely 'increasing with', instead of 'equal to'.)

(15) The situation described here is a formal consequence of the bare idea that degree of corroboration is a report of tests and of their severity. Yet (v) restates the intuitive remarks made under (1) in section 27 concerning the theories of Maxwell and Fresnel; for we may take a to be Maxwell's theory, b to be Fresnel's, and d to be a test (such as Hertz's) of the electromagnetic part of Maxwell's theory, or the conjunction of a test of this kind and one which applies to both theories (such as double refraction, or Fizeau's experiment).

(16) The conditions so far discussed—except for the two conventions mentioned under (3) and (6)—all derive from our fundamental principle that degree of corroboration is a report on the severity of the tests which a theory has passed. But now we add a new idea.

Take, as before, a hypothesis a from which a weaker hypothesis b is deducible. (We may again take a to be Maxwell's and b to be Fresnel's theory.) Now assume that we have tested b by the test x (Fizeau's experiment, or double refraction) but not that part of a which goes beyond b. In this case we shall say that $C(a,x)<C(b,x)$, in spite of the greater corroborability of a. For a contains a part, not yet tested, which may be refuted at the first test we undertake. But if x tests this latter part also (as does Hertz's experiment) then we have $C(a,x)>C(b,x)$, provided the test was successful, and sufficiently severe.

We can put this consideration in several other ways. For example, we can say that if b is well corroborated by x, and we add to b a completely new c which is in no way tested by x, then the resulting $a = bc$ will be less well tested by x than b, and thus less well corroborated—even though its corroboration will rise beyond that of b as soon as a few severe tests are made of c, that is to say, of that part of a which goes beyond b. We can express all this by the demand:

If x is explained by b alone in at least the same degree as it is explained by $a = bc$, then $C(a,x) < C(b,x)$. Here 'explained' may be replaced by 'made probable'. We thus come to the formulation:

(vi) If b follows from a; if $0 \leqslant p(a) < p(b) < 1$; and if $p(x,b) \geqslant p(x,a)$, then

$$C(a,x) < C(b,x).$$

It is clear that this last demand is, like (iii), one of those in which degree of corroboration resembles probability. For the following analogous formula is derivable from the calculus of probability.

(vi)$_p$ If b follows from a; if $0 \leqslant p(a) < p(b) < 1$; and if $p(x,b) \geq p(x,a)$, then

$$p(a,x) < p(b,x).$$

In view of this new demand, the solution of our two problems (of section 30) becomes even more urgent.

These are all the demands we need to consider. I first found a definition of $C(a,b)$ which satisfies them all with the help of a highly intuitive definition of the *explanatory power* of a with respect to b. (See footnote 1 to the present section.) Later I found a simpler definition, as follows. (See definition D in section 31.)

(vii) Let a be a consistent[4] hypothesis and let b be the description of the result of all our attempts to refute a, so that $p(b) \neq 0$. Then

$$C(a,b) \quad = \quad \frac{p(b,a) - p(b)}{p(b,a) - p(ab) + p(b).}$$

With the help of the sole assumption that p satisfies the calculus of probability, it can be shown that all our demands are satisfied; that the definition, and therefore the set of demands, is consistent; and that $C(a,b)$ is not the same as $p(a,b)$. For $C(a,b)$ does not, like $p(a,b)$, satisfy the rules of the calculus of probability.

[4]We need not demand consistency if we are prepared to make our formula a little more complicated by inserting ' $-p(\bar{a},a)$ ' in both numerator and denominator, *i.e.*, by writing $C(a,b) = p(b,a) - p(\bar{a},a) - p(b)/(p(b,a) - p(\bar{a},a) - p(ab) + p(b)$. As a result, $C(a,b) = -1$ whenever a is inconsistent. A corresponding remark applies to the relativized definition (viii).

Thus both the problems are solved. Yet there seems to be not only room but a real need for two further improvements of our definition (vii).

First, we can generalize our definition by introducing a third variable, c say, which denotes our background knowledge.[5] (This was, of course, represented in section 31 by 'b'.) So we are led to

(viii) Let a be a (consistent)[6] hypothesis; let b be the result of genuine attempts to refute a; and let c be our (consistent)[7] background knowledge consisting of theories not under test, and also of initial conditions. Then

$$C(a,b,c) \ = \ \frac{p(b,ac) \ - \ p(b,c)}{p(b,ac) \ - \ p(ab,c) \ + \ p(b,c).}$$

This corresponds precisely to the definition (D) of section 31, which was arrived at by a somewhat different argument.

A technical remark may be made regarding these two definitions, (vii) and (viii). They are based upon $p(b,a)$—called the likelihood of a with respect to b by R. A. Fisher—rather than upon $p(a,b)$. This is an important point, because they thus retain their full meaning even if $p(a) = 0$; a case which, as shown in sections 80 and 81 of *L.Sc.D.*, may be generally assumed to hold if a is a universal theory.[8] For a universal theory may be considered, *in an infinite universe* (infinite with respect to either space or time), as an infinite product of singular statements, and therefore 'infinitely improbable'; or more

---

[5]This presupposes the relativization of $Ct(a)$: we shall put

$$Ct(a,c) \ = \ C(a,a,c) \ = \ 1 \ - \ p(a,c).$$

[6]See footnote 4.

[7]See footnote 4.

[8]See also *L.Sc.D.*, Appendix *vii. That all universal laws have zero probability and therefore zero degree of confirmation (if degree of confirmation is defined as probability) is also a result of Carnap's. (*Logical Foundations of Probability*, 1950, pp. 570 f.) In consequence, universal laws are non-confirmable in Carnap's sense. For this reason, Carnap replaces (as I do) probability or degree of confirmation by another idea—though he does it only *ad hoc*—when it comes to universal laws. But the new function, which he calls 'qualified instance confirmation', leads to contradictions, as I pointed out in *Brit. Journ. Phil. Science* 7, 1956, pp. 252f. In fact, one can obtain, with the help of formula (17) on p. 573 of Carnap's book, vastly different 'degrees of qualified instance confirmation', *on the same evidence, for two laws which differ merely in their verbal formulations.*

precisely, as having zero probability. This causes no difficulty to our theory owing to the fact that I have developed a generalization of the customary calculus of probability which gives meaning to $p(x,y)$ even if $p(y) = 0$. (See *L.Sc.D.*, Appendix *iv.) As a consequence, the expressions '$p(b,a)$', or '$p(b,ac)$' and '$p(b,c)$', which occur in the definiens of (vii) and (viii) respectively, will be meaningful even if a (and c) are laws of zero probability. If in the definiens of (vii) we were to operate with '$p(a,b)$' instead of '$p(b,a)$' then, for any universal law a, we should obtain $C(a,b) = 0/0$; for in this case $p(a,b) - p(a)$ would be equal to zero, and so would $p(a,b) + p(a)$.[9] This point is of considerable importance, and it can of course be generalized by the introduction of c, that is to say, of 'background knowledge', and so extended to (viii).

Our second improvement is this. We can do a little more towards making explicit our informal requirement—the condition that b must consist of genuine attempts to falsify a: as stated in section 31, we can partly formalize it by demanding that our empirical test statements should be unexpected or improbable in the light of our background knowledge; that is to say, their probability, given the background knowledge, should be (considerably) less than $1/2$.

[9]The fact that $p(a,b) - p(a)$ equals zero if a is universal while $p(b) > 0$, is of considerable interest in connection with Carnap's concept '*confirmed*' (*Logical Foundations*, pp. 463 f.). For Carnap defines '$Co(a,b)$', that is to say '*a is confirmed by b*', with the help of an expression which is equivalent to '$p(a,b) - p(a) > 0$', in our notation; that is to say, equivalent to '$p(a,b) > p(a)$'. As a consequence, no evidential statement b 'confirms' a universal law a, if $p(a) = 0$ for universal laws (as indeed it does in Carnap's theory). *Thus laws cannot be 'confirmed'*, in Carnap's sense. Moreover, Co contradicts Carnap's concept of degree of confirmation— which is a probability, $p(a,b)$, say—in the presence of the tautology 'If x is not warm and y is warm then it is never the case that x is warmer than y, or equally warm as y'; or rather, 'If x is not confirmed by z and y is confirmed by z then it is never the case that x is better confirmed by z than y is, or equally well confirmed by z as y is.' For this may be expressed by 'If non-$Co(x,z)$ and $Co(y,z)$, then not $p(x,z) \geqslant p(y,z)$'; from which we obtain at once, by substituting according to the equivalence for Co mentioned above:

(1)   If $p(x,z) \leqslant p(x)$ and $p(y,z) > p(y)$ then $p(x,z) < p(y,z)$
(2)   If $p(x,z) = p(x)$ and $p(y,z) > p(y)$ then $p(x,z) < p(y,z)$ (by 1)
(3)   If $p(x,z) = p(x)$ and $p(y,z) > p(y)$ then $p(x) < p(y,z)$ (by 2).

But this is absurd; for we may choose $p(x)$ as near to 1 as we like, or equal to 1; and if we then choose $p(y,z)$ equal to $1/2$, say, and $p(y)$ equal to $1/4$, say, we get a contradiction; also, more generally, if we choose $y = xz$, and assume x and z to be independent. (See *Brit. Journ. Phil. Science* 7, 1956, pp. 254 f.; also my introductory remarks to Appendix *ix, *L.Sc.D.*)

It seems to me that we cannot avoid an intuitive element in our definition, amounting to the condition that the '*b*' in '*C(a,b)*' denotes a description of the results of *genuine tests*. For *C(a,b)* is intended to denote something like the acceptability of a theory—a useful, applicable, and convincing appraisal of it. In other words, we are dealing with an *interpreted* rather than with an uninterpreted formalism; and our informal or intuitive element only comes in as a warning: 'if you wish to interpret *C(a,b)* or *C(a,b,c)* as degree of corroboration (or acceptability) of a theory *a*, then you must make sure that *b* describes the results of genuine tests.'[10]

In summing up, I wish to point out two things. First, our concept of degree of corroboration takes (automatically, as it were) full account of the *weight of evidence* which causes so much trouble to probability theories.[11] Secondly, our definition gives the desired comparative results, not only in the case of Maxwell's and Fresnel's theory but also, for example, in the case of Einstein's and Newton's. And it even allows us fully to express both the somewhat doubtful character of the eclipse results and the fact that they are more favourable to Einstein than to Newton. For if a_1 is Newton's theory, a_2 Einstein's, and *b* the eclipse measurements, we certainly get

$$p(b,a_1c) < p(b,a_2c),$$

because the *b* are *much* further from Newton's prediction than from Einstein's. We see from this at once that *b* corroborates Einstein much better than it does Newton—in agreement with common sense, and with what is generally accepted.

This is due not to any particular subtlety of our theory but to the simple fact that our definition satisfies our requirements. It contains a straightforward assessment of the severity of the tests which the hypothesis has passed, and of how well it has passed them—or how badly it has failed them.

To repeat, I do not believe that my definition of degree of corroboration is a contribution to science except, perhaps, that it may be useful as an appraisal of statistical tests (*cf.* Appendix *ix to

[10]It should be remembered that there is a parallel case in probability theory. If $p(a,b)$ is to be interpreted as the degree of rational belief in *a*, then *b* must denote the total evidential knowledge in the possession of the believer.

[11]*Cf.* my 'Third Note' in Appendix *ix to *L.Sc.D.*

L.Sc.D.). Nor do I believe that it makes a positive contribution to methodology or to philosophy—except in the sense that it may help (or so I hope) to clear up the great confusions which have sprung from the prejudice of induction, and from the prejudice that we aim in science at high probabilities—in the sense of the probability calculus—rather than at high content, and at severe tests.

33. *Humanism, Science, and the Inductive Prejudice.*

There is no probalistic induction. Human experience, in ordinary life as well as in science, is acquired by fundamentally the same procedure: the free, unjustified, and unjustifiable invention of hypotheses or anticipations or expectations, and their subsequent testing. These tests cannot make the hypothesis 'probable'. They can only corroborate it—and this only because 'degree of corroboration' is just a label attached to a report, or an appraisal of the severity of tests passed by the hypothesis.

But new theories of induction are published almost every month. For there is considerable intuitive force in the assertion that the probability of a law increases with the number of its observed instances. I have attempted to explain this intuitive force by pointing out that probability and degree of corroboration have not been properly distinguished. Whether or not my explanation is satisfactory, the present superabundance of untenable theories of induction must be highly unsatisfactory even to an inductivist.

In view of the situation, I wish to address myself to the authors of any future theory of induction who may claim for their theories anything like scientific status. Remembering Kant's famous appeal to those interested in metaphysics, I will now try *to convince all those who find it worth their while to take an active interest in the theory of induction: that it is unavoidable, and indeed necessary, that they should desist from any further effort for the time being; take everything that has been done as undone; and raise the question, first of all, whether such a thing as a probabilistic theory of induction is at all possible.*[1]

To be more specific, I challenge anybody who believes that it is

[1]*Cf.* Kant's *Prolegomena to Any Future Metaphysics that Will be Able to Claim Scientific Status,* 1783. The words put in italics are Kant's, except that Kant's term 'metaphysics' has been replaced by 'theory of induction'.

possible to increase the probability of a theory by means of some inductive procedure, to explain four things:

(i) Why scientists invariably prefer a highly testable theory whose content goes far beyond all observed evidence to an *ad hoc* hypothesis, designed to explain just this evidence, and little beyond it, even though the latter must always be more probable than the former on any given evidence. How is the demand for a high informative content of a theory—for knowledge—to be combined with the demand for a high probability, which means lack of content, and lack of knowledge?

(ii) How to avoid obtaining probabilities equal to 1 for all not yet refuted universal laws, considering that they are all instantiated almost everywhere; for both the laws 'All swans are white', that is 'There is no non-white swan', and the law 'All swans are non-white', that is 'There is no white swan', are instantiated in every region in which there is no swan—that is, according to our present knowledge, almost everywhere in the universe.

(iii) How to avoid, in an infinite universe (or in a practically infinite one) obtaining the probability zero for all universal laws, considering that a universal law about an infinite universe can always be expressed as an infinite product of singular statements. (For example, 'All swans are white' can be expressed by '*Everything has the property P*' (where 'having the property *P*' is defined by the phrase 'either being white or not being a swan'.)

(iv) How do they answer my objections raised in *L.Sc.D.*, Appendix *ix, especially p. 390?

This is my challenge.

The main argument against induction is the same as that against idealism: induction is too cheap. But in the case of induction—as opposed to the case of idealism—the argument can be used to show the absurdity of the theory. For since Bacon, all induction consists in the collection and (statistical) tabulation of instances, especially of confirming instances.

But confirming instances, 'verifications', can be had for the asking. *In fact, any 'instance' whatever confirms every universal theory, with the sole exception of a falsifying instance.* Hence confirming instances are not worth having. (See, for example, section 28 of *L.Sc.D.*, especially note *1, and section 80, especially note *4.)

256

That instances are so very cheap is closely connected with the law of excluded middle. This can be seen very easily if we put our theories in the form of a there-does-not-exist-statement, that is to say, of a negated existential statement (as can be done with every universal theory; see section 15). The statement 'There does not exist a non-white swan' is indeed confirmed by anything whatever in the world, with the exception of a non-white swan; for it is equivalent to the statement 'Everything has the property of not being a non-white swan'. Similarly, the statement 'There does not exist a perpetual motion machine' is equivalent to the statement 'Everything has the property of not being a perpetual motion machine'. Therefore, everything is a verifying instance of this statement with the sole exception of an instance which is indeed a perpetual motion machine.[2] This is the main reason why only genuine attempts to refute a theory can count.

[2]The assertion made here (*cf.* sections 28 and 80 of *L.Sc.D.*) is closely connected with what C. G. Hempel has called the 'paradox of confirmation'. Hempel seems to think (*Mind*, N.S. **55**, 1946, p. 79) that the 'paradoxes' are only 'peculiarities', and 'upon closer analysis . . . prove to be reasonable', so that we should indeed accept the fact that a white handkerchief or a black raven *confirms* the statement 'All swans are white'. This may be so. But I think that all these 'confirming' or 'verifying' instances are clearly too cheap to be accepted as *corroborations*, because they are not in general the results of genuine attempts to refute the theory 'All swans are white'.

Hempel's concept of 'confirmation' or 'confirming instance' (*Journal of Symbolic Logic* **8**, 1943, pp. 122–143) is 'verificationist': the idea is that the statement *x* confirms the statement *y* if and only if *x* describes a verifying *instance* of *y*. But owing to the inadequacy of Hempel's 'logical requirements', his definition has the following clearly unintended consequences. Let *a* be the name of an individual thing. Then the singular statement '*a* is a white swan' not only confirms, as intended, 'All swans are white', and 'All non-white things are non-swans', but it also confirms (which is hardly intended) 'All things are white'; 'All things are swans'; 'All things are white swans'; and except for his *ad hoc* exclusion of identity (about which see below) 'There is only one white swan'; 'There exists one and only one thing, and this is a white swan'. Moreover, our singular statement confirms *only such statements as are compatible with the statements here quoted;* so that it confirms 'All swans are white' only in the Pickwickian sense in which it means 'All swans are white because—apart from the *a* we have seen—there aren't any'. This shows that Hempel's definition is clearly both *too narrow and too wide*. It confirms statements which are not intended to be confirmed, and it fails to confirm those which are intended to be confirmed. For example, the two statements 'Every human being has a human mother' and 'Every female human being has a human mother' are both non-confirmable by human beings (although they are confirmed by '*a* is not human'), according to Hempel's theory (since they have no finite 'development', in Hempel's sense, among human beings); but the statement 'Every

'Nothing even in mathematical science can be more certain than that a collection of scientific facts are of themselves incapable of leading to discovery,' wrote Brewster over a hundred years ago—expressing, perhaps, a more mature attitude towards science than our own.[3] For today, we are told that universal laws and theories are not really needed in science—that science can infer its predictions (with probability) from a mere collection of singular facts.[4]

Two attitudes or tendencies that are at times found together foster a belief in induction. One is the wish for a super-human authority—the authority of science, far above human whims, and exemplified in the 'exact' science of mathematics, and in the natural sciences, so far as they are based, firmly and squarely upon fact: verified, confirmed fact. The other is the wish to see in science not the work of an inspiration or revelation of the human spirit, but a more or less mechanical compilation which in principle might be performed by machines. (For what else are we but machines?) At bottom, the two tendencies may be one: the tendency to debunk man.

Now a little debunking may do us a lot of good, especially if it is done with the good grace and the good humour of Bertrand Russell. 'Put in a nutshell,' he wrote in the preface to *Mysticism and Logic*, 'the change in my outlook comes to this, that I no longer regard solemnity as a means of attaining truth; observation of life shows one that solemn people are generally humbugs, and solemn moods also contain some humbugging quality.'

But there are other debunkers, and among them quite solemn ones. And as opposed to those who honour Science with a capital 'S', because it is verified, or (since I have debunked verification)

male human being has a human mother' is confirmable, for example, by the statement 'Mary is the mother of Edward (and Mary is a female human being and Edward is a male human being)'. As to Hempel's exclusion of identity from his model language, I do not think that a definition can be made more adequate by the attempt to prevent us from expressing in our model language consequences which are palpably counter-intuitive. (*Cf.* also J. W. N. Watkins, 'Between Analytical and Empirical', *Philosophy*, 1957, pp. 116–123 and pp. 125–127.)

Incidentally, I have been criticized for not giving a 'precise' definition either of the *'instantiation of a law'* or of a *'refuting instance of a law'*. Yet 'semantic' definitions of these concepts are given in the last footnote of my 'Note on Tarski's Definition of Truth', now in my *Objective Knowledge*, pp. 335–340.

[3]Sir David Brewster, *Memoirs of the Life, Writings and Discoveries of Sir Isaac Newton*, second edition, 1860, vol. ii, p. 328.

[4]See text to notes 10 and 11 to section 13 above.

'confirmed', or 'exact', there are not a few who believe that though it is made 'firm' by confirmation,[5] it is not deep; that since the world has no depth, science does not need any theories but only a collection of facts from whose frequency it can induce the probability of their future repetitions.

I see science very differently. As to its authority, or confirmation, or probability, I believe that it is nil; it is all guesswork, *doxa* rather than *epistēmē*. And probability theory even 'confirms' me in this, by attributing zero probability to universal theories.

But seen as the result of human endeavour, of human dreams, hopes, passions, and most of all, as the result of the most admirable union of creative imagination and rational critical thought, I should like to write 'Science' with the biggest capital 'S' to be found in the printer's upper case.

Science is not only, like art and literature, an adventure of the human spirit, but it is among the creative arts perhaps the most human: full of human failings and shortsightedness, it shows those flashes of insight which open our eyes to the wonders of the world and of the human spirit. But this is not all. Science is the direct result of that most human of all human endeavours—to liberate ourselves. It is part of our endeavour to see more clearly, to understand the world and ourselves, and to act as adult, responsible, and enlightened beings. 'Enlightenment', Kant writes, 'is the emancipation of man from the state of self-imposed tutelage . . . from a state of incapacity to use his own intelligence without external guidance. Such a state of tutelage I call "self-imposed" if it is due not to any lack of intelligence but to the lack of courage or determination to use one's own intelligence instead of relying upon a leader. *Sapere aude*! Dare to use your own intelligence! This is the maxim of the enlightenment.'[6]

Kant challenges us to use our intelligence instead of relying upon a leader, upon an authority. This should be taken as a challenge to reject even the scientific expert as a leader, or even *science itself*. Science has no authority. It is not the magical product of the given, the data, the observations. It is not a gospel of truth. It is the result of our own endeavours and mistakes. It is you and I who make science, as well as we can. It is you and I who are responsible for it.

[5]Cf. *Philosophy of Science* **12**, 1945, note 2 on p. 99.
[6]I. Kant, *Was ist Aufklärung?* (What Is Enlightenment?), 1785.

Science, one might be tempted to say at times, is nothing but enlightened and responsible common sense—common sense broadened by imaginative critical thinking. But it is more. It represents our wish to know, our hope of emancipating ourselves from ignorance and narrow-mindedness, from fear and superstition. And this includes the ignorance of the expert, the narrow-mindedness of the specialist, the fear of being proved wrong, or of being proved 'inexact', or of having failed to prove or to justify our case. And it includes the superstitious belief in the authority of science itself (or in the authority of 'inductive procedures' or 'skills').

The nuclear bomb (and possibly also the so-called 'peaceful use of atomic energy' whose consequences may be even worse in the long run) have, I think, shown us the shallowness of the worship of science as an 'instrument' of our 'command over nature' or the 'control of our physical environment': it has shown us that this command, this control, is apt to be self-defeating, and apt to enslave us rather than to make us free—if it does not do away with us altogether. And while knowledge is worth dying for, power is not. (Knowledge is one of the few things that are worth dying for, together with liberty, love, kindness, and helping those who are in need of help.)

All this may be trite. But it has to be said from time to time. The First World War destroyed not only the commonwealth of learning; it very nearly destroyed science and the tradition of rationalism. For it made science technical, instrumental. It led to increased specialization and it estranged from science what ought to be its true users—the amateur, the lover of wisdom, the ordinary, responsible citizen who has a wish to know. All this was made much worse by the Second World War and the bomb. This is why these things have to be said again. For our Atlantic democracies cannot live without science. Their most fundamental value—apart from helping to reduce suffering—is truth. They cannot live if we let the tradition of rationalism decay. But what we can learn from science is that truth is hard to come by: that it is the result of untold defeats, of heartbreaking endeavour, of sleepless nights. This is one of the great messages of science, and I do not think that we can do without it.

But it is just this message which modern specialization and organized research threatens to undermine; and the re-emergence of Bacon's naive views of induction—of the belief that science is the collection and tabulation of instances, and especially of confirming instances—must be combated by those who believe in human reason.

As I recall my own experience at the meeting of the Aristotelian Society (which I recounted in section 1), I now see that my attempt to explain my views was bound to fail. There was too much to it, and too much against it. I may have shown, by now, the insufficiency of the beautiful method of stating one's case clearly and simply, and leaving it at that, without discussing all those 'isms' which may soften the impact of any new ideas. I did indeed try this method: I stated my solution of the problem fully and briefly in sections 2 and 3. In a way, no more than this should have been needed, and it is possible that some readers may have felt the full force of the argument. But I do not think that the long way we have travelled since then has been unnecessary. If anything, I doubt whether I have gone far enough into all the assumptions and prejudices which beset this simple and straightforward logical problem, and which stand in the way of impartial examination of any simple and straightforward solution.

Of course, while writing this, I am inclined to believe that this time (in contrast to that meeting of the Aristotelian Society) I may have succeeded in getting my point across. But when I try to take a more detached view of the matter, I rather doubt it. People not only hear what they expect to hear but they only read what they expect to read. And since I always say that we can grasp things only with the help of our theories, I should be the last to complain about this fact. No doubt, there are two possible ways of interpreting while reading: a reader may be uncritical or critical towards his own interpretations. Yet I fear that a time and a place where readers habitually try to refute their own interpretations and expectations of what they are reading are only in a writer's dream.

Addendum: Critical Remarks on Meaning Analysis

The reason why I do not wish to present my definition of degree of corroboration as a goal attained (see section 30 above) is that—in opposition, I fear, to most philosophers—I feel convinced of the truth of the following three propositions.

(1) *What-is? questions*, such as *What is Justice?* or *What is degree*

of corroboration? are always pointless—without philosophical or scientific interest; and so are all *answers* to what-is? questions, such as *definitions*. It must be admitted that some definitions may sometimes be of help in answering other questions: urgent questions which cannot be dismissed: genuine difficulties which may have arisen in science or in philosophy. But what-is? questions as such do not raise this kind of difficulty.

(2) It makes no difference whether a what-is? question is raised in order to inquire into the essence or into the nature of a thing, or whether it is raised in order to inquire into the essential meaning or into the proper use of an expression. These kinds of what-is? questions are fundamentally the same. Again, it must be admitted that an answer to a what-is question—for example, an answer pointing out distinctions between two meanings of a word which have often been confused—may not be without point, provided the confusion led to serious difficulties. But in this case, it is not the what-is? question which we are trying to solve; we hope rather to resolve certain contradictions that arise from our reliance upon somewhat naive intuitive ideas. (The first example discussed below—that of the ideas of a derivative and of an integral—will furnish an illustration of this case.) The solution may well be the *elimination* (rather than the clarification) of the naive idea, and its replacement by a totally different one.[1] But an answer to an essentialist question (a what-is? question) is never fruitful. For words, or concepts, or notions, are never more than mere instruments, useful for formulating our theories. (Much as I am opposed to the instrumentalist interpretation of scientific *theories*, I am an instrumentalist as regards *words* or *concepts* or *notions*.[2])

In consequence, it *cannot* be the main aim of philosophy (or indeed of any rational or critical enterprise) to clarify or define ideas or concepts or notions or meanings, or to replace some given ideas or concepts or meanings by more exact ones.

(3) The problem, more especially, of replacing an 'inexact' term by an 'exact' one—for example, the problem of giving a definition in

[1] [*Cf.* the discussion of 'dialysis' in *Unended Quest*, section 7. Ed.]

[2] I have in various places described this position as methodological nominalism, in opposition to both methodological essentialism, which asks and answers *what-is?* questions, and to metaphysial nominalism, which asserts that meaningful words *are* nothing but names of things (or of memory images of things) and which thereby give an essentialist answer to an (implied) essentialist question—and a wrong answer to boot.

'exact' or 'precise' terms—is a pseudo-problem. It depends essentially upon the inexact and imprecise terms 'exact' and 'precise'. These are most misleading, not only because they strongly suggest that there exists what does not exist—absolute exactness or precision—but also because they are emotionally highly charged: under the guise of scientific character and of scientific objectivity, they suggest that precision or exactness is something superior, a kind of ultimate value, and that it is wrong, or unscientific, or muddle-headed, to use inexact terms (as it is indeed wrong not to speak as lucidly and simply as possible). But there is no such thing as an 'exact' term, or terms made 'precise' by 'precise definitions'. Also, a definition must always use undefined terms in its definiens (since otherwise we should get involved in an infinite regress or in a circle); and if we have to operate with a number of undefined terms, it hardly matters whether we use a few more. Of course, if a definition helps to solve a genuine problem, the situation is different; and some problems cannot be solved without an increase of precision. Indeed, this is the only way in which we can reasonably speak of precision: the demand for precision is empty, unless it is raised *relative* to some requirements that arise from our attempts to solve a definite problem.

These three propositions are to be defended in the present section.

They run counter, I am afraid, to the main current or trend or drift of contemporary philosophy—including the philosophy of science—which tends to combine an instrumentalist interpretation of *theories* with an essentialist interpretation of *concepts*. This tendency is illustrated by a characteristic passage of Schlick's in which he expounds certain ideas of Wittgenstein's. Schlick tells us that the famous what-is? questions in Plato's dialogues, such as the question 'What is Justice?' posed by the 'Socrates' of the *Republic*, shows that Socrates's philosophy was devoted to 'what we may call "the pursuit of meaning". He tried to clarify', Schlick writes, 'the meaning of our expressions and the real sense of our propositions . . . Let me state shortly and clearly', Schlick sums up his position, 'that I believe that science should be defined as the *pursuit of truth*, and philosophy as the *pursuit of meaning*.'[3]

[3]M. Schlick, 'The Future of Philosophy' (*Publications in Philosophy*, edited by the College of the Pacific, 1932), quoted here from Schlick, *Gesammelte Aufsätze 1926–1936* (published 1938), p. 126. Schlick says that, *qua* scientist, the scientist is

Schlick's definition of the essential nature of philosophy is still very influential. It will be obvious from my first two theses in this section that I do not think such a definition can tell us anything worth telling (definitions never convey information—except to those in need of a dictionary) or that it comes anywhere near to clarifying the task or aim of most philosophers (who, if worth listening to, are searchers for truth). Incidentally, I do not believe that Schlick correctly represents the view of the historical Socrates. Yet I do agree that Plato (and the Platonic 'Socrates' of the *Republic*[4]) was much preoccupied with what-is? questions. Indeed, the belief that philosophy should analyze the meaning of words or of concepts, that it has to answer what-is? questions, that it should give definitions, derives from Platonic and Aristotelian metaphysics. But it goes back to still deeper roots: to animism. (This may be one of the reasons why it is so hard to eradicate.[5]) As shown by our

interested only in truth—in verification—and not in meaning. But how, Schlick asks, can a scientist test the truth of a statement whose meaning he does not understand? Especially if the meaning of a statement is the method of its verification? And on p. 127, Schlick writes: 'In so far as the scientist does find out the hidden meaning of the propositions which he uses in his science he is a philosopher.' And he mentions, as examples, Newton's discovery of the concept of mass, and 'Einstein's analysis of the meaning of the word "simultaneity" as it is used in physics'. I discussed and dismissed the latter example in my *Open Society*, chapter II, section ii. The former, Newton's concept of mass (which Newton defined as the space-integral of density), is an unfortunate example which would have embarrassed Newton greatly; indeed, the example is incomprehensible except on the assumption that what Schlick had in mind was not Newton's definition of 'mass' but rather the idea of matter—together with the essentialist interpretation of Newton's theory of gravity, which Newton himself rejected. I have discussed this interpretation in my 'Three Views Concerning Human Knowledge', section iii, reprinted in *Conjectures and Refutations*.

[4]For a discussion of the Socratic problem (that is to say, the problem of the relationship between Platonism and the historical Socrates), see my *Open Society*, note 56 to Chapter 10, and Richard Robinson in *The Philosophical Review* 60, 1951, especially pp. 494f.

[5]The animistic belief in the *power of words*—in word magic—is, I conjecture, involved in the very process by which the child learns to speak; a process which is linked with his experience of controlling his environment by pre-linguistic noises. I think that this belief is involved in what we mean when we speak of the 'meaning of words'; which is perhaps one reason why some people feel that meaning-analysis is important. An acknowledgment is due to Freud for his deep understanding of animism and its belief in the 'omnipotence of thoughts', symbols, and words. Yet Freud never suspected the animist basis of his own essentialist approach. This becomes clear in one of his last papers (*Collected Papers* 5, 1952). There he speaks of the problem of getting 'nearer to the *nature*, or, as people sometimes say, the

passage from Schlick, this belief was never overcome by the nominalist and positivist opponents of Platonic idealism, and it is present in Berkeley, Wittgenstein, and their followers—just as much as in neo-Aristotelianism. It inspires what Edmund Husserl has called 'pure phenomenology' or 'the intuition of essences'; what G. E. Moore calls 'philosophical analysis'; and what Rudolf Carnap calls the 'explication of concepts'.[6] However profound the differences between these schools may be, they all take it to be the task of philosophy to clarify the meaning of ideas or of concepts or of words; and for this reason they may be dealt with together here.

I do not wish to repeat here what I have said in other places about essences, meaning analysis, and definitions.[7] Instead, I will attempt to defend the three theses formulated at the beginning of this section by analyzing a number of *prima facie* counter examples; that is to say, a number of cases which, it may be claimed, refute my three

---

essence of the mental' (p. 377) and he declares that 'being conscious cannot be the essence of what is mental', and further (with an acknowledgment to T. Lipps) 'that the mental is in itself unconscious and that the unconscious is the truly mental' (p. 382). Perhaps the most interesting point in this somewhat pointless discussion (pointless because asking for the essence of the mental is like asking for the spirit of brandy, or for the anima of the soul; and if the unconscious is the essence of the mental, then mental energy may be the essence of the unconscious, and libido the essence of mental energy, . . .) is Freud's realization (p. 378) that a physicist would *not* ask 'what is the nature of electricity?'. This, Freud sees, raises an objection, because 'psychology, too, is a natural science'. Freud's dismissal of this objection does not follow from his somewhat ambivalent arguments. And when he suggests that, in psychology, the 'case is different' because 'everyone behaves as . . . an amateur psychologist', and because every amateur psychologist believes in essences, he comes pretty close (but not quite close enough) to saying that essentialism is a pre-scientific attitude, and thus to uncovering its magical and animistic roots.

[6]Hume speaks of 'the accurate explication' of words (such as 'power'; *cf.* his footnote to part ii of section iv of the *Enquiry*) by which he means 'to fix the precise meaning' of a word 'and thereby remove part of that obscurity which is so much complained of in . . . philosophy' (*op. cit.*, section vii).

Carnap, in the first chapter, 'On Explication', of his *Logical Foundations of Probability*, 1950, refers to Kant, Husserl, Moore, and Langford. He assumes, explicitly, that what-is? questions are relevant problems; and he criticizes only those who 'start to look for an answer'—that is, to a what-is? question—without first making sure that 'the terms of the question' are sufficiently clear or exact (p. 4). The task of '*explicating*' some '*explicandum*' is defined as transforming the *explicandum* into (or replacing it by) an exact concept, the *explicatum*.

[7]See note 2 to section 19 of *L.Sc.D.*, and section ii of Chapter 2 of *The Open Society* (see also section 10 of *The Poverty of Historicism*). [*Cf. Unended Quest*, section 7. Ed.]

theses by establishing the importance of definitions which clarify or explicate or make precise the meaning of some important concepts. My first example will be the definition of the derivative (the differential quotient or measure of the slope of a curve) as a limit of a quotient of differences. This example will be discussed at some length. My second example will be Russell's 'theory of descriptions'—a kind of standard example in this field. My third and fourth examples will be the set-theoretical definition of dimension, and Tarski's definition of truth.

In each of these cases I am going to assume, without further discussion, that the examples might be claimed, by devotees of meaning analysis or of explication, to refute my three theses. My aim will be to show that, on the contrary, they support them: that they are not examples of clarifying or 'explicating' meanings, but something different: attempts to solve concrete problems, such as the elimination of contradictions; and that they achieve their aim (at least in the first two cases) by *abandoning* the attempt to clarify, or make exact, or 'explicate' the intended or intuitive meaning of the concepts in question.

My first example is most instructive. The modern history of the problem of the foundations of mathematics is largely, it has been asserted, the history of the 'clarification' of the fundamental ideas of the differential and integral calculus. The concept of a *derivative* (the slope of a curve or the rate of increase of a function) has been made 'exact' or 'precise' by defining it as a limit of the quotient of differences (given a differentiable function); and the concept of an integral (the area or 'quadrature' of a region enclosed by a curve) has likewise been 'exactly defined'. The example is of crucial historical importance for our problem. Attempts to eliminate the contradictions in this field constitute not only one of the main motives of the development of mathematics during the last hundred or even two hundred years, but they have also motivated modern research into the 'foundations' of the various sciences and, more particularly, the modern quest for precision or exactness. 'Thus mathematicians', Bertrand Russell says, writing about one of the most important phases of this development, 'were only awakened from their "dogmatic slumbers" when Weierstrass and his followers showed that many of their most cherished propositions are in general false.

Macaulay, contrasting the certainty of mathematics with the uncertainty of philosophy, asks who ever heard of a reaction against Taylor's theorem? If he had lived now, he himself might have heard of such a reaction, for this is precisely one of the theorems which modern investigations have overthrown. Such rude shocks to mathematical faith have produced that love of formalism which appears, to those who are ignorant of its motive, to be mere outrageous pedantry.'[8]

It would perhaps be too much to read into this passage of Russell's his agreement with a view which I hold to be true: that without 'such rude shocks'—that is to say, without the urgent need to remove contradictions—the love of formalism is indeed 'mere outrageous pedantry'. But I think that Russell does convey his view that without an urgent need, an urgent problem to be solved, the mere demand for precision is indefensible.

But this is only a minor point. My main point is this. Most people, including mathematicians, look upon the definition of the derivative, in terms of limits of sequences, as if it were a definition in the sense that it analyses, or makes precise, or 'explicates', the intuitive *meaning of the definiendum*—of the derivative. But this widespread belief is mistaken.

We may illustrate two senses (out of many) of the word 'definition', by looking at a few possible definitions of the number 23.

We may write:

(a) $\qquad x = 23$ if, and only if, $x = 20 + 3$
(b) $\qquad\qquad\qquad\qquad\qquad x = 2 \times 10 + 3$
(c) $\qquad\qquad\qquad\qquad\qquad x = 2^4 + 2^2 + 2^1 + 2^0$
(d) $\qquad\qquad\qquad\qquad\qquad x = 3^3 - 2^2$
(e) $\qquad\qquad\qquad\qquad\qquad x = 2018 - (15 \times 133).$

Most people will say that the equivalences (a) and (b) may define '23', and perhaps also (c), if we want a definition in terms of powers of 2; but neither (d) nor (e) will be a 'reasonable definition' of '23', although they are of course true equivalences; for in no way do they analyse or elucidate or explicate the meaning of '23'. Nobody who

[8]Bertrand Russell, *Mysticism and Logic*, 1918; *cf.* the penultimate paragraph of Chapter v.

understands the meaning of '23' would explain or 'explicate' its meaning by (d) or (e).

Now the point I wish to make is this. Not only are (d) and (e) perfectly legitimate definitions in the general sense of the term 'definition' which is at present accepted by most mathematicians and logicians;[9] but (d) and (e) may serve very well to illustrate just that sense of the word 'defined' in which the derivative is defined as a limit of the quotient of differences; a sense which, I am afraid, is obscured by the customary way of treating the subject.

Newton and Leibniz and their successors did not deny that a derivative, or an integral, could be *calculated* as a limit of certain sequences—roughly in the sense in which (d) or (e) may be used to calculate *x*. But they would not have regarded these limits as possible definitions, because they do not give the *meaning*, the *idea*, of a derivative or an integral.[10]

For the derivative is a measure of a velocity, or a slope of a curve. Now the velocity of a body at a certain instant is something real—a concrete (relational) attribute of that body at that instant. By contrast the limit of a sequence of average velocities is something highly abstract—something that exists only in our thoughts. The *average* velocities themselves are unreal. Their unending sequence is even more so; and the limit of this unending sequence is a purely mathematical construction out of these unreal entities. Now it is intuitively quite obvious that this limit must numerically coincide with the velocity, and that, if the limit can be calculated, we can thereby calculate the velocity. But according to the views of Newton and his contemporaries, it would be putting the cart before the horse were we to *define* the velocity as being identical with this limit, rather than as a real state of the body—at a certain instant, or at a certain point, of its track—to be calculated by any mathematical contrivance we may be able to think of.

The same holds of course for the slope of a curve in a given point. Its measure will be equal to the limit of a sequence of measures of certain other average slopes (rather than actual slopes) of this curve. But it is not, in its proper meaning or essence, a limit of a sequence: the slope is something we can sometimes actually draw on paper, and construct with compasses and rulers, while a limit is in essence

[9]See A. Tarski, *Logic, Semantics, Metamathematics*, 1956, p. 299.
[10]*Cf.* R. Courant and H. Robbins, *What is Mathematics?*, 1941, p. 433.

something abstract, rarely actually reached or realized, but only approached, nearer and nearer, by a sequence of numbers.

Similarly, a circle can of course be approximated by a sequence of inscribed polygons, and we can *calculate* its area or 'quadrature' in this way, *approximately* (to any desired degree of approximation); but it would be sheer nonsense, from the point of view of an understanding of its meaning or essence, to say that a circle *is* the (never realized) limit of this sequence of inscribed polygons. For it is the circle which defines this sequence—as the sequence of polygons inscribed in this circle—and not the other way round. Similarly, it is clear that the area of the circle will equal the limit of the areas of the inscribed polygons; but it would be sheer nonsense to explicate it in this way; just as it would be sheer nonsense to describe the body of my cat (his shape at the present instant) as the limit of the inscribed polyhedra. My cat is what he is, and only because he is there, in his own right, can I speak of inscribed polyhedra, and understand that these may be made more and more closely to approach his shape; in which case, obviously, their volume will also approach that of my cat. Thus I may use the sequence of these abstract polyhedra for purposes of calculation; but since this sequence is defined as inscribed in my cat, we cannot use these abstract sequences to *define* either the shape of the cat, or its volume—*in the old sense of 'define'*.

Or as Berkeley puts it ' . . . however expedient such analogies or such expressions may be found for facilitating the modern quadratures, yet we shall not find any light given us thereby into the *original real nature* of fluxions; or that we are enabled to form from thence just ideas of fluxions considered in themselves.'[11] Thus mere means for facilitating our calculations cannot be considered as explications or definitions.

This was the view of all mathematicians of the period, including Newton and Leibniz.

If we now look at the modern point of view, then we see that we have completely given up the idea of definition in the sense in which it was understood by the founders of the calculus, as well as by Berkeley. We have given up the idea of a definition which explains the meaning (for example of the derivative). This fact is veiled by

[11]See Berkeley, *The Analyst*, 1734, section 47. (The italics are mine.)

our retaining the old symbol of 'definition' for some equivalences which we use, not to explain the idea or the essence of a derivative, but to *eliminate* it. And it is veiled by our retention of the name 'differential quotient' or 'derivative', and the old symbol dy/dx which once denoted an idea which we have now discarded. For the name, and the symbol, now have no function other than to serve as labels for the definiens—the limit of the sequence.

Thus we have given up 'explication' as a bad job. The intuitive idea, we found, led to contradictions. But we can solve our problems without it, retaining the bulk of the technique of calculation which originally was based upon the intuitive idea. Or more precisely, we retain *only* this technique, as far as it was sound, and eliminate the idea with its help. The derivative and the integral are both *eliminated*; they are replaced, in effect, by certain standard *methods* of *calculating* limits.

But this change was made surreptitiously, and it remained little understood by the observer (especially by the essentialist observer) because the old term 'definition' and the old names and symbols for the derivative, etc., were retained. This retention is quite in order: only an essentialist could object to it.

It is fascinating to consider that this whole admirable development might have been nipped in the bud (as in the days of Archimedes) had the mathematicians of the day been more sensitive to Berkeley's demand—in itself quite reasonable—that we should strictly adhere to the rules of logic, and to the rule of always speaking sense.

We now know that Berkeley was right when, in *The Analyst*, he blamed Newton as well as 'the foreign mathematicians' (such as Leibniz) for obtaining their mathematical results in the theory of fluxions or 'in the *calculus differentialis*' by illegitimate reasoning. And he was completely right when he indicated that their symbols were without meaning. 'Nothing is easier', he wrote, 'than to devise expressions and notations, for fluxions and infinitesimals of the first, second, third, fourth, and subsequent orders, $\dot{x}$, $\ddot{x}$, $\dddot{x}$, $\ddddot{x}$, . . . , etc. These expressions indeed are clear and distinct, and the mind finds no difficulty in conceiving them to be continued beyond any assignable bounds. But if . . . we look underneath, if, laying aside the expressions, we set ourselves attentively to consider the things themselves which are supposed to be expressed or marked thereby, we shall discover much emptiness, darkness, and confusion . . . , direct impossibilities, and contradictions.'[12]

But the mathematicians of his day did not listen to Berkeley. They got their results, and they were not afraid of contradictions as long as they felt that they could dodge them with a little skill. This was fortunate. For the attempt to *analyse the meaning*' or to *explicate*' their concepts would, we know now, have led to nothing. Berkeley was right: all these concepts were meaningless, in his sense *and* in the traditional sense of the word 'meaning': they were empty, for they denoted nothing, they stood for nothing. Had this fact been realized at the time, the development of the calculus might have been stopped again, as it had been stopped before. It was the neglect of precision, the almost instinctive neglect of all meaning analysis or explication, which made the wonderful development of the calculus possible.

The problem underlying the whole development was, of course, to retain the powerful instrument of the calculus without the contradictions which had been found in it. There is no doubt that our present methods are *more* exact than the earlier ones. But this is not due to the fact that they use 'exactly defined' terms. Nor does it mean that they are exact: the main point of the definition by way of limits is always an *existential* assertion, and the meaning of the little phrase 'there exists a number' has become the centre of disturbance in contemporary mathematics. Whatever one may think about the problems of the theory of the continuum, and of Brouwer's intuitionism, there can be no doubt that we were only awakened from our dogmatic slumbers when Brouwer brought home to us the difference between constructive and non-constructive proofs; a difference whose significance is admitted today even by his opponents.[13] This illustrates my point that the attribute of exactness is not absolute, and that it is inexact and highly misleading to use the terms 'exact' or 'precise' as if they had any exact or precise meaning.

My second example of a definition or explication is Russell's famous 'theory of description'. Its apparent task was to give a definition or explication of the meaning of a phrase like 'the (such and such)', such as, for example, 'the smallest prime greater than

[12]*The Analyst*, section 8.
[13][See *Objective Knowledge, op. cit.*, pp. 128–40. Ed.]

19'. But there were other problems which led to the task of defining this phrase. One of them, as Russell tells us, was his dissatisfaction with Meinong's views concerning the mode of existence of non-existent entities.[14] Another problem—in a way a more important one—was to show how statements containing an expression of the form 'the such and such' could be logically deduced from statements which did not contain an expression of this form; for example, how the statement '23 is the smallest prime greater than 19' could be deduced from, '23 is a prime, and there is no prime between 19 and 23, and $19 < 23$'.

Russell's answer to the problem was, no doubt, based on an analysis of the intuitive meaning of statements containing the phrase 'the such and such'. But his resulting definition was, exactly as in the case of the definition of the derivative, a means of *discarding or eliminating* the intuitive definiendum. The symbol was retained: but it became a shorthand symbol for the definiens. The question whether its intended meaning was properly rendered by the definiens thereby loses its point: the defined symbol has no meaning other than to be a shorthand symbol for the definiens.

It is for this reason that most criticisms of Russell's definition (I have in mind G. E. Moore's contribution to P. A. Schilpp's volume, *The Philosophy of Bertrand Russell*) miss the point, even if they are perhaps more sensitive than Russell to the intuitive meaning of the term analyzed.

---

[14]See *The Philosophy of Bertrand Russell*, ed. by P. A. Schilpp, 1944, p. 13. It may be noted that G. E. Moore, in his essay on Russell's 'Theory of Descriptions' (in the same volume, pp. 177–225) nowhere mentions any of the problems which Russell's theory was intended to solve and which it did solve—although Moore takes it as one of his own main problems to discover what reasons F. P. Ramsey might have had for describing Russell's theory as a *'paradigm of philosophy'*. In other words, Moore wants to evaluate Russell's theory. But nobody can even begin to understand the value of a theory unless he relates it to the *problems* it was designed to solve. I greatly admire Moore as a realist and as a defender of common sense. But I cannot admire him as an analyst. And this essay of his—now famous as a *paradigm of analysis*—with its various little criticisms and improvements of Russell's theory, all irrelevant to the *problems* which this theory was designed to solve, has had, I fear, a devastating influence. It has encouraged others to emulate it, and thus to ignore the maxim that we should always criticize a theory in its strongest possible form—rather than attack minor weaknesses of its presentation—if we want our criticism to be worthwhile. It has thus contributed to a lowering of the all too precarious standards of philosophical discussion.

A third example is the definition of *dimension* (developed by Poincaré, Brouwer, Urysohn and Menger) which plays some part in *L.Sc.D.*[15] The main point of this definition is that it helps to solve a problem—the problem of the topological invariance of dimensionality—which had first become urgent through Cantor's proof (of 1878) that the cardinal numbers of the points of a one-dimensional continuum and of a continuum of more than one dimension are the same. Of course, the definition of dimension in set-theoretic terms led to the solution of many other problems. Considering the kind of problem it was designed to solve, the salient point of the definition was that the definiens employed *set-theoretic terms*. Note that I say 'set-theoretic terms', and *not* 'exact terms'. In fact, the exactness of the intuitive set theory upon which the definition was based was somewhat dubious at a time when there already existed a more exact (axiomatic) set theory. But it so happens that the particular problem which this definition was designed to solve is not touched by those difficulties which led to the more exact developments; moreover, the definition can be taken over without change into the various more exact forms of axiomatic set theory.

My last example is Tarski's definition of the concept of truth. Tarski says that his sole intention was to define truth, in its absolute sense; or more precisely, the predicate '. . . is true' in the sense of '. . . corresponds to the facts' or '. . . corresponds to reality'. Yet it is clear that this intention grew out of the need to solve some very serious problems.

On the one hand, the idea of truth appeared to have an important function. Philosophers wished to say such things as, for example: 'The search for a true theory is not the same as the search for a useful instrument.' And logicians wished to say such things as, for example: 'If the conclusion of a valid inference is false, then the premises cannot all be true.' In other words, there was a real need for a concept such as truth.

On the other hand, it was asserted, perfectly correctly, that the phrase 'It is true that . . .'—for example, 'It is true that $3 + 5 = 8$'—was always logically redundant, and that '$3 + 5 = 8$' asserted just as much; and it was pointed out (a little less correctly)

[15]See *L.Sc.D.*, sections 32 and 38 *ff.*

that '. . . is true' in a phrase like ' "3 + 5 = 8" is true' was also redundant. (I say 'less correctly' because this phrase is short for 'The statement "3 + 5 = 8" is true'; and here 'is true' is obviously not redundant.) In addition to the suggested redundancy of the concept of truth, it was pointed out that the concept of truth led to such paradoxes as the paradox of the liar. And finally, it was pointed out that the idea of a correspondence with the facts, or with reality, was at the same time intuitively inescapable and hopelessly obscure. For what do we mean when we talk of the *facts*, or of *reality*? And what do we mean when we talk of a *correspondence* between a statement, or a proposition, and a fact? Wittgenstein had suggested, in the *Tractatus,* that this correspondence was a similarity of structure—that it was like the correspondence between a melody and the vibrations in the phonographic needle. But this mistaken suggestion only contributed to the obscurity.

This was, roughly, the problem situation, or rather, part of the problem situation; and Tarski's work succeeded, in the most straightforward way, in clearing up all the problems and obscurities mentioned here. It is by this success that his work must be judged. In addition, his theory turned out to be surprisingly fruitful in the more technical field of mathematical logic, and even in mathematics itself.

As to the intuitive ordinary everyday concept of truth, Tarski's work did provide, I think, something like a valuable meaning analysis. Not, however, by providing us with a definition of this concept. On the contrary, Tarski showed that if we introduce, without special artificial precautions, a definition of truth into ordinary language (that is to say, into a language which is 'universalistic' in the sense that we can talk in it about everything), then this language becomes inconsistent. In this sense, then, an 'exact' definition of the intuitive idea of truth turns out to be impossible. Nevertheless, we can of course use the concept of truth quite consistently within ordinary language if we learn a lesson which can easily be drawn from Tarski's work. The lesson, in brief, is this.[16] We must distinguish between *use* and *mention*; that is to say, between the

[16]See also my dialogue, 'Self-Reference and Meaning in Ordinary Language', *Mind* **63**, 1954, pp. 162 *ff.*; now also in my *Conjectures and Refutations.* [See Popper's discussion of Tarski in *Conjectures and Refutations,* Chapter 10, and *Objective Knowledge,* Chapters 2 and 9. Ed.]

language in use and the parts of the language which are talked about; and we must remember that if we assert *of* a statement that it is true, then this statement belongs to the language mentioned, as distinct from the language in use—the language *in* which we assert that a statement is true. But this lesson is not, of course, the same as a 'precise definition'.

A precise definition, Tarski showed, can be given only with respect to an *interpreted formalized* language—an artificial language system. (The salient point is that a precise definition of a 'well formed formula' must precede the definition of a 'true statement'.) With respect to artificial formalized languages, Tarski has given a *method* of defining truth rather than a definition; and it is the knowledge of how to define truth for such languages, and the knowledge that a definition can be given, together with a knowledge of the conclusions which could be drawn with its help, which make this theory important. This knowledge, embodied in Tarski's theory, is far more important than the definition which he actually gives, as an illustration of his method, with respect to a certain very simple language.[17]

Many more examples could be given. But I hope that these four, together with the example of our own problem of defining degree of corroboration, will help to make my main thesis clear: problems of definition, or 'analysis', or 'explication' are, by themselves, without any significance; they can be significant only relative to other and more serious problems. It can never be a rational enterprise to replace a term by a more 'exact' one—this really is not a serious problem; but it may become a serious task in connection with some serious problem whose solution might be made easier if we had a definition, or if we cleared up some terminological ambiguities first.[18]

[17]Tarski's work has encouraged the view that we have to construct an artificial model language whenever we wish to give an 'exact' definition. I do not think that this view properly interprets Tarski's work which refers to *all* languages of a very comprehensive set.

[18]A useful definition—one which helps us to solve real problems—will always be an *eliminating* definition, rather than an *explicating* one, as indicated above; though it should be stressed that our inquiry will here, as always, be guided by intuition. But the intuition of meanings which may guide us to propose a certain definition is our private affair; and the acceptance of the proposal will depend on its fertility—on the help it renders us in the solution of our problems. This distinction between (a) eliminating and (b) explicating definitions corresponds closely to what

I am ready to admit that an increase of precision can be of great importance: not for its own sake, but because it may increase the *testability* of a theory. (*Cf.* section 37 of the *L.Sc.D.*) In other cases, it may help us to make just those distinctions which are needed to solve a serious problem. But precision should never become a fetish. Nor should it ever be treated as a value in itself—like clarity or lucidity.

Incidentally, there is a whole category of problems that look as if their solution might be furthered by a definition while in fact definitions would merely rob them of their empirical character, by turning a factual problem into a verbal one. For example, if we wish to solve the admittedly very vague but possibly factual problem 'Are all works of art beautiful, or are there works of art which are not beautiful?' we should beware of the temptation to define either 'work of art' or 'beautiful', but rather try to remember things that might reasonably be described both as works of art and, at the same time, as non-beautiful. In other words, we should look for examples. Schönberg's *Pierrot Lunaire;* Orwell's *1984;* some caricatures (even by Leonardo); and Gulliver in Brobdingnag, may perhaps qualify. This list is obviously as sketchy as it is subjective, and the solution based upon it is therefore dubious. But it might be improved. Moreover, the vagueness and dubiousness of the solution perfectly fits the vagueness and dubiousness of the problem. Now if, on the other hand, we define, say, 'work of art' so as to include beauty among its properties, we can safely answer our problem with 'precision'. But in doing so we substitute (perhaps unwittingly) a *purely verbal problem* for our factual one, and no reflection on actual works of art can contribute any longer to its solution. My list will no longer be sketchy and subjective but it will now be completely irrelevant; we can say *a priori* that every one of its items must be either beautiful, contrary to my suggestion, or else no work of art, also contrary to my suggestion. And we can assert this even if we do not know any of the works mentioned in this or any similar list. (In the scientific field, a very similar transformation of factual problems into purely verbal ones is achieved by conventionalism, and by all conventionalist interpretations or stratagems; *cf.* sections

I have elsewhere described as (a) right-to-left definitions and (b) left-to-right, or essentialist, or Aristotelian definitions; see my *Open Society,* Chapter 11, section ii. [*Cf. Unended Quest,* p. 31. Ed.]

19 and 20 of my *L.Sc.D.* In philosophy, the transformation of philosophical or other problems into verbal problems is simply a curse—especially since the days when definition and 'explication' became the declared aim of 'analytic' philosophy.)

My examples may help to emphasize a lesson taught by the whole history of science: that absolute exactness does not exist, not even in logic and mathematics (as illustrated by the example of the still unfinished history of the calculus); that we should never try to be more exact than is necessary for the solution of the problem in hand; and that the demand for 'something more exact' cannot in itself constitute a genuine problem (except, of course, when improved exactness may improve the testability of some theory).

As far as meaning-analysis for its own sake is concerned, it is *always* pointless. Where two or more senses of a word are habitually confused (as for example, in the case of 'probability') it is advisable to draw attention to this fact by showing that there are statements which hold true for the one sense or the other and which become incompatible if the senses are not *distinguished*. But otherwise, questions such as 'What is corroboration?' or 'What does "corroboration" mean?' are empty.[19]

[19]I think that this remark clears up the so-called 'paradox of analysis'; for in the light of my remark, this 'paradox' can be interpreted as showing that an analysis (or a definition, other than one that introduces a short label for a long story) is *always pointless*—except if the *kind of term to be admitted* in the 'analysans' (or definiens) is indicated *by a definite problem*—one towards whose solution the analysis is supposed to make a contribution.

The paradox of analysis usually illustrated by the analytic definition, 'a brother is a male sibling', is quite a good example, and one that has the approval of Moore himself. I have found some examples that are perhaps more instructive and entertaining, yet are in intention strikingly similar, in a widely distributed pamphlet, edited by the Metropolitan Police Driving School and published, for the Commissioner of Police of the Metropolis, by Her Majesty's Stationery Office, in 1955. Its title is *Roadcraft; A Manual of Driving Instruction for Students of the Motor Car Wing*, and it is much concerned with explication and meaning analysis, as the following fairly representative quotation may show:

'5. Concentration may be defined as the full application of mind and body to a particular endeavour, to the complete exclusion of everything not relevant to that endeavour.' (p. 7.)

Moral philosophers may be interested in the analysis of *'good judgement'* contained in the following paragraph:

'8. With the aid of *VISION, HEARING, GENERAL FITNESS* and *CONCENTRATION* a driver will be able to exercise good *JUDGEMENT*, which is the ability to distinguish between right and wrong, good and bad, and safe and unsafe.' (pp. 7 *f.*)

The following two quotations are examples of straightforward definitions:

'15. Effort in this sense may be defined as the exertion of strength.' (p. 11.)

'18. Skidding may be defined as follows: *Involuntary movement of the car due to the grip of the tyres on the road becoming less than a force* or forces acting on the car.' (p. 18.)

I believe that no serious student of the history of ideas would question my assertion that it is philosophy, and philosophers, who are ultimately responsible for this sort of empty verbalism, rather than the Commissioner of Police.

PART II

THE PROPENSITY INTERPRETATION
OF PROBABILITY

CHAPTER I

OBJECTIVE AND SUBJECTIVE PROBABILITIES

I N this chapter of the *Postscript,* I propose to develop at considerable length the distinction (made in section 48 of *The Logic of Scientific Discovery*) between objective and subjective interpretations of the probability calculus.[1]

The subjective theory of probability springs from the belief that we use probability only if we have *insufficient knowledge.* This theory was criticized in many places in *The Logic of Scientific Discovery*: but, as might have been expected, it survived my criticism. Even among physicists, the theory is still very popular. Yet in my opinion it has led, within physics, to a great deal of confusion and invalid reasoning. For this reason I am going to renew my attack on it, with the aim of showing that the theory should be scrapped, in spite of its *prima facie* plausibility.

[1][With the material presented in this chapter and the following, compare the following publications by K. R. Popper: 'Two Autonomous Axiom Systems for the Calculus of Probabilities', *The British Journal for the Philosophy of Science* 6, 1955, no. 21, pp. 51–7 (see also 6, no. 22, p. 176, and no. 24, p. 351, where some misprints are noted); 'The Propensity Interpretation of the Calculus of Probability, and the Quantum Theory', in *Observation and Interpretation, Proceedings of the Ninth Symposium of the Colston Research Society, University of Bristol,* ed. Stephan Körner in collaboration with M. H. L. Pryce (London, 1957), pp. 65–70 and 88–89 (for errata see *British Journal for the Philosophy of Science* 8, 1958, no. 32, p. 301, n. 1); 'Probability Magic, or Knowledge out of Ignorance', *Dialectica* 11, 1957, no. 3/4, pp. 354–372; 'On Mr. Roy Harrod's New Argument for Induction', *British Journal for the Philosophy of Science* 9, 1958, no. 35, pp. 221–224; 'The Propensity Interpretation of Probability', *British Journal for the Philosophy of Science* 10, 1959, no. 37, pp. 25–42; 'Probabilistic Independence and Corroboration by Empirical Tests', *British Journal for the Philosophy of Science* 10, 1960, no. 40, pp. 315–318; 'Creative and Non-Creative Definitions in the Calculus of Probability', *Synthese* 15, no. 2, June 1963, pp. 167–86 (for errata see *Synthese* 21, no. 1, 1970, p. 107).

See also Popper's 'On Rules of Detachment and So-called Inductive Logic', and his 'Theories, Experience, and Probabilistic Intuitions', in I. Lakatos, ed.: *The Problem of Inductive Logic,* 1968, pp. 130–144, and pp. 285–303. Ed.]

The attack upon the subjective interpretation is the main topic of this chapter and the ones that follow. But these chapters are preceded and concluded by a plea for a new objective interpretation of probability which I call the 'propensity interpretation'. It marks perhaps the most significant change in my views since 1934, and it will be very fully discussed and applied also in Volumes II and III of the *Postscript*.[2]

1. *The Meanings of Probability.*[1]

It is a mistake to believe that the word 'probability' has only one important meaning, or at most two. The word has many different meanings, some of which have little in common. I propose to use the word 'probability' here, and in other places, for all and only those meanings that satisfy the well known *mathematical calculus of probabilities*, and to introduce alternative terms—such as 'acceptability', 'degree of corroboration' or 'degree of confirmation', etc.—in all those cases in which the word 'probability' may have been used in perfect agreement with philosophical or ordinary usage, but in which *it has been questioned* (for example by myself in *L.Sc.D.*) *whether this usage is, in fact, in agreement with the laws of*

---

[2][When Popper first presented his propensity theory, in a paper read for him in Bristol, in his absence, by his pupil Paul K. Feyerabend (see S. Körner, ed., *Observation and Interpretation, op. cit.*), R. B. Braithwaite compared Popper's idea of propensity, interestingly, with C. S. Peirce's 'would-be' or 'habit'; and ever since then Popper's theory has been attributed to Peirce. But there are of course important differences. For example, just as a field of force may be physically present even when there is no (test) body on which it can act, so a propensity may exist for a coin to fall heads even though it falls only once, and on that occasion shows tails. There may indeed be a propensity without any fall at all. The most important difference from Peirce's theory, stressed by Popper in 1957, is his *relational* theory of propensity. It may be mentioned that the only criticism advanced by D. H. Mellor (*A Matter of Chance*, 1971), who adopts Popper's terminology, is that he does not agree with Popper's relational theory (apparently for reasons of language habits) and so returns to Peirce's view. But the relational view is clearly a considerable improvement: the death risk of a sick person may be greatly reduced (or greatly increased) by the invention of a new therapy. And the propensity (half-life) of an atomic nucleus to disintegrate may be radically increased by, say, bombarding it with slow neutrons. Ed.]

[1]The fundamental ideas of this and the next sections go back to *L.Sc.D.*, section 48. Most of the innovations were introduced in my note in *Mind*, 1938, reprinted in Appendix *ii of *L.Sc.D.*

the calculus of probabilities. (For the laws of that calculus, see especially the Appendices *iii to *v of my *L.Sc.D.*)

Degree of corroboration will not be discussed again here (see Part I, Chapter 4, above). Only such probability concepts will be discussed as unquestionably satisfy the formal calculus of probability. (But not all such concepts will be discussed.)

If my terminological proposal is accepted, then a relative frequency is, for example, rightly called a 'probability'; for on the basis of the definition given in *L.Sc.D.* (section 52 and Appendices *ii and *vi), relative frequencies satisfy tautologically all the laws of the calculus of probability.

But relative frequency is by no means the only interpretation which satisfies this calculus. There are many more interpretations. Of these, the following are of major importance: the set-theoretic interpretation; the interpretation of probability as a (weighted) measure of possibilities, and more especially, the propensity interpretation which takes these possibilities to be physical propensities (which may be considered to be physically as real as are physical forces); and the logical interpretation.

Now where I have changed my mind since 1934—that is, since publishing *L.Sc.D.*—is, in the main, over the following point. While previously I believed that the frequency interpretation is fundamental for the understanding of probabilistic *physical theories* (and of theories of games of chance), I now believe that the propensity interpretation is more adequate.

However, before explaining this change of mind, I am going to discuss again two very simple formulae, in order to make quite clear what kind of formalism it is which we have to interpret.

2. *Relative and Absolute Probabilities.*

There are only two formulae which have to be referred to in a discussion of the interpretations of the probability calculus; the first is

(R) $$p(a,b) = r,$$

or in words 'the (relative) probability of *a* given *b* is equal to *r*', where *r* is some fraction between 0 and 1 (these limits included).

The second formula is

(A) $$p(a) = r,$$

in words 'the (absolute) probability of a is equal to r'.

The relative probability of a given b has sometimes also been called 'conditional probability of a under the condition b'; and the absolute probability of a has sometimes been called the 'prior' or the 'initial' or the *a priori* probability of a.

If we interpret (R) in the sense of the *frequency interpretation*, discussed at length in *L.Sc.D.*, then

$$p(a,b) = r$$

becomes 'the relative frequency of a within the reference class b (or the reference sequence b) is equal to r'.

On the other hand,

$$p(a) = r$$

can hardly be interpreted in frequency terms (except in a trivial way—by taking the reference class b as 'understood'). For there is no point in saying 'sparrows occur frequently' unless 'among European birds', or something like this, is taken as understood. But

$$p(a) = r$$

can be interpreted without difficulty in other than frequency terms, for example within what I call the 'logical interpretation' of probability. Here we take the letters 'a', 'b', etc., to be names of statements. The absolute logical probability of a—i.e., $p(a)$—is then what I described in *L.Sc.D.* as 'logical probability': its value r is the greater the less the statement a says. Or in other words, the greater the content of a, the smaller is the value of its absolute logical probability. This is justified by the fact that the assertion a (it will rain tomorrow) is clearly more probable than the assertion ab (it will rain tomorrow and it will be sunny on Saturday week), provided b does not follow from a (which it does if, for example, it is a tautology).

It has sometimes been said that the idea of absolute probability is meaningless; and indeed, it is difficult to give it a useful meaning within the frequency interpretation, as we have seen. Nevertheless, $p(a)$ is hardly to be described as meaningless, even within the frequency theory. For the meaning of absolute probability can be defined with the help of relative probability.

In order to show this, we may remember that all theories of

probability operate with the idea of the conjunction or the product or the intersection of a and of b, here denoted by 'ab' (read 'a and b'), and with the idea of the negation or the complement of a, here denoted by $\bar{a}$ (read '$non\text{-}a$'). If these two ideas are introduced, then $a\bar{a}$ becomes a self-contradictory or empty element (event, or class, or statement, etc.) of the system; and the negation of $a\bar{a}$, i.e., $\overline{a\bar{a}}$, becomes a 'full' or a tautologous element—for example, a tautological statement. We can agree to write 't' (i.e., tautology) instead of $\overline{a\bar{a}}$. Then we can define absolute probability as follows:

(DA) $$p(a) = p(a,t).$$

Since we have now defined absolute probability in terms of relative probability, it must be 'meaningful'. And if we assume a finite universe of discourse—say, with n elements—the meaning of

$$p(a) = r \qquad (0 \leqslant r \leqslant 1)$$

becomes perfectly clear from the point of view of relative frequency: $p(a) = r$ asserts that among the n elements of the universe, nr elements have the property a. The belief that $p(a)$ is meaningless in the frequency theory is due to the fact that we do not, as a rule, assume that the universe in which we actually live is finite. (At least we do not as a rule assume that it is finite in time.) And in an infinite universe, the meaning of '$p(a) = r$' becomes indeed highly obscure; for if it is neither 1 nor 0, it will in general be indeterminate, in the mathematical sense.

In a finite universe, we can also define relative probability in terms of absolute probability, by way of the definition

(DR) $$p(a,b) = p(ab)/p(b).$$

This definition can be extended to an infinite universe, but only under the condition that $p(b) \neq 0$. (In a finite universe we can put $p(a,b) = 1$ whenever $p(b) = 0$.)

In what follows I am going to operate in the main with relative probabilities; that is to say, I am going to investigate various interpretations of

$$p(a,b) = r.$$

If absolute probabilities are considered, we may take them as defined by relative probability, in the manner of (DA).

I do not assume here any knowledge of the actual calculus. An exposition of its axioms and elementary derivations will be found in Appendices *iv and *v of *L.Sc.D.*

3. *The Propensity Interpretation. Objective and Subjective Interpretations.*

In this section I intend to introduce what I have called the propensity interpretation without, however, discussing it in detail, or even giving my main arguments in its favour. (These are given in section 20 below.)

The propensity interpretation may be considered in connection with the classical interpretation, which defines probability as the number of favourable cases divided by the number of possible cases. This suggests that we may interpret probability as a *measure of possibilities*. If and only if the number of possibilities can be counted, and if these possibilities are all equal, only then do we have a case to which the classical definition clearly applies. Thus the interpretation of probability as a measure of possibilities is a natural generalization of the classical definition.

Now the propensity interpretation is very closely related to the interpretation which takes probability as a measure of possibilities. All that it adds to this is a physical interpretation of the possibilities, which it takes to be not mere abstractions but physical tendencies or propensities to bring about the possible state of affairs—tendencies or propensities to realize what is possible. And it assumes that the relative strength of a tendency or propensity of this kind expresses itself in the relative frequency with which it succeeds in realizing the possibility in question.

Thus relative frequencies can be considered as the results, or the outward expressions, or the appearances, of a hidden and not directly observable physical disposition or tendency or propensity; and a hypothesis concerning the strength of this physical disposition or tendency or propensity may be tested by statistical tests, that is to say, by observations of relative frequencies.

This, in brief, is the propensity interpretation. It will be asked why I propose to introduce hidden propensities behind the frequencies? My reply is that I conjecture that these propensities are physically real in the sense in which, say, attractive or repulsive forces

286

may be physically real. Apart from this I shall try to show that we may with great advantage operate in physics with these propensities (and even with fields of propensities which are analogous to fields of forces). I now regard the frequency interpretation as an attempt to do without the hidden physical reality—an attempt which was worth making, and which can be carried far with great success, but which is not in all fields completely successful. I believe that it will have to give way, in the end, to the propensity interpretation.

Formally considered, the difference between the two interpretations is this. The frequency interpretation can explain *singular probability statements*—such as 'There is a probability ½ that at the next toss with this coin heads will turn up'—as merely grammatically (or 'formally') singular, as I have tried to explain in *L.Sc.D.*, section 71. It attributes to the single event a probability *merely* in so far as this single event is an element of a sequence of events with a relative frequency.

As opposed to this, the propensity interpretation attaches a probability to a single event as a representative of a *virtual or conceivable sequence* of events, rather than as an element of an actual sequence. It attaches to the event *a* a probability *p(a,b)* by considering the *conditions which would define this virtual sequence*: these are the conditions *b*, the conditions that produce the hidden propensity, and that give the single case a certain numerical probability. Only if we wish to *test* the ascribed numerical probability we shall have to realize a segment of the virtual sequence long enough to make it possible for us to apply to it a significant statistical test.

Thus the main difference between the frequency interpretation and the propensity interpretation lies in the status of singular probability statements. They play a peripheral role in the frequency theory but a central role in the propensity interpretation which sees, as it were, every single case as the outcome of a propensity, or perhaps of contesting propensities, even though these can be tested only statistically.

I shall have much to say later on the propensity interpretation. In the present context, I merely wanted to introduce the idea. All I need to add at present is a brief remark (in the next section) on the way it fits into my general scheme of proposing hypotheses and testing them by experiments.

In order to have a common name for the propensity and the

frequency interpretations (and perhaps for similar interpretations which may yet be proposed), I shall call them *'objective'* interpretations, thus somewhat extending the scope of a term introduced in *L.Sc.D.*, section 48. The objective interpretations assume that the probability of tossing heads depends solely upon physical or similar conditions, and *not* upon the state of our knowledge. Whereas I call those interpretations *'subjective'* which interpret the probability of tossing heads as being dependent upon the state of our (subjective) knowledge, or perhaps upon the state of our beliefs.

4. *Experimental Tests and their Repetition: Independence.*

Every scientific theory implies that under certain conditions, certain things will happen. Every test consists in an attempt to realize these conditions, and to find out whether we can obtain a counter-example even if these conditions are realized; for example, by varying other conditions which are not mentioned in the theory. (This shows, by the way, that a *ceteris paribus* clause, like 'all things being equal', must not be added to a theory since it would destroy its testability.)

This fundamentally clear and simple procedure of experimental testing can in principle be applied to probabilistic hypotheses in the same way as it can be applied to *non-probabilistic* or, as we may say, for brevity's sake, *'causal'* hypotheses.

I emphasized this point even when I championed the frequency interpretation of probabilistic (physical) hypotheses. I explained then that probabilistic hypotheses make assertions about the statistical distribution of certain properties of events within *classes or sequences* of events, and that these hypothetical assertions can be tested in a way analogous to the testing of causal hypotheses.

I am now inclined to say that the relation between probabilistic and causal hypotheses, and the manner of testing them, is even closer than I previously thought: from the point of view of the propensity interpretation, causal hypotheses can easily be interpreted as hypotheses asserting a propensity equal to 1 ('certainty').

This interpretation is in keeping with the character of causal hypotheses, which give rise to *singular* predictions; for the propensity interpretation attributes probabilities also to 'singular events', that is, to 'occurrences' (in the sense of *L.Sc.D.*, section 23).

Thus we test hypotheses, whether probabilistic or causal, by creating certain conditions and seeing whether they produce the predicted (singular) effect if other conditions are varied. The probabilistic hypothesis predicts that the singular event has a certain propensity to be realized. This prediction can be tested by *repeating* the experiment under the conditions prescribed, and noting the frequency distribution in repeated experiments. But the causal hypothesis is also tested by repeated experiments: and here corroboration demands that the relative frequencies of the results within the sequence of repeated tests are equal to unity (or in practice very nearly so).

Testing by experiments thus has two aspects: variation of conditions is one; and keeping constant the conditions which are mentioned as relevant in the hypothesis is another—the one aspect which interests us here. It is decisive for the idea of *repeating an experiment.*

The experiment to be repeated is defined, or described, by these conditions. It is therefore absolutely essential for a repetition of an experiment that each repetition occur under the same stated conditions. But this means that in an experimental set-up, *the earlier experiments must not affect the later ones.* For otherwise the later ones operate under new conditions. In other words, it is an essential property of a sequence of repeated experiments that *there must be no after-effect of earlier experiments upon the later ones.* The experiments must be *independent.* This is part of the idea of repetition, and has nothing whatever to do with the question whether the hypothesis under test is a probabilistic hypothesis or a causal one.

To put this point in practical terms: let our experiment end with the reading of a meter consisting of a needle ranging over a scale. If this apparatus is not properly oiled, it may get stuck, and a repetition of the experiment may, for this reason, end with the same reading as the previous experiment. Clearly, we have here a case of 'after-effect' or *dependence* (non-independence) of the experiments.

Accordingly, if we are interested in experiments which involve after-effects (such as Brownian motion, or rote learning), then we shall consider, from the point of view of testing our theory, the experiments which show after-effects together with the original experiment as part of one *single* experiment, rather than as repetitions of the experiment; and we shall consider as a genuine repetition only an experiment which starts right from the beginning, and which includes in its turn the study of the after-effects.

In brief, it is essential for a sequence of repeated experiments that each be conducted under the same stated conditions, and that, for this very reason, the experiments be *independent,* which is merely another way of saying that the sequence must be free from after-effects.

Tests of the simplest probabilistic hypotheses involve such sequences of repeated and therefore independent experiments—as do also tests of causal hypotheses. And the hypothetically estimated probability or propensity will be tested by the frequency distributions in these independent test sequences. (The frequency distribution of an independent sequence ought to be 'normal' or 'Gaussian'; and as a consequence it ought to indicate clearly whether or not the conjectured propensity should be regarded as refuted or corroborated by the statistical test.)

Now independence, in the sense mentioned, has for a long time been one of the fundamental concepts of probability theory. It is explained in this theory as follows: the later experiments are independent of the former ones if and only if the probability of obtaining a certain result is uninfluenced by the former experiments; or, in other words, if the probabilities remain the same for every experiment throughout the sequence.

This clearly indicates the idea that the probabilities depend upon the experimental conditions, the experimental set-up. What is assumed, in this theory of independence, is that if the experimental set-up of the later experiments is the same as that of the former experiments, then the probabilities of obtaining results will also be the same; and *vice versa*: if the probabilities are unchanged after each experiment, then the experiments have no after-effect: they do not change the conditions.

To be more exact I should have said 'they do not change the *relevant* conditions'. For clearly, the fact that another experiment has been conducted before, and had a certain result, is part of 'the conditions', just like the fact that a rainstorm was raging in Borneo or a brainstorm in Alexandria, while the coin was tossed in London. The point is that these conditions have no effect upon the probability of tossing heads, that they are *irrelevant*: they do not influence the set of *specified and relevant* conditions, and thus do not influence the probability.

An experiment is thus called '*independent*' of another, or of certain conditions, or not affected by these conditions, if and only if they do not change the probability of the result. And conditions which in this way have no effect upon the probability of the result are called '*irrelevant*' conditions.

The definitions of these terms with the help of the formal calculus is very easy. Let a be the result in whose probability we are interested, and b the relevant experimental conditions. Then

$$p(a,b) = r$$

will be the assertion that under the conditions b, the result a has the probability r. (In the propensity interpretation, it is the assertion that the conditions b produce a propensity r to realize the result a.)

Now let c be some additional condition—for example, the condition that the experiment in question is carried out after two earlier experiments (which both, say, yielded the result a). Then we say that these conditions are irrelevant, or that the result a is independent of them, if the probability of a remains the same under the conditions bc (that is the conjunction of both conditions) as it was under condition b alone. That is to say, if and only if

$(*)$ $$p(a,bc) = p(a,b).$$

To be more exact, we have to mention b in our verbal formulations also: we say in this case that a is independent of c, given b; and that c is irrelevant to a, given b.

Independence of experiments or events, and irrelevance of conditions, are, as will now be seen, a correlative set of terms which express *one* idea. This fact is emphasized by the following result of the probability calculus: If $p(a,b) > 0$ and $p(c,b) > 0$ then $p(a,bc) = p(a,b)$ if and only if $p(c,ba) = p(c,b)$ or in words, a is independent of c, given b (and c, irrelevant to a, given b) if and only if c is in its turn independent of a, given b (and a irrelevant to c, given b). For this reason, there is no need to distinguish between independence and irrelevance of conditions.

5. *The Logical Interpretation.*

It is best to consider a formula such as

$$p(a,b) = r$$

as a statement or an assertion about the objects a and b. The symbols 'a' and 'b' are thus (variable) *names* of the objects we are talking about. These objects may be, for example, *events*, or classes or sets of events; or they may be, for example, *statements* (or perhaps sets of statements, such as deductive theories).

There is a fairly direct kind of transition from one interpretation which takes a and b to be events to another which takes a and b to be statements: we can simply choose those statements which describe the events. We thus proceed from one interpretation to another in a way which does not affect what we really wish to assert if we assert '$p(a,b) = r$'. Only our method of asserting it, our method of expression, will have changed. And if the statement '$p(a,b) = r$' was a factual statement to start with, for example, a hypothetical probability estimate, rather than a logical or tautological one, then it will of course retain its character after this re-interpretation.

But if we take our arguments, 'a', 'b', 'c', . . . to be names of statements, then $p(a,b) = r$ may also be interpreted very differently. In what I have called (in *L.Sc.D.*, section 48) the *logical interpretation of probability*, 'a' and 'b' are interpreted as names of *statements* (or propositions) and

$$p(a,b) = r$$

as an assertion about the *contents* of a and b and their *degree* of *logical proximity*; or more precisely, about the degree to which the statement a contains information which is contained by b. If b says all that is said by a, so that a follows from b, then $p(a,b) = 1$. If b is consistent and contradicted by a, then $p(a,b) = 0$, while of course $p(a,ab) = 1$. If a neither follows from b nor contradicts b, then the value of $p(a,b)$ will lie somewhere between 0 and 1 (these end points included); and it will be close to 1 if a says only a little more than b says, and close to 0 if a says things very different from those that b says.

This 'logical interpretation' takes the probability calculus as a generalization of ordinary logic, as it were. It can be justified either on intuitive grounds or, more fully, on formal grounds.

The intuitive justification runs as follows. Let a be the statement 'Socrates is mortal' and b the statement 'All men (or 100 per cent of all men) are mortal and Socrates is a man'; then we shall say that $p(a,b) = 1$, because a follows from b; and indeed, given b, we may consider a as certain.

But let a be the same statement as before, and b the statement '92 per cent of all men are mortal, and Socrates is a man', then a will not be *certain* on the information b, but highly *probable*; and we may indeed say that the probability which on the information b is attached to a will not be far from 0.92; that is to say, $p(a,b)$ will be about 0.92. Thus one statement b may make another statement probable. (This is sometimes called 'the simple rule of induction'.[1])

This argument is intuitively fairly convincing, but it does not in itself justify the claim that the calculus of probability may be interpreted, in the way indicated, as a generalization of propositional logic (or of deductive logic); for our intuitive argument does not make sure that there will not somewhere be a clash between the laws of the formal calculus and these intuitive ideas. A full justification, however, is provided in Appendix *v, *L.Sc.D.*, where Boolean algebra (and therefore the theory of statement composition, that is to say, propositional logic) is derived from the axioms of the calculus of probability. This shows that a logical interpretation of the calculus of probability is admissible.

If we interpret the calculus in the logical sense, then all its theorems obtain a status similar to that of the theorems of the calculus of propositions. We may therefore call them 'analytical' or 'tautological'. Although we may not know of many probabilistic formulae whether they are true or false, they will in fact all be tautological or contradictory. Keep, say, both a and b as in our last example (the one with 92 per cent), and let us ask for the value of $p(b,a)$ rather than $p(a,b)$. We may be unable to answer this question; but the correct answer should nevertheless be tautological, and all other

[1][*(Added 1980) The simple rule of induction has been shown, by David Miller, in 1965, to be paradoxical. There have been a number of replies to Miller, but to my knowledge, none of them are tenable, and they were all in their turn replied to by Miller, who clearly had the best of the debate. Nevertheless, quite a number of authors on this subject have continued as if nothing had happened. See David Miller: 'A Paradox of Information', in *The British Journal for the Philosophy of Science* **17**, 1966, pp. 59–61; and my 'A Comment on Miller's New Paradox of Information', same issue, pp. 61–69.]

answers contradictory. We see from this that the logical interpretation is very different from either the frequency or the propensity interpretation which both interpret a formula '$p(a,b) = r$' in general as *a hypothesis which can be experimentally tested*.

Moreover, let us take some fixed statements a and b for which we obtain

$$p(a,b) = r$$

in the logical interpretation. The value of $p(a,b)$, for the same statements a and b, will then in general differ from r in the frequency or propensity interpretation. This is clear, first, because in choosing a certain value of $p(a,b)$—say, the value s—we are making a *hypothetical estimate* and we are clearly free to conjecture any estimate we like. (Of course, we should afterwards *test* it.) But apart from our freedom of choice, there is not the slightest reason to believe that the r of the logical theory will represent a *good* estimate from the point of view of the frequency or of the propensity interpretation.

To show this, let a be the statement (function) 'x is an ellipse', and b the statement (function) 'x is a planetary orbit'. Then we can (if we wish to express it probabilistically) express Kepler's first law—a bold conjecture—by the probability assertion

$$p(a,b) = 1$$

since this will mean: given that x is a planetary orbit, the probability that x is an ellipse equals 1, which is the same as saying '100 per cent of all planetary orbits are ellipses'. On the other hand it is quite clear that from the point of view of the logical interpretation, $p(a,b)$ will be either equal to zero (considering the infinity of possibilities, i.e., the infinity of possible shapes of planetary orbits) or very nearly so. And even if we incorporate into b the evidence provided by Tycho on which Kepler worked, the situation will be essentially the same. Not only because this evidence referred in the main to one planet only; but mainly because it was fragmentary even with respect to this planet.

This example shows that a very 'good' hypothetical estimate of probability may differ widely from the value of the corresponding logical probability.

6. *Comparing the Objective and the Subjective Interpretations.*

Our main interest, I may remind the reader, is the status of probabilistic hypotheses in the sciences, especially in physics. The logical interpretation does not enter here, simply because it takes probability statements as analytic or tautological, and therefore as *untestable*. It comes in indirectly, however, by way of the subjective interpretations.

As has been explained, I call both the frequency and the propensity interpretations '*objective*': both of them take probabilities as properties of certain physical systems—experimental set-ups, for example. I call those interpretations '*subjective*' which take these same probabilities as measures of our own imperfect knowledge as to what happens in these physical systems.

The subjective interpretation is mainly the result of the plausible and widely-held view that whenever probability enters our considerations, this is due to our imperfect knowledge: were our knowledge perfect, we should not need probability, for we should always have certainty.

This view is profoundly mistaken. Its aim is to elucidate and to analyze the meaning of probability statements in physics, and in the theory of games of chance. Statements such as 'The probability of tossing heads equals $1/2$', are interpreted to mean: 'The statement that heads will be tossed and the alternative statement that heads will not be tossed are both equally well supported by the insufficient evidence which constitutes our knowledge.' In other words, the subjective interpretation tries to replace the interpretations which I have called 'objective' by another one, believed to have certain advantages. Yet the subjective interpretation clashes with some of the simplest, most important and best confirmed consequences of the objective theory.

I do not, of course, have any objection to the translation of an objective theory into a subjective language. If anybody prefers to say, 'I have reason to believe that the sun is shining', 'I have knowledge that it is daytime' (etc.), instead of, 'The sun is shining' and 'It is daytime', I shall not waste my time trying to convert him. Similarly, if anybody wishes to replace an objective statement of the form '$p(a,b) = 1/3$' by something like: 'If I were to bet against a then I should be ready, in the light of my knowledge of *the conditions b*, to offer odds of 2:1 against it', I shall again not argue against him. A subjectivist re-interpretation of this kind which mirrors the objective theory would not be very interesting and would not enlighten us much, but it might be harmless.

It is not, however, the objective theory—in which b represents the (experimental) *conditions*—which the subjective interpretation attempts to mirror, but the *logical interpretation*, of which we have already briefly shown that it will at times lead to probability values that differ from those to which the objective theory leads.

Thus the status of the subjective theory is hopeless: (a) its aim is to analyze and to explain objective probabilities; it is (b) bound to take the logical theory as its basis (as I shall show in this section); but we have seen (c) logical and objective probabilities may give rise to conflicting values, owing to the fact that logical probability statements are *untestable tautologies* (or something of this kind) while objective probabilities are freely chosen *testable hypotheses*.

First, I shall only show (b)—that the subjective theory is bound to base itself on the logical interpretation.

This is so because the subjective theory tries to interpret $p(a,b)$ as that degree of belief in a which may be rationally justified by our actual total knowledge b. But this is really the same problem (only expressed in subjective language) which the logical interpretation tries to answer; for the logical interpretation tries to assess the degree to which the statement a is logically backed, or supported, or justified, by the statement b. Thus if we interpret the logical interpretation subjectively—that is, in terms of our knowledge or nescience—then $p(a,b)$ becomes, precisely, the degree to which our actual total knowledge b rationally justifies a dubious or hypothetical a.

7. The Objectivist and Subjectivist Interpretations of 'b' in 'p(a,b)'.

My last paragraph points towards a small but significant difference between the two approaches. The subjectivist takes a as his hypothesis and $p(a,b)$ as our degree of belief in it, while the objectivist takes '$p(a,b) = r$' as his hypothesis. (He may or may not believe in it.) We shall return to this point later. At the moment I wish to stress the difference between the objectivist and subjectivist views as to the part played by 'b' in '$p(a,b)$'.

The objectivist view of b is that b states the repeatable conditions of a repeatable situation. As we have seen, the results of previous experiments are not part of the information b; and if they were part of b, they could be omitted without loss; for they must be irrelevant, since we assume the repeatability of the conditions b; and repeatability entails independence of predecessors.

The subjectivist view of b is very different. Here b must contain all our relevant knowledge; and observation of past results of the experiment in question will be highly relevant. It is only from them, according to the subjectivist, that we can judge the value of $p(a,b)$. If our past knowledge—that is b—tells us that a has occurred often in the past, then $p(a,b)$ will have a much greater value than if b tells us that a has been a very rare occurrence.

This plausible view which, indeed, is the fundamental intuition underlying the subjective approach, will be elaborated in the next section. Here I wish only to show that

(i) the b of the objective theory is not in general the same as the b of the subjective theory,

(ii) even where it is the same, or roughly the same, the b of the subjective theory will contain information which, according to the subjective theory, must be highly relevant, though it must be irrelevant according to the objective theory.

As to (i), both the objective and the logical theories can choose their b with complete freedom. The logical theory is interested in $p(a,b)$ for any a and b. Similarly, the objective theory may raise the question of the probability of a, given any set of conditions b. Of course, the actual choice of a and b for the objective theory will largely depend upon the scientific problem situation. But there is no reason whatever to restrict b to, say, already observed or otherwise known situations.

As opposed to this, the subjective theory looks upon $p(a,b)$ as the degree of our belief in a if b represents our actual total knowledge. In all applications b will be interpreted by the subjective theory as the sum total of what 'we' know. In all applications, the subjective theory is therefore not free in the choice of b: it is not a constant, because our knowledge changes all the time; but it is not a variable in the sense that we can choose it freely; on the contrary, it is something completely beyond our control: our knowledge is what is 'given' to us.

As opposed to this, the objective theory looks upon b, even in applications, as open to free choice. It is I who can freely ask what the probability of heads will be if I bend the coin in a certain way, so that heads is on the concave side. How to find out the new probability is another question. The conditions are not given, they are chosen or selected or constructed by us when we choose our problem.

Thus there is no reason whatever why in any actual problem of determining the probability of a, the two theories should operate with the same b.

But, the subjectivist will answer, you are naive, and a victim of superficial formulations. If you say that you 'choose' b, you only mean that you can choose to bend a coin. Who denies this? If you wish to find the probability of heads in the case of a bent coin, you will simply have to use past experience if you wish to get any reasonable result: you will use all your knowledge, and the fact that you have bent this very coin will be an important and relevant part of it. To look at it as the only constituent of b is naive. It may be the most conspicuous part of b—so conspicuous as to make you and others like you forget the rest and take it for granted. But the logician must always ask for suppressed premises—for those premises which are implicitly assumed in an argument. And these suppressed premises are, in this case, all our past knowledge. Not all of it will be relevant, no doubt. But previous experiences with coin tossing, and also with the influence of asymmetries on the frequencies of coins and dice, and in fact our whole knowledge of physics, may all be relevant. And how will you find out the value of $p(a,b)$—of tossing heads with this coin—if not by observing the relative frequency in a set of tosses? After a hundred or a few hundred tosses, you may be able to estimate $p(a,b)$.

I have tried to present the case of the defender of the subjective theory convincingly; nevertheless, what he says here is all, I believe, mistaken and muddled. But I will challenge here only those of his points that pertain to the problem of the status of b.

But first I must note that the subjectivist is interested, as I am, in the problem of actually determining the probability of heads turning up in tossing this particular bent coin. In this we agree; this is the common ground, the firm basis of our discussion.

Now our subjectivist asserts that the true or best or proper value of $p(a,b)$ will be strongly influenced by the results of earlier experiments.

Here we may stop and consider. The objective theory says that the results of earlier experiments are irrelevant: otherwise the experiments would not be a sequence of repetitions. The subjectivist says that they are relevant. Who is right? Feelings tend, I suppose, to support the subjectivist here. I postpone the answer to the following sections and ask now instead: is this merely a difference of opinion or is there more behind it? Irrelevance and independence are not, after all, philosophical or epistemological playthings but concepts which both parties define by the same well-known mathematical formulae (the special multiplication law).

The objectivist can point out that his theory of independence or irrelevance is one of the most important parts of the applications of the calculus. For example, it is equivalent to the law of the excluded gambling system (discussed in *L.Sc.D.*, sections 49 *ff.*), which has been well tested statistically. Independence can be and has been tested, and tests are sensitive enough to detect, for example, that the usual methods of shuffling cards are not very good.[1]

Now the common ground between the objective and the subjective theory was that both wished to explain probability theory as used in physics or in games of chance; and the claim of the subjective theory is that it can do this as well, or better, than any objective theory. But here there is a conflict between the subjective theory and all applications: these require independence from previous experiments and their irrelevance, where the subjective theory (no doubt very plausibly) demands dependence and relevance.

Thus the objectivist can say that even if he adopts the allegedly suppressed premises (the results of previous experiments) explicitly as constituent parts of *b,* his mathematical theory tells him that he may nevertheless ignore these constituents as irrelevant. He therefore does not overlook his past experience naively, but does so in following the demands of a highly successful theory—in fact, the same theory that the subjectivist proposes to explain.

The difference about *b* remains, because the subjectivist believes that parts of *b* are highly relevant which the objectivist claims must be dismissed as irrelevant; whether the objectivist or the subjectivist is right in fact, their theories are completely different, and the subjectivist's claim to understand and interpret the theory as used in physics and in the theory of games must be dismissed, simply because of the status of *b*.

[1] See W. Feller, *Introduction to Probability Theory*, 1950, p. 336, esp. note 9; see third edition, 1968, pp. 406 *f.*, esp. note 16.

To sum up, the objectivist holds that under constant objective physical conditions b (constant ratio of black balls to white balls in the urn; constant thorough mixing), the probability of each draw from the urn remains constant. The subjectivist disagrees: the probability changes after each draw of whose result we are informed even if we are also informed of the constancy of the physical conditions.

Historical Note: Keynes has the great merit of having noticed (*A Treatise on Probability*, 1921, pp. 342 *f.*) that the applicability of some of the essential parts of the classical doctrine (like the binomial and Bernoullian laws) can neither be explained nor even upheld from the point of view of his subjective theory. For this applicability assumes independence or irrelevance—a condition that could hardly ever obtain from the point of view of his subjective theory. Later subjectivists have as a rule overlooked the point, operating with the classical theory of independence as if it presented no great problem within their theory. An exception is Carnap who endorses Keynes's warning (*Logical Foundations of Probability*, 1950, p. 499) and whose subjectivism is perhaps the purest of all. Although he distinguishes the subjective theory of degrees of rational belief ('probability$_1$') from the objective theory of relative frequency ('probability$_2$'), he explains the estimates of objective relative frequency as determined by subjective probability (the optimal estimate of the value of objective probability he indeed takes as identical with the subjective probability value); and this is of interest to us here—he holds that our b (he calls it 'e', or 'evidence') of the objective probability is always the same total knowledge which enters into the subjective formulae. As to the binomial law, Carnap also says, like Keynes, that 'some traditional uses of it are not admissible' *(loc.cit.)*. But these are precisely the uses to which I have referred above as characteristic of the theory of games of chance, and of the law of the excluded gambling systems. Accordingly, the binomial law is inapplicable also to those cases in which its applicability is asserted by Carnap (see *op.cit.*, p. 500, 'the experiment described with replacement of each ball') because our b—that is, Carnap's e—changes constantly with every new draw, since every result changes our relevant experience. And this must be so even if our evidence involves definite information concerning the composition of the urn.

CRITICISM OF PROBABILISTIC INDUCTION

8. *The Simple Inductive Rule.*

The subjective interpretation does not succeed in analyzing, or explaining, the traditional theory of games of chance. It fails in the analysis of chance-like or independent sequences. Nor can it explain the applicability of this theory. On the contrary, the assertions of the subjective theory are bound to clash with the classical results of Bernoulli and his successors. Thus the subjectivists are mistaken over a very important point: in analyzing the classical theory, they show, by their programme, that they fail to understand it.

Serious as this criticism is, it does not settle the question: who is right? Who has the correct theory, the objectivist or the subjectivist? 'Correct' would mean here: the theory applicable to physical reality—for example, to games of chance. It is conceivable that the classical theory may be wrong, in spite of its successes, and that the subjective theory offers a better approximation.

I do not believe that this is so. On the contrary, I believe that the subjective interpretation is completely mistaken—that it is an obvious misinterpretation.

But the subjective interpretation is extremely influential among physicists; not so much as a clear-cut and coherent theory, but rather as a convenient second line of defence to fall back upon in case any difficulty arises elsewhere. In this way a kind of double-talk has arisen in physics, both in statistical mechanics and in quantum theory; and at times we find objective physical facts 'explained' by our lack of knowledge.[1]

Since this is the situation, it seems important seriously to consider the reasoning which leads to the subjective interpretation. The main arguments are undoubtedly these.

The idea of probability enters into our considerations when we

[1][See Volume III of the *Postscript: Quantum Theory and the Schism in Physics.* Ed.]

do not know enough to be certain: if we knew whether heads or tails would turn up, there would be no need to speak about the 'probability' of heads or of tails.

Ignorance, however, is not enough. If we compare coin tossing and playing with a die, we find that *our knowledge* that there are now six possibilities instead of two does influence the probabilities. And if we consider, further, betting on possibly biased coins or loaded dice, then the probabilities we assume will depend very largely upon past results. It is difficult to calculate probabilities for a loaded die, but easy to estimate them if we have time to throw the die say a hundred thousand times: everybody will admit that the relative frequencies obtained in such a long sequence will be excellent estimates of the probabilities.

Thus our past experience, our knowledge of past events, clearly determines our probability estimates in this case; and there is every reason to believe that it always plays a powerful role in determining the best estimates of the probabilities in question. How else should we determine probabilities if not in the light of our past experience, especially in non-symmetrical cases like loaded dies? And how should we know that a die is not loaded except in the light of past experience?

These considerations suggest the following *simple inductive rule*: if in a very large number of repetitions of an experiment we find that the result *a* occurs with the frequency m/n, then the best estimate of the probability of *a*, with respect to this experimental evidence, is equal, or approximately equal, to m/n.

I have purposely given this rule a form which is not very definite, and therefore not very strong. In the form given, it applies only to *'large'* numbers of repetitions; and it does not claim that m/n is the best estimate: it only claims that the best estimate will be approximately equal to m/n.

Since the rule, in this form, is weak, all its stronger variants will be destroyed with it, if we can refute it, as indeed we can.[2]

For the simple inductive rule is false. The reasons have already been indicated in the preceding section. A simple example will show that it may lead to utterly absurd results where the condition of *independence* (in the sense of the objective theory) is not satisfied.

---

[2] The variants refuted include countless numbers of theories of induction (among them the whole of Carnap's so-called 'Continuum of Inductive Methods', of 1952).

On the other hand, the subjective-inductive theory is unable to combine its 'simple rule' with the demand for independence. The very rule itself makes the results of past experiments tell upon future probabilities. These results are far from irrelevant: the rule makes them the most relevant information available, especially in the case of long sequences. So the subjectivist cannot demand independence.

A striking example by which we may refute the simple inductive rule is the game 'Red or Blue'.

A gambler plays heads or tails on credit terms with his banker: a book is kept in which he is credited 1 shilling every time he wins, and debited 1 shilling every time he loses. If, after any toss, he owes money to the bank, we say that we have observed the event 'Red'. If he does not owe any money, the observed event is 'Blue'. The game I call 'Red or Blue' consists in betting on the occurrence of one of these two observable events, 'Red' or 'Blue'.

Now from the point of view of the *objective theory*, it is quite clear that:

(i) the probability of Red = the probability of Blue = $1/2$, or very nearly so. (There is a slight but negligible asymmetry in the conditions, in so far as 'Red' means 'debit', while 'Blue' means 'credit or zero balance'.) Thus we may say, from the objective point of view, that

$$p(a,b) = p(\bar{a},b) = 1/2,$$

where a is Red, $\bar{a}$ is Blue, and b are the conditions of the game (of the experimental set-up).

(ii) The sequence of a and $\bar{a}$ is not independent. The calculations[3] show that this fact leads to very unexpected results. If we arrange for a colossal experiment, tossing the coin every second for a whole year, then the following holds: there will be a probability of about 0.9 that the *difference* between the two observed frequencies—that of a and that of $\bar{a}$—will exceed $1/6$ (that is 2 months). There will be a probability of more than 0.5 that the difference between the two observed frequencies will exceed $2/3$ (that is 8 months). Thus it will be more probable than not that the observed frequencies will differ like $1/6$ and $5/6$, while the actual probabilities are $1/2$ and $1/2$.

[3]All the results are taken from W. Feller, *An Introduction to Probability Theory and its Applications*, third edition, Chapter III, section 4, esp. pp. 82 f. The original papers are quoted there in footnotes on p. 252.

In other words, in this game it is objectively extremely improbable that estimates according to the simple inductive rule will succeed, and very probable that the deviations between the inductive estimate and the value $1/2$ will be very great indeed; and this will be so even if we take tremendous numbers of observations—more than could possibly be observed in a lifetime.

Thus we have here a case in which the simple inductive rule will almost certainly fail in spite of its plausibility. The reason is that the successive results of the game 'Red or Blue' are not independent, and that the simple inductive rule is likely to lead to results which agree with good objective estimates only in cases of *objectively independent* observations, when we can apply Bernoulli's methods. However, these explanations themselves belong essentially to the objective theory. It is impossible to translate them into the language of the subjective or inductive theory because, upon translation, objective independence would become subjective irrelevance. But not only is subjective irrelevance incompatible with the simple inductive rule, but the relevance-status of observations in 'Red or Blue' is, from the subjective point of view, exactly the same as in 'Heads or Tails'. From the objective point of view, the repeated observations of 'Red' and 'Blue' are *in no sense repeated experiments*: the after-effect shows that the conditions are not reproduced. (In order to repeat the experiment, we have to start a Red or Blue game, of definite length, all over again from the beginning.) But the subjective theory has no means of thus distinguishing between repeated observations and repeated experiments, since this fundamental distinction is based upon a theory of independence which it cannot formulate.

The failure of the simple inductive rule in the game 'Red or Blue' may be illustrated by contrasting the behaviour of an *inductive gambler* with that of a *rational gambler*. Let us assume the observations so far show 100,000 occurrences of Red and 10 of Blue. Then the inductive gambler will be prepared to offer odds of something like 10,000 to 1 on Red, whatever the last occurrence may have been. The rational gambler (who takes account not merely of the observed results of the game but also of its mathematical structure) will, however, know that, in view of the after-effect inherent in the game, a 1:1 bet on Blue would be highly advantageous to him— provided that the last occurrence was one of Blue.

The game 'Red or Blue' is of course only a particularly striking example of the breakdown of the simple inductive rule if used with the aim of assessing an objective probability with a reasonably good approximation. If used with this aim, it works of course very well in the case of ordinary coin tossing, which is independent. But it would work very badly even in a coin-tossing game, if used as a *gambling system*. (The induced probabilities would sometimes deviate from $1/2$ so that in these cases objectively unfavourable odds would become acceptable to the gambler.)

To sum up, the specious plausibility or 'self-evidence' of the simple inductive rule is merely a misinterpretation of the objectively independent examples which are usually considered. And the difference which we found to hold between the examples involves the idea of objective independence and thus transcends the subjective-inductive theory.[4]

9. *How to Interpret the Simple Inductive Rule Where It Works.*

There is no doubt that the 'simple inductive rule' works at times: if we find, in a few thousand throws with a die, that the side six comes up, say, with a frequency of just over $1/4$, instead of $1/6$, then (it could be argued), we shall apply the simple inductive rule: we shall say that the probability is nearer to $1/4$ than to $1/6$.

I am prepared to admit this, up to a point; but it has to be interpreted.

(i) First of all, we ought not to draw this kind of inference unless we have good reason to believe that our sequence is *independent*. If

[4]Take a sequence such as, for example,

0011000011110000000011111111100000000000000000001111111111111111...

in which 2^n zeros are followed by 2^n ones to be followed by 2^{n+1} zeros . . . , so that the frequencies of the zeros and ones have the limit $1/2$. The probability of accidentally hitting, in this sequence, upon an approximately representative segment approaches zero for every chosen length of the segment (even if the degree of approximation required is very poor). This seems to me a clear refutation of Carnap's assertion (*Logical Foundations of Probability*, p. 500) that the binomial theorem holds precisely, *without independence condition*, 'likewise for an *infinite population* . . . with respect to a given serial order of the elements' of this population. [*I should repeat here that in 1965 David Miller showed that the simple inductive rule is logically paradoxical. See his 'A Paradox of Information', *British Journal for the Philosophy of Science* 17, 1966, pp. 59–61.]

the corresponding situation arises in a game of cards, we should probably first suspect the shuffling. But if the hypothesis of independence logically precedes the application of the inductive rule, then it cannot be based in its turn on the inductive rule: it must be a genuine conjecture, made in view of an assessment of the total objective situation; perhaps testable, but certainly not a result of the simple inductive rule.

(ii) Under the assumption of independence, we may use Bernoulli's argument that most large samples (or long sequences) will be representative. Thus we shall take our sequence of throws with the die in question as representative, and conclude that the probabilities of its six sides are not equal, and thus not symmetrical: we shall conjecture that the die is loaded.

(iii) However, if we are really interested in the case, we shall in several ways check our new hypothesis that the die is loaded. One way to do this is to make new statistical tests. But we shall also try to find the position of its centre of gravity by direct physical investigation (including perhaps X-ray tests, etc.).

(iv) Should we find, in this investigation, that the die is, physically, highly homogeneous and symmetrical, while statistics continue to give the same results, then we shall be faced with a riddle. In this case, we have the choice of whether to consider our statistics or our direct physical measurements as 'influenced by an unknown systematic error' (every experimentalist knows that such 'occult effects' do happen; *cf. L.Sc.D.*, section 8); or whether we should attribute to chance the statistical deviations from symmetry.

(v) If, stimulated by our statistics, we discover that the die is physically loaded with a piece of lead, then we shall replace our hypothetical *b* ('this die is homogeneous') by a new *b* ('this die is loaded to such and such an extent'). That is to say, our statistics may influence us to *revise* our 'knowledge' of the conditions. It may make us realize that the conditions *b* in which we believed as if they were certain, were in fact mistaken. *But our statistics do not themselves become part of the conditions,* that is to say, of our new *b*.

The point is of fundamental importance: our experience here is without doubt relevant to the question 'What is the appropriate *b* in the case of tosses with the die before us?' But it is not relevant, in the probabilistic sense, to the question: 'What is the probability of *a*, in the presence of *b*?' The subjective theory fails to distinguish these two questions. Its failure is a necessary consequence of its subjectivism, of regarding probability as describing our state of knowledge; a view expressed by Peirce who says: 'I cannot make a valid probable inference without taking into account whatever knowledge I have (or, at least, whatever occurs to my mind) that bears upon the question.'[1] Certainly, we may obtain in this way something like a tautological statement of the logical probability of *a* in the light of our total knowledge. But as our argument shows, this is completely different from (a) the actual objective probability and (b) the best available estimate of this actual objective probability.

What our objective theory—the theory of Bernoulli, in fact—tells us is this: in case of a very large number of independent experiments (or independent observations), it is extremely probable that (a), the actual objective probability, will lie very close to the observed frequency. Thus the observed frequency will *probably* be a very good (b), that is to say, a very good estimate of the objective probability.

This result, which explains when[2] the simple inductive rule may be expected to work, and why it may be expected to work, is part of the purely objective theory. Thus the objective theory *explains* the working of the rule; and the fact that the rule works (sometimes) is no argument whatever in favour of the subjective theory.

But this, of course, is understating the seriousness of the situation. We have shown that the probabilistic theory of induction is incompatible with independence, and that independence is a condition for the legitimate employment of the simple inductive rule. Thus the legitimate employment of the simple inductive rule turns out to be incompatible with the probabilistic theory of induction.

---

[1] C. S. Peirce, 'A Theory of Probable Inference', *Studies in Logic* by members of the Johns Hopkins University, 1883, p. 161; also in *Collected Papers*, vol. ii, p. 461. (*Cf.* also Carnap, *op. cit.*, p. 212.)

[2] I do not say that independence is a *necessary* condition for a successful application of the simple inductive rule, for obviously there are trivial cases like the (non-independent) sequence 01010101 . . . in which the simple inductive rule would yield a correct frequency estimate (and even correct predictions of every single case). But for a large class of cases independence is a necessary condition; and even the above sequence may be re-interpreted as a sequence 22222 . . . where '2' stands for '01'; and a transformation of this kind would bring it, together with all causal sequences, into the class of independent sequences, as the formula 'If $p(a) = 1$ then $p(ab) = p(a)p(b)$' indicates.

10. *Summing up of the Status of 'b' in 'p(a,b)'.*

Fundamental to all objective interpretations of probability is the view that we can *explain* certain statistical (physical or biological, etc.) effects by way of a probabilistic hypothesis of the form

$$p(a,b) = r$$

and that we can test this hypothesis by statistical tests. Here the task of b is to state the conditions of the effect in question; and these conditions are sometimes far from manifest (as in the case of a loaded die): the hypothesis that b truly states all the objectively relevant conditions is often in need of revision, and statistical tests may play an important part in connection with the revision of our hypothesis. The upshot of it all is that the statement

$$p(a,b) = r$$

claims to describe some objective state of affairs: to say that the probability of throwing 'three' with this die is slightly less than $1/6$ is interpreted as a statement about the outcome of an experiment with a certain set-up; or as a statement of the conditions of an experiment, and of the results liable to be produced by these conditions. Thus '$p(a,b) = r$' is a hypothesis, exactly like other physical statements; and its hypothetical character has no particular connection with its probabilistic character.

This is the view shared by the various objective interpretations.

As opposed to this, the subjective interpretations do not take

$$p(a,b) = r$$

as the hypothesis in which they are interested; for they take '$p(a,b) = r$' to be an analytic statement, rather than an empirical hypothesis. But they take the statement

$$a$$

as the hypothesis under consideration; and the probabilistic formula '$p(a,b) = r$' they consider as informing us about the reliability of, or the degree of our rational trust in, the hypothesis a, in the light of the knowledge b.

Now if this interpretation is to be used in practice, then we need some rule telling us that if

$$p(a,b) = r$$

is analytic, and if b is indeed our total present accumulated knowledge, then we may assert, (since b is known) that, *for the present,*

$$p(a) = r \; ;$$

that is to say, a rule telling us that, under the conditions stated, our present rational belief in a would be equal to r. This rule is analogous to the *modus ponens* of classical logic; for both of them allow us to assert (with or without specifying a probability) some statement a unconditionally or absolutely which we first asserted only conditionally—under the condition b; provided always we *know* that the condition b is actually fulfilled. We may call a rule of this kind—one that allows us to assert some statement a absolutely rather than conditionally and so to absolve or free it from the fetters of its conditions— a '*rule of absolution*'.

There are two significant differences between the classical logical rule of absolution (*modus ponens,* sometimes called the rule of detachment) and its probabilistic counterpart: first, the result of the operation of absolution is in one case the assertion of a, or of the truth of a; in the other case the assertion that a has a certain probability (degree of rational belief). Secondly, the basis of applying the rule is in the one case the assertion of b or of the truth of b (where b is the condition or antecedent of a); in the probabilistic case, it is insufficient to assert that b is true: we must indeed assert that b is true, or that we know b to be true: we must also assert that b comprises *all* we know at present: it is the sum total of our knowledge. Only if we assert that b is at present our total knowledge can we derive what the present probability of a is, or how much we can rely on a.

Now my point is that this analysis shows why the subjective theory must make the probability of a absolute. Indeed, we cannot *act* upon a *relative* logical probability any more than upon conditional information: both need the application of the respective rules of absolution before they can furnish us with that kind of information which is a useful basis of practical action. And we cannot act upon a relative *logical* probability either—no more than upon a tautology. But the absolute statement

'the probability of a is (at present) equal to r'

which results from applying the rule of absolution is no longer analytic: it is a synthetic statement, a statement based upon b, upon our experience, and changing with our experience; for the value of r will in general change with every change of b. *This is the way we learn from experience—at least, according to the subjective theory.*

This is the line the subjective theory must take. But it makes it difficult if not impossible to accept the view that the subjective theory has anything to do with such problems as the probability that Mr. Andrews will survive the next five years (in the sense in which a life insurance office may bet on it), or the probability of throwing five with this die.

In the practice of *insurance* we can fairly clearly distinguish between business of two types which may be characterized as (i) fair betting, or rational betting, and (ii) gambling. The first type is based on large numbers and rational statistics, and the premium charged is highly competitive and in so far reasonably 'fair'. The second type is very different. It is insurance against rare or even unique risks such as, for example, the risk that a certain dictator will start a war within a certain period of time, or that the income tax will be decreased at the next budget. Statistics do not help here as the basis of computing risks because the events in question are rare if not unique.

Now the point I wish to make is this: ordinary life insurance which is of the type (i), and based upon rational statistics, is very far from taking into account, in every case, all the relevant knowledge available. Instead, it is based upon a rough questionnaire method which aims at subsuming every case under a very large class. If for some reason this cannot be done, the case becomes one of type (ii).

To put it in a different way, enough is known in every case to place the person seeking insurance into a class by himself. (*Proof*: he must be individually identifiable on the basis of his proposal.) And more and more information about the person might be acquired: some people would think his horoscope relevant, others his reading habits, others whether he habitually eats yogurt. But if the insurance company treats the case as belonging to type (i), then it will not take too many questions into account. Only age and average good health are considered fundamental. This is so because type (i) cases are assumed to be *repetitive,* and to belong to a huge class of repetitions. The information sought has to define the conditions of the experiment, and if these conditions make the experiment unique, or even rare, then the statistical method of type (i) becomes inapplicable. For there are two conditions basic for applying Bernoullian methods: independence of the repetitions (which, as we have seen, are not to be considered as repetitions if they are not independent); and *large numbers*.

Too much relevant information will, if considered, always make the case unique: it will take it out of the 'rational' or 'fair betting' cases of type (i) and put it into the irrational gambling cases of type (ii).

Thus the rational and statistical methods are based upon conscious neglect of available relevant information. This is due to the fact that *b* is considered as defining repetitive experimental conditions, rather than as a summary of our total relevant knowledge. What the insurance office tries to do is to find a reasonably stable *r* for a not too specific *b*. This procedure contrasts very sharply with the subjective theory according to which *b* constantly grows and *r*, consequently, constantly changes.

Type (ii) cases may perhaps be said to resemble more closely what the subjective theorist has in mind, for here the insurer will make use of the experience of a life-time: his whole knowledge of men, of politics, etc., will enter into his calculations. This is true; and yet, type (ii) cases are neither statistical nor rational. The simple inductive rule cannot be applied here, and there is clearly no rational method which would make it possible to base the insurer's estimate of his risk upon anything resembling logical inference.

If we now consider the types (i) and (ii) in connection with the probabilistic rule of absolution, then we find that we operate with this rule only in cases of type (ii)—that is, irrational gambling. Gambling upon unique events is indeed a kind of absolute act: once we have made up our mind (on the basis of our total experience), we have embarked on our adventure, and from this moment on we need no longer consider our total experience *b*.

Type (i) is very different and does not give rise to an application of the probabilistic rule of absolution. There is clearly no point whatever in saying that, given the truth of the information that Mr. Andrews belongs to the class *b*, we may say absolutely that Mr. Andrews's expectation of life is equal to *r*. On the contrary, it is only in so far as he is a member of the class defined by *b* that his expectation of life equals *r*; as seen from the fact that we may know that the same Mr. Andrews also truly belongs to the classes determined by b' (his horoscope), b'' (his reading habits), etc., and that his expectation of life *qua* member of these classes is different (simply because *any* change of the reference class may produce a different r). Frequency theorists (and I myself when writing *L.Sc.D.*) might have expressed this by saying that we cannot define the probability of an event, only the probability of an event *qua* member of a class of events. Today I should say that those objective conditions which are conjectured to characterize the event (or experiment) *and its repetitions* determine the propensity, and that we can in practice speak of the propensity only relative to those selected repeatable conditions; for we can of course in practice never consider *all* the conditions under which an actual event has occurred or an actual experiment has taken place.

Thus in any explanatory probabilistic hypothesis, part of our hypothesis will always be that we have got the relevant list of conditions, b, characteristic of the kind of event which we wish to explain. The subjectivist takes b as given, as unalterable data, while his hypothesis is a, as we have seen. The objectivist as a rule takes not a, but only the value of r as a hypothesis.

As to a, the attitude of the two is again completely different. The subjectivist may find an interesting hypothesis a—the hypothesis of his choice, as it were—and in asking for $p(a,b)$ he is asking whether a can be counted on (given b). The objectivist, on the other hand, whenever he is considering a formula $p(a,b) = r$, will be no more interested in the particular event a than in the event non-a. He does not look upon a as his chosen hypothesis, but as one of the possible events; and what he is interested in is not a itself but the probability of a. Thus the insurance actuary who is responsible for estimating survival probabilities is no more interested in the fact a (that a client has survived) than he is in the fact *non-a* (that a client has died): both are equivalent material for him. (This is so, even though his office will indeed be 'interested' in a rather than in non-a, in the sense of sincerely wishing that all their clients reach a ripe old age.)

Turning to the probability of throwing a 'five' with a die, we may take note of further discrepancies between the subjective theory and the objective theory and practice.

According to the subjective theory, we should, in a game of chance, determine (by some form or other of the simple rule) the subjective probability of the event—say 'five' in throwing a certain die. We should do so, for example, by observing games with dice in general and with one die in particular, and by compiling some kind of statistics. Having so determined the correct or rational degree of belief, it is rational to act upon it, that is to say, to accept the corresponding odds, *absolutely.* That is to say, we again apply the rule of absolution.

But this does not describe at all the ordinary attitude of a better. He always bets *assuming that the conditions*—i.e., *b*—are satisfied; for example, that the die is not loaded. Even his bet is not absolute: if he finds later that the die is loaded, he feels that he has the right to cancel his bet. Or if he finds that any other 'trick' was played on him implicitly forbidden by the accepted conditions of the game, he will again consider that he is not bound. Thus he does not bet upon *a* (as the subjectivist asserts) but he bets upon *a*, provided some *objective* conditions *b* are satisfied.

It may be said in passing that the subjectivist would be mistaken if he believed that he could interpret this situation by saying that we do not bet that *a* will happen but rather upon a conditional statement. For the probability of a conditional is very different from a conditional (or relative) probability, as may be seen as follows. Let *b* again be our total knowledge, *c* the conditions of the game; then, he may suggest, we do not bet upon *a* (given *b*) but upon 'if *c* then *a*' (given *b*); and after applying the rule of absolution, upon 'if *c* then *a*', absolutely. This interpretation is not compatible with the laws of the probability calculus, since 'if *c* then *a*' will have a higher probability than *a*; unless, indeed, *c* is part of *b*, in which case the condition *c* loses its force; that is to say, we have in this case $b = bc$, and as a consequence, p (if *c* then $a,b) = p(a,b)$ and there is no reason why, after applying the rule of absolution, we should obtain the present probability of 'if *c* then *a*', rather than the present probability of *a*.

Thus the role of the evidence *b* in the subjective theory is fundamentally different from the role of the experimental conditions *b* in the objective theory; and it does not appear that the subjective theory comes anywhere near to treating those facts and those problems in which the objective theory is interested.

11. *The Diminishing Returns of Learning by Induction.*

Closely connected with these considerations is the following difference between the objective and the subjective interpretations.

The objective theorist conjectures that, by keeping constant the conditions *b*, he will obtain a fairly stable *r*. This he may estimate if he has a very large sample, because Bernoulli tells him that the sample will very probably be *representative*, i.e., that it will show a frequency close to that *r* which we try to determine.

The subjective theorist, on the other hand, works with a changing *b* and consequently with a changing *r*. He may say that, with his statistical experience accumulating, he will in fact get a more and more stable *r*; and he will attempt to identify the objectivist *r* with this ever more stable *r* of his.

From the various difficulties created by this view, the following may be singled out here for discussion.

In the subjectivist view, the changes in *r* reflect precisely *the way in which we learn from experience*. And the fact that *r* becomes ever more stable reflects the way in which our increased learning stabilizes our rational beliefs.

The fact of the stabilization itself cannot be doubted: according to the simple inductive rule, *r* will be approximately equal to m/n, where *n* is the number of our total relevant observations, and *m* the number of observations in which the same property as expressed by *a* was observed (or something favourable to *a*); here again *a* is the hypothesis we are considering, and *b* the evidence from which we obtain the statistics expressed by m/n; and we have

$$p(a,b) = r \approx m/n.$$

Now if *n* grows very large, then any particular new observation will have very little influence upon *r*; for if *n* is very large, $(m + 1)/(n + 1)$ or $m/(n + 1)$ will be almost the same as m/n.

This simple fact means that *r* must become comparatively stable when our experience has accumulated, and *n* has grown large.

But this fact also has a different and less favourable aspect. It shows that the subjective theory of learning attributes an immense authority to our past. After a lifetime of learning, there can be no hope of learning more: the authority of our past experience makes a revision practically impossible. In other words, the older we get, the more rusty, the more sluggish we get, being tied down by the past. And the same must hold for science.

Even if the facts described were true, as a matter of psychological and historical fact (and although they may be true for some individuals, they are clearly not true for the historical development of science), they clearly represent a strangely unsatisfactory theory of learning. It is a theory of diminishing returns.

There is something like a very obvious reply to this critical attack. It is this. The increasing stability of our beliefs is simply the result of the fact that our beliefs are better and better founded in experience. They are becoming ever more reliable. If we have progressed in our endeavour to know, we must not complain if, the further we progress, the nearer we approach something like a state of finality.

My rejoinder to this defence is this. Actual finality is of course not reached in this way, nor was it implied in the critical attack. If learning means changing the value r of the probability, and if these changes are improvements, then there must exist something like a 'true' probability value, a true value of r which we approach by way of our inductive values. What the subjectivist apologist who replied to me had in mind was something like this: by throwing this die very often, our subjective probability values will, in time, approximate more and more closely to the true value.

But what is a true probability value if not the objective value, the value of the objective theorist? And was not the replacement (or elimination) of this 'true' or 'objective' value one of the major aims of the subjective theory—an aim chosen because of the basically unsatisfactory character of the objective theories of probability?

By an appeal to anything like a 'true' or 'objective' value of probability, subjectivism would commit suicide: it would simply give the show away.[1]

Yet apart from this, the argument of diminishing returns has not really been answered at all. Diminishing returns are, as I showed, the simple arithmetical result of accumulating experience—whether

---

[1] I cannot see anything except an appeal to objective probability in the following lines of Carnap's 'Inductive Logic and Science', (*Proceedings of the American Academy of Arts and Sciences*, vol. 80, no. 3, March 1953, p. 191): 'The answer to the question: "How long then shall we make the series of throws with the die in order to determine the probability?" is the same as the answer to the question: "How fine a thermometer should we use to measure the temperature?" In both cases the answer depends, on the one hand, on the time and money available and, on the other hand, on the desired degree of precision. More specifically it depends on the theoretical or practical advantages to be expected from higher precision. The finer the thermometer and the longer the series of throws, the higher the precision which is achieved. In neither case is there a perfect procedure.'

or not this accumulation was successful in approaching nearer and nearer to some 'true' value. We have no reason whatever to believe that the increasing sluggishness of learning is a result of a better approximation; indeed, we can see clearly that it is quite independent of it. (Remember the game 'Red or Blue'.)

12. *The Paradox of Inductive Learning.*

The criticism of the subjective-inductive theory of learning takes us more closely to the heart of the subjective interpretation than any of my previous remarks. For in the work of most subjectivists, we find the following argument, which I shall dub the 'transcendental' argument because of its close resemblance to the argument in which Kant appealed to the *fact* that pure science, or knowledge that is valid *a priori,* actually exists.

The transcendental argument of the subjectivists may be put as follows.

Nobody knows better than we subjectivists that the theory of induction is extremely difficult. This is why we work on it. The difficulties presented by you neither shock nor surprise us: they are our bread and butter. They may be very real, and we may not be able to solve them here and now. But *we subjectivists know one thing—that they must be soluble.* For we know that we learn from experience: this is a fact; and no sceptical quibbles will shake our confidence in this fact.

I have no intention of denying this fact; and although I am fortunate in never having been so hard pressed by sceptical attacks or doubts as to have recourse to a transcendental argument, I gladly admit that this argument is one which deserves respect. So much so that I am prepared to adopt it here in order to turn the tables against the subjectivists: I shall try to show that, if we adopt the subjective-inductive theory of learning, *learning becomes impossible.*

We first consider what will happen if we choose, for the b of a logical-subjective formula $p(a,b) = r$, that information which gives us precisely the objective conditions of the experiment, and nothing else; more especially, if we do *not* include, in b, any information about previous outcomes (the outcome, a, or $\bar{a}$) of the experiment, and about their frequencies. We assume that the experiment is not one with a perfect symmetrical gambling device (a die, or a coin, etc.), but more of the nature of an ordinary physical experiment, say, one which tests the statistical theory of gases. It is clear that in this case the value r of the logical (or of the subjective) probability

$$p(a,b) = r$$

will not in general correspond to the 'objective' value r; otherwise all our objective knowledge would be logically true, or analytic; and no induction would be necessary, or even possible: *we could not learn from experience.*

Thus we do get a paradoxical result if we assume that we may identify the b of the subjective theory with information about objective conditions. But we get similarly paradoxical results if we abandon this assumption. And indeed, we must abandon it; for the b of the subjective theory must include knowledge not only of the objective conditions, but also of past *results* of the experiment.

The subjective-inductive theory tells us that we learn by applying some form of the simple rule of induction. It tells us that we can learn, by applying this rule, to expect 'heads' with a probability of $1/2$ if we see that a penny has been tossed, in the past, and that it fell heads uppermost with a frequency of $1/2$.

But this is precisely what we cannot learn, from the point of view of the subjective theory. For it is impossible for this theory to operate with the idea of the *objective conditions of an experiment.*

As we have seen, the subjective theory must interpret b, in

$$p(a,b) = r$$

as a summary of our total knowledge—or of our relevant total knowledge. This is a consequence of its basic view that '$p(a,b) = r$' is to be interpreted as a statement of the degree of our rational belief, *rational in view of the evidence in our possession.*

Thus, within the subjective theory, b cannot be a statement of the experimental conditions. These conditions may admittedly form part of b since we may be informed of them or may have observed them. But as subjectivists, we cannot separate them from the rest of our knowledge. The objectivist separates them with the help of his idea of relevance. But the idea of relevance leads the subjectivist to a very different b; for the results of previous experiments are most relevant to him, while to the objectivist they are irrelevant, since the new results must be independent of the old ones.

As a consequence, the subjectivist cannot express the idea of the *repetition of an experiment.* For every experiment will proceed under essentially different relevant conditions (not objective conditions, but conditions of knowledge). The eighteenth experiment will proceed under conditions different from those of the seventeenth; for among its conditions will be our knowledge of the result of the seventeenth experiment.

Thus we cannot apply the simple inductive rule, since we can never repeat the conditions of an experiment.

We can put the paradox in this way.

Assume that our knowledge grows, in accordance with the subjective theory, if and only if we observe a repetition of an experiment. Then it cannot grow; for since its growth would alter the *known* conditions of any experiment, no experiment can ever be repeated.

In other words, the assumption that the new experiment is a repetition of the old one is contradictory, from the subjective point of view. For if it is a repetition, then the simple inductive rule applies which makes all previous instances highly relevant conditions, so that it must be a case essentially different from these previous cases. Thus no experiment can ever be repeated.

(Zeno might have said: if it is the same, then it is not the same. Therefore it cannot be the same.)

Repetition of an *experiment* thus turns out to be a concept foreign to the subjective theory. This theory, however, may employ the idea of the repetition of an *observation,* or of an observed fact. There is no difficulty for it in asserting that we have observed fifty tosses giving tails. What it cannot assert is that in these fifty cases the conditions were equal, or largely equal, or mainly equal; for according to the presuppositions of the inductive rule they were, in every case, compared with any of the other cases, different in the most relevant sense.

Against this way of arguing it may be objected that I have cunningly mixed up two senses of the word 'condition', and of the word 'knowledge'. It will be said that we must distinguish (i) our knowledge of the objective conditions of the experiment and (ii) the psychological or subjective conditions existing when we observe the new experiment (which include, of course, our knowledge of previous results). My reply is that the subjective theory cannot make this distinction. All it can do is to distinguish between relevant and irrelevant conditions; and the only thing of which we can be certain is that the knowledge of previous results must be highly relevant, from the point of view of the subjective theory.[1]

Thus learning by experience is impossible according to the subjective-inductive theory.[2] But we know that we learn by experience.

Thus the subjective-inductive theory must be false.

This negative use of the transcendental argument is its only proper use. It should not be used, as it so often is, to argue in favour of any particular theory of learning; for it is an argument which always amounts to the more or less explicit claim that the theory in question is the only possible one. (It is surprising to find how many incompatible theories have been claimed to be the only possible ones.)

13. *An Inductive Machine.*

In spite of my objections to the inductive theory of learning, I believe in the possibility of inductive machines—of a certain kind.

In order to see the possibility of an inductive machine, we consider a *simplified universe* which the machine will have to investigate. The universe may consist, for example, of 'individuals' or 'individual events', with a certain limited number of properties. We choose for our individuals balls of one inch diameter. An individual event may consist in the appearance of a ball at the end of a pipe; it then rolls down a groove and disappears into a hole. There is to be such an individual event every second.

The balls may have certain *properties*. They may be of steel or of brass, and they may be painted in a number of different colours. We now assume that our artificial universe may be operated in accordance with a number of 'natural laws'. For example, we may so adjust

---

[1]See, for example, R. Carnap, *The Continuum of Inductive Methods*, 1952, p. 14, Rule C9, which asserts that 'the probability (in the logical sense) that the next throw of a given die will yield an odd number is *generally regarded (sic)* as depending *merely* on the number of odd results and the number of even results obtained so far with this die'. (Italics mine.)

[2]Compare with this argument my analysis of Hume's views on induction in sections iv and v of my paper 'Philosophy of Science: A Personal Report', reprinted in *Conjectures and Refutations*, Chapter 1; and also Appendix *x of *L.Sc.D.*

it that steel balls always come in sequences of three, and after these a longish but varying sequence of copper balls; or we may so adjust it that, out of each hundred steel balls, 10 balls are blue, 20 green, 30 red, and 40 yellow: this would be a statistical 'law'. Other 'laws' can be arranged.

So much concerning our simple universe. Now to the induction machine. This can be so constructed that, in a reasonable period of time, it will 'discover' the laws which are valid during this period in its universe. And it will show its capacity by finding, if the laws of its universe are changed, the new set of laws. (Of course, sufficient time must be allowed for these discoveries, and the 'laws', statistical and otherwise, must be kept constant during this time.)

The induction machine will have to be equipped with a detector (perhaps a magnetic needle) which allows it to distinguish between steel and copper balls, and another detector for colours. Moreover, it will have to be equipped with counting devices. It will have to be able to draw up statistics of the various distinguishable occurrences, and to calculate averages. If it is further adjusted to the *kind* of law (not to the actual laws) which the universe may exhibit—laws of succession, general or conditional frequencies of a certain stability, etc.—then it may be so constructed as to be quite efficient, for example, in formulating hypotheses and in testing and eliminating them. Thus it may detect simple regularities in the universe without any difficulty: *it may thus learn from experience.*

It is useful to discuss some of the more primitive stages in the evolution of induction machines in a little more detail.

(i) The most primitive stage in the evolution of the induction machine may be described as follows.

The machine notes the properties (or predicates) of every single event, i.e., whether steel or copper, whether blue or green, etc. It *counts* the events, and it also counts the events falling under any one of the various properties, and those falling under two or more properties ('steel and blue'). Moreover, it forms at once the corresponding quotients, notes them symbolically, and corrects them after every new event, so that at any moment it can give us the relative frequencies of the occurrence of the various properties, and combination of properties.

If we have built our 'world' so that these frequencies are reasonably stable (or if we know it is so constructed) then the frequencies calculated by the machine will, of course, approach more and more closely to the 'true' or 'objective' frequencies which we have built into our 'world'; and we can then say that the machine can answer at any moment a question such as: 'What is the probability of the next event possessing the property P, or perhaps the set of properties $\{P, Q, R\}$ with a reasonable approximation to the objective value?'

At the stage just described, the machine uses the simple inductive rule (as I have called it) in its most primitive form. Against this form the objection has been raised that the early results—of the initial steps of the induction—are unreasonable. For example, if the first ball is blue and of steel, then the machine would have to attribute to the predicates 'blue' and 'steel', and also of course to the predicate 'blue and steel', the probability 1, and to the others the probability zero. The objection raised here appears to have little significance; for if we wish to avoid such early results, we can always construct our machine so that it starts issuing probabilistic predictions only after the 1000th event, say, or after any other number n which we may choose, bearing in mind the number of different properties in our 'world'. (The problem is so trivial that it is not worth making any effort to solve it systematically; for we know, after all, that applications of the simple inductive rule will never give us more than increasingly good approximations.[1])

[1]In his *Logical Foundations of Probability*, 1949, and its sequel, *The Continuum of Inductive Methods*, 1952, Carnap is concerned with constructing 'inductive methods' which correspond closely to what I have here described as the most primitive stage (i) in the evolution of an induction machine; and the various inductive methods contained in his *Continuum* can be shown to be so many attempts to solve the problem I have just described as trivial; the problem, that is, of how to avoid the counter-intuitive initial results of the simple inductive rule in its most primitive form. See for example, *Logical Foundations*, p. 227, where Carnap discusses what he calls the '*Straight Rule*', which is our simple rule, interpreted as giving exact results rather than approximate results. (This is why Carnap raises *two* objections—one in connection with the initial results of the rule, and the other in connection with the fact that we should hardly at every moment of time ascribe an extreme probability such as zero to every event which up to that moment has not been observed; yet this second objection disappears if we replace 'zero' by 'not far from zero'.) Carnap's 'continuum of inductive methods' simply consists of various methods of obtaining not too implausible results even below the number which, in the text, I have called 'n', and of choosing that number n. Since he believes that the theory of induction must be a logical theory, he attempts in his *Logical Foundations* (pp. 563ff.) to give a kind of logical justification for choosing a particular solution of this problem. This attempt, it seems, is given up in his *Continuum*, where the variety of the possible solutions is stressed. (Thus Carnap's

(ii) In a second and slightly less primitive stage, the induction machine may, in a similar way, formulate the statistics of consecutive pairs, triplets, etc., of events. It may discover, in this way, that the probability of a copper-steel pair to be followed by a copper event, or by a steel-copper pair, or by any triplet except a steel-steel-copper triplet, is zero. (Which means that the machine has discovered the 'law' that steel events tend to occur, if at all, in a succession of exactly three.)

(iii) Further stages in its evolution may now be introduced by enabling the machine to frame hypotheses and to test them. I shall not go deeper here into the interesting but difficult problem of how a machine might do this. Instead, I turn to the question:

If a machine can learn from experience in this way, that is, by applying the simple inductive rule, is it not obvious that we can do the same?

Of course I never said that we cannot learn from experience. Nor did I ever say that we cannot successfully use the simple inductive rule—if the objective conditions are appropriate. But I do assert that we cannot find out by induction whether the objective conditions are appropriate for using the simple inductive rule.

This may be illustrated by our machine, in its stage (i).

If our machine is successful in its stage (i), then it is so because we have constructed our 'world' in such a way that the simple inductive rule can be successfully applied to it. That this is so becomes obvious if we remember our game 'Red or Blue'. Nothing is easier than constructing a 'world' which incorporates this game, or a generalized version of it. But, as we have seen, this 'world' will *in all probability* defeat our machine. It can be made to do so for any probability we may choose (short of unity); for any period of time which we may allow to the machine for its induction; and for practically any criterion of 'defeat' we may adopt.

In constructing an induction machine we, the architects of the machine, must decide what constitutes its 'world'; what things are to be its individual events; what constitutes a property, or a relation; or, in other words, *what constitutes a repetition.* And it is we who must decide what kind of questions we wish the machine to answer.

But this means that all the more important and difficult questions

number λ and Kemeny's number κ are something like highly sophisticated substitutes for the number I have here called '*n*'.)

322

were already solved by us when we constructed the 'world', and the machine.[2]

14. *The Impossibility of an Inductive Logic.*

The inductive machine described in the preceding section—with all the fundamental faults and limitations which make it unacceptable as a model of induction—is a simplified and improved version of what inductive logicians seem to have in mind. About its comparative simplicity there can be no doubt, if we compare it with the heavy volumes—heavy in every respect—which have been written on the subject. As to the machine's being an improvement, I should like to mention an argument in favour of discussing inductive theories in terms of machines rather than of languages; an argument which should appeal to all inductivists. It is that cats and dogs assuredly learn from experience, but without a symbolic language. Hence a logical or a linguistic theory of induction will always remain a highly unconvincing artifice, as compared with the discussion of possible inductive machines.

But the main improvement brought about by our machine is that it breaks with the traditional view that induction, and more especially some form or other of the simple inductive rule, is part of an inductive logic that, in its turn, is a generalization of deductive logic.

I have said in an earlier section that there exists a logical interpretation of the probability calculus which makes logical derivability a special case of a relation expressible in terms of the calculus of probability. Thus I assert the existence of a probability logic which is a genuine generalization of deductive logic.

But I deny the possibility that this probability logic can ever be interpreted as an inductive logic. More especially, I deny the possibility of linking this probability logic with any form of the simple inductive rule.

The existence of the link is usually taken for granted. This seems to be due to a tacit or explicit argument on the following lines.

---

[2]Compare with the last three paragraphs the end of section v of my paper 'Philosophy of Science: A Personal Report', cited in previous section, note 2, and Appendix *x of *L.Sc.D.*

Inductive reasoning is in some respects similar to deductive reasoning, although not quite as conclusive. This situation may be explained by assuming that a conclusive (or an almost conclusive) argument corresponds to the probability 1, an inconclusive argument to a lower probability. Thus inductive reasoning will be part of probability logic; and so will some form or other of the simple inductive rule which is the simplest and most elementary form of all inductive reasoning.

This argument is completely mistaken, and the simple inductive rule has nothing whatever to do with logic—neither with logic in its usual deductive form, nor with probability logic. And nobody has ever developed an argument leading from probability logic to any form of the inductive rule. Most writers on the subject have taken this link for granted; and the few who saw that there is a problem here (especially Carnap in his *Logical Foundations of Probability*) tried to bridge the gap by some form or other of the argument which I have described, in the section before the last, as 'transcendental': there had to be such a link because otherwise we could never learn from experience, they believed.

It is easy enough to prove that the simple inductive rule must be independent of anything like logic. The independence proof proceeds, like all such proofs, by the construction of a model. We merely take for our world the events of the game 'Red or Blue': in this world, 'logic' remains (objectively) valid, as it does in all logically possible worlds. But the inductive rule becomes (objectively) invalid, in the sense that it will most probably fail—with a probability as high as we choose.

We now see why our inductive machine is so simple, compared with almost all attempts to construct an inductive logic. The reason is this. We have not tried to establish the validity of our machine by any inductive logic, or to establish a link between the machine and probability logic. We have not done so because it cannot be done. An inductive logic is precisely an attempt to establish the impossible. It is this which makes books on inductive logic so complicated and loosely reasoned, and which is responsible for the introduction of endless and complicated substitutes for simple argument.

15. *Probability Logic vs. Inductive Logic.*

So far I have attacked our problem from the side of the inductive rule, by establishing its lack of validity. I am now going to attack from the other side, the side of probability logic.

Probability logic is supposed to be a generalization of deductive logic, and its formulae are supposed to be logically true, or analytic, or tautologous, as are those of all logic.

Keeping this in mind, we may look at a few formulae of the calculus of probability which make this interpretation possible.

(1) $$p(a,a) = 1.$$

This formula may be interpreted as 'a follows from a'.

(2) $$p(a,ab) = 1 = p(a,ba) = p(a,bac).$$

This formula may be interpreted as 'a follows from any conjunction of which a is a component'.

(3) $$p(ab) = p(a)p(b) \text{ if and only if } p(a,b) = p(a).$$

This we may interpret as 'a and b are (absolutely) independent in the sense of probability logic if and only if the information b leaves the absolute probability of a unchanged; or in other words, if b is irrelevant to a'.

(4) $$\text{If } p(ab) \neq p(a)p(b) \text{ then either}$$
$$\text{(i) } p(a,b) > p(a), \text{ or}$$
$$\text{(ii) } p(a,b) < p(a) \text{ and } p(\bar{a},b) > p(\bar{a}).$$

This we may interpret as 'if a and b are not logically independent in the sense of probability logic, then either (i) b is favourable to a, or supports a, or (ii) b is unfavourable to a, or undermines a by supporting non-a'.

This shows that probabilistic-logical independence is to be interpreted as logical neutrality: a and b are logically independent if and only if b favours neither a nor $\bar{a}$.

The following theorem may now be formulated which throws a great deal of light upon the tautological character of this logic.[1]

[1] I am discussing this problem in a simple form, relying upon the possibility of analysing contents in terms of conjunction. A more formal discussion might have to introduce conjunctions of relative atomic statements (whose logical independence would then be the main point of the argument).

(5) Let a be a conjunction of x and y, i.e., $a = xy$, with $p(y,x) \neq 1$. (The contents of both x and y are assumed $\neq 0$.) Let $p(xb) = p(x)p(b)$, so that x and b are (probabilistically) independent. Let $b = yz$. Then

$$p(a,b) = p(x,b) = p(x) > p(a).$$

That is to say, if a is the conjunction of an x that is (probabilistically) independent of b, and of a y that follows from b, then $p(a,b)$, although equal to the absolute probability of x, will nevertheless (provided $p(y,x) \neq 1$) be greater than the absolute probability of a.[2]

We are thus led by (5) to the following explanation of the tautological character of this extension of deductive logic: b favours a whenever a is equivalent to a conjunction xy such that y is not a tautology and follows from b, provided x (the part of a that is not entailed by b) is independent of b or not too strongly undermined by b: it need in no way be supported by b. If b is independent of x and thus neutral with respect to x, the probability of a given b will be equal to that of x given b; and therefore to the absolute probability $p(x) > p(a)$.

Or in other words, b supports a if it entails part of its content and is neutral (or not strongly unfavourable) towards the rest.

But if this interpretation is correct—and it seems to me clear that it is—*then probability logic cannot be inductive logic*. For the theory of induction tries to justify the attempt to obtain more from the

---

[2][*(Added January 1981.) David Miller has some important comments on this, two of which may be mentioned here.

(i) Miller points out that it would be interesting to see whether any of the various concepts of (non-probabilistic) *logical* independence (such as complete independence, or maximal independence: see A. Tarski, *Logic, Semantics, Metamathematics*, 1956, p. 36) can be related to probabilistic independence. And he actually constructs a link between the latter and maximal independence: two statements a and b are (according to H. M. Sheffer) maximally independent if $a \vee b$ is a tautology so that $p(a \vee b) = 1$; and Miller proves that, if $p(a \vee b) = 1$ (and $p(b) \neq 0$), it is not possible that b supports a: it can only undermine a or be (at best) neutral to a. Miller's proof is indirect: he shows that if $p(a,b) > p(a)$ then $p(a \vee b) < 1$. [From $p(a,b) > p(a)$ and $p(b) \neq 0$ we get $p(ab) > p(a)p(b)$ and so $p(a \vee b) < p(a) + p(b) - p(a)p(b) = p(a)(1 - p(b)) + p(b) \leq 1 - p(b) + p(b)$. Thus $p(a \vee b) < 1$.]

(ii) Miller uses this proof in order to establish the following theorem (6): If b supports a without entailing it then there always exist two statements x and y such that $a = xy$ and such that y sums up all the content of a entailed by b while b is at best neutral to x. (Proof: put $x = a \vee \bar{b}$; $y = a \vee b$. Now b cannot support x, since $x \vee b$ is a tautology. Incidentally, x and y may be said to be maximally independent; and they may be called 'maximally independent relative to a'.)]

premises than there is in them: to go with the conclusion *a* beyond the premise *b*, even at the price of a loss in certainty. Yet logical probability says that if the content of *a* goes beyond the premise *b*, then only that part of *a* which follows from *b* is made certain, and the rest remains exactly as probable or improbable as it was without *b*. The increase in the probability of *a* which *b* may procure is thus entirely due to the fact that part of *a* is logically entailed by *b*; it is not due to any effect of *b* upon that part of *a* which does not follow from *b*. [*Thus there is in probability theory no ampliative probabilistic inference. The apparent ampliative inference is due to the fact that part of the content of *a* is deducible from *b*.]

It is this situation which gives probability logic its tautological character. Any other interpretation—say one which would enable *b* to be favourable to *a* without logically entailing part of it—would make the resulting formulae non-analytic.

Probability logic thus generalizes derivational logic by considering not only conclusions which are entailed by the premises, but also partial conclusions which are only partially entailed by the premises (and several further possibilities also). But it does not, in doing so, allow us to proceed from the known premise—with less than certainty—to an unknown conclusion; in fact, what goes beyond the premise remains as probable or as improbable as it was before.

Induction, of course, is essentially an attempt to extend our knowledge: to proceed from the known to the unknown, by increasing at least the probability of something that is unknown. My argument here is that whatever we may think of induction it cannot be identified with probability logic.

16. *The Inductivist Interpretation of Probability.*

What, then, is 'inductive logic'? There is (i) a logical interpretation of the probability calculus, or 'probability logic', which has nothing to do with induction; and there is (ii) an inductivist interpretation of the probability calculus, which has nothing to do with logic—except that it may be so formulated that probability logic becomes a zero case of it. But it is *not a generalization* of probability logic because in so far as it is inductive, it is in direct conflict with the logical interpretation; and if we take its zero case, then we exclude any inductive argument. To talk of inductive logic is therefore highly misleading. But may we perhaps call the probability calculus if interpreted so as to yield the simple rule of induction, the '*inductive calculus*', and the probabilities calculated by it '*inductive probabilities*'?

The 'inductive calculus' is, formally considered, the probability calculus plus a *rule* asserting the following:

Any observed instance of a property P tends to increase (if perhaps only slightly) the probability of that property P. Consequently, any observed instance of the complementary property $\bar{P}$ tends to decrease the probability of P (since it tends to increase that of $\bar{P}$).

Carnap[1] calls this the 'Theorem of *Instantial Relevance*', and formulates it as follows: '. . . one instance of a property is positively relevant to the prediction of another instance of the same property'. (He continues: 'This seems a basic feature of all inductive reasoning concerning the prediction of a future event.')

The rule means, in practice, that, for example, the observation of green things increases the probability that the next will be green; or that the observation of objects which are clever increase the probability that the next object observed will be clever. As Carnap shows, it is a form of what I have here called '*the simple inductive rule*'.

I do not want to discuss here the plausibility or implausibility of this rule, nor to re-open the issue of its validity or invalidity. I only wish to show that, even had I not proved, by way of a counter-example, that it is not part of logic,[2] the whole character of the rule would preclude its acceptance as part of probability logic.

For there is nothing illogical (in the sense of 'self-contradictory') in the opposite rule. I mean the famous *gambler's fallacy* of waiting for a block of 'heads' and then betting on 'tails'—'because it is now time for a change'. Although this is a fallacy as far as the usual games

[1] R. Carnap, 'On the Comparative Concept of Confirmation', *Brit. Journ. Philos. Science* 3, p. 315. (The rule itself is formulated as T1 on p. 314, and A4 on p. 316.)

[2] Apart from the game 'Red or Blue', I have also constructed, in *L.Sc.D.*, Appendix *vii, note 9, an example of property B which contradicts the '*Theorem of Instantial Relevance*'. Admittedly, B can only be constructed in a language richer than Carnap's; but a rule which, if extended to a richer language, leads to a contradiction, can certainly not be said to be a logical principle, or analytic; especially if there is no argument in its favour (except the 'transcendental' argument; *cf.* note 3 to Appendix *vii).

of chance are concerned, we can easily construct a sequence (say, with a reduced frequency of blocks) in which the gambler's fallacy would lead to success.

But the formulation of the inductive rule just given shows its non-logical character not only by being in conflict with the gambler's fallacy, or with our game of 'Red or Blue'; it is in conflict with any normal random sequence. For interpreted objectively, it postulates an after-effect which in a normal random sequence does not occur. The rule thus conflicts, for example, with the estimate

$$p(a,b) = 1/2,$$

where a is throwing heads with a coin that is homogeneous, as b tells us.

This is obscured by what I have called above the 'law of diminishing returns' which ensures that the probability, calculated by the simple inductive rule, stabilizes itself very close to the 'true' or 'objective' value of the probability of a random sequence, as we have seen, thus leading ultimately to 'success'. But even after being stabilized in this way, the induced probability may still flutter a little if an unusually long block (it need not be a pure one) should come along—even if this block is somewhat overdue, from an objective point of view.

The fact that all forms of the simple inductive rule are in conflict with objective probability, even for the ordinary coin-tossing game, deserves to be well stressed. The conflict may be a minor one from the start, and it is bound to die down completely if the sequence of games becomes infinitely long. But this does not change the fact that if higher stakes are involved, great losses must be expected, according to the objective theory, by any gambler who uses a form of the simple inductive rule as his *'guide of life'*, or in other words, as his *gambling system*. For the probability calculated according to the simple inductive rule will again and again deviate from the objective probability (which we put at 1/2 since we are considering coin-tossing). Whenever this deviation occurs, the inductive gambler should be prepared to accept odds which from the objective point of view are unfavourable. But although this need not necessarily lead to a loss, the (objectively) *expected* or *probable* loss can be made as big as we like: a clear indication that inductive probabilities cannot be tautologous.

Our discussion establishes, I believe, two things: (i) the consistency of the (interpreted) inductive calculus, i.e., of inductive probability[3]; and (ii) its non-logical character. One really ought not to argue the latter point any longer, 216 years after Hume's *Treatise*; but the relevant passages have clearly not yet received the attention they deserve. Hume argued that there can be no valid demonstrative (or logical) argument which would allow us to show '*that those instances, of which we have had no experience, resemble those, of which we have had experience*'. Consequently, '*even after the observation of the frequent or constant conjunction of objects, we have no reason to draw any inference concerning any object beyond those of which we have had experience*'. For 'should it be said that we have experience'—experience teaching us that objects constantly conjoined with certain other objects continue to be so conjoined—then, Hume says, 'I wou'd renew my question, *why from this experience we form any conclusion beyond those past instances, of which we have had experience*'.[4] In other words, an attempt to justify the practice of inductive logic is untenable, and the attempt to justify it by an appeal to experience must lead to an *infinite regress*.

Thus Carnap's rule establishing the mutual dependence, or relevance, of any two instances of the same property is a non-analytic inductive principle; and if it is upheld (and if Hume's infinite regress is to be avoided), it could only be upheld as a synthetic proposition, that is, as valid *a priori*.

---

[3][*It is consistent only if it is applied to a stabilized (sufficiently long) sequence of objectively independent occurrences; but it is inconsistent in so far as it asserts the relevance (and therefore the opposite of independence) of these occurrences. But if the sequence is stabilized, *this* theoretical inconsistency does not matter in applications. (Of course, the game 'Red or Blue' shows that it must not be applied to objectively dependent sequences.) See also David Miller, 'A Paradox of Information', *British Journal for the Philosophy of Science* 17, 1966, pp. 59–61.]

[4]The italics are all Hume's. *Cf.* Hume's *Treatise of Human Nature*, 1739–40, book i, part iii, sections vi and xii. (Selby Bigge's edition, pp. 89, 139, and 91.) Hume was very clear that an appeal to probability would not alter the situation. Thus he writes in his *Abstract:* ' 'Tis evident that *Adam*, with all his science, would never have been able to *demonstrate*, that the course of nature must continue *uniformly* the same . . . Nay, I will go farther, and assert, that he could not so much as prove by any *probable* arguments, that the future must be conformable to the past. All probable arguments are built on the supposition, that there is this conformity betwixt the future and the past, and therefore can never prove it.' *Cf. An Abstract of a book lately published entitled a Treatise on Human Nature*, 1740, ed. by J. M. Keynes and P. Sraffa, 1938, p. 15. (The italics are Hume's.)

How then can we argue in its favour? By the transcendental argument of course: this is now most fashionable among all kinds of inductive logicians who scorn Kant as an *apriorist* and transcendentalist. (I am a great admirer of Kant, but I think that the positive use of the transcendental argument in support of principles which he believed to be valid *a priori* was his gravest mistake. Yet there were strong mitigating circumstances in those days, long before Einstein.)[5] The argument may be found in connection with various probabilistic theories of induction in Russell, Jeffreys, and Reichenbach. Carnap uses it too, but with a certain hesitancy which in my eyes does him great credit. For he writes about his choice of the function *m** (which amounts to the choice of a certain special form of the simple inductive rule):

'The preceding considerations show that the following argument, admittedly not a strong one, can be offered in favour of *m**. Of the two *m*-functions which are most simple and suggest themselves as the most natural ones, *m** is the only one which is not entirely inadequate.'[6]

Choosing the other *m*-function, which is 'entirely inadequate', would amount to choosing what I have called *logical probability*. Its 'inadequacy' consists, of course, in the fact that it does not imply any form of the simple inductive rule. But from my point of view, this cannot be otherwise, *since probability logic is not inductive; since inductive probability is not logic; and since an inductive logic cannot exist.*

But although an inductive logic cannot and does not exist, inductive logicians do. Indeed, they have been fairly abundant ever since Bacon (if not since Aristotle). And they are not so easily refuted. This is so because inductive probabilities—that is to say, applied inductive calculi—may be consistent, even though they are not tautological. However, they are consistent *only* as long as inductive probability is clearly distinguished from both logical and objective probability. For the probability values of inductive probability are in general incompatible with both the logical probability and the objective probability of even a symmetrical and independent game such as tossing pennies, as we have just seen. I hardly need add that some or all of these different kinds of probability are, as a rule,

[5][See *Conjectures and Refutations*, p. 27 and pp. 190–1. Ed.]
[6]*Logical Foundations of Probability*, p. 565.

mistakenly identified in the transcendental arguments of the inductive logicians.[7]

17. *The Redundancy of Theories.*

The method of hypothesis which is defended in *L.Sc.D.* is very old indeed. What may be new is the attitude towards predictive tests. These have at times been considered as verifications or even as proofs of a theory. Later they have been considered as 'merely' capable of making the theory probable. I consider them as attempted falsifications; and only if, in spite of being sincere and ingenious attempts, they do not succeed in refuting the theory do I consider them as capable of corroborating it. By corroborating theories, they can make them seem 'probable'—but not in any of the many senses of the word 'probable' which satisfies the calculus of probabilities: for this reason I prefer to speak of 'corroboration' (or 'confirmation' or 'acceptability').

The importance of the method of hypothesis, or of the hypothetico-deductive method, is now generally recognized, and it is fully admitted even by inductivists that a theory of induction which does not give full weight to this method would be preposterous. Some of the leading inductivists such as Jeffreys or Carnap have gone out of their way to emphasize this point. Jeffreys puts the question of universal theories and of their simplicity at the centre of his theory of induction. And Carnap emphasizes, in the introductory part of his book, that by 'induction' he does not mean a method which collects data and generalizes them, but a method of assessing the 'degree of confirmation' which a freely-invented theory obtains by being empirically tested.

'If, for instance,' Carnap writes, 'a report of observational results is given, and we want to find a hypothesis which is well confirmed and furnishes a good explanation for the events observed, then there is no set of fixed rules which would lead us automatically to the best hypothesis or even a good one. *It is a matter of ingenuity and luck for the scientist to hit upon a suitable hypothesis; and, if he finds one,*

---

[7] I wish to make it quite clear, on this occasion, that my 'degree of corroboration' (or confirmation, or acceptability) can be defined in terms of *logical* probability, which thus turns out to be perfectly adequate for this purpose. But it is not, of course, identical with logical probability or any other function satisfying the probability calculus.

he can never be certain whether there might not be another hypothesis which would fit the observed facts still better even before any new observations are made. This point, the impossibility of an automatic inductive procedure, has been especially emphasized, among others, by Karl Popper . . . who also quotes a statement by Einstein . . .'[1]

A little later, he writes:

'Given: a sentence e as evidence; wanted: a hypothesis h which is highly confirmed. . . . There is no effective procedure for solving these problems; that is the point emphasized by Einstein and Popper, as mentioned above.'

The problem which inductive logic should try to solve is, according to the programme of Carnap, not the discovery of hypotheses, but the determination of their degree of confirmation:

'Given: two sentences e and h. Wanted: the value of . . . the degree of confirmation of h on the evidence e.'

Now this programme clearly takes due account of the method of hypotheses; and I fully agree with everything that has been said by Carnap in the passages quoted (except, of course, that I could not accept the interpretation of 'degree of confirmation' as probability in the sense of the calculus of probability, and least of all as an 'inductive probability').

But this well-defined programme clashes in the most astonishing way with its execution. First, in the course of his book, Carnap introduces a method which allows us to 'discover', or rather to calculate, the *best* confirmed hypothesis, i.e., the most probable hypothesis, on any given evidence. But this means, of course, that we are no longer free to invent a hypothesis; or rather, that there never was any point in racking our brains, trying to find a good hypothesis, since we could simply have calculated the best hypothesis.

Thus the method of hypothesis becomes completely redundant, in direct contradiction to Carnap's programme. And without explicitly rejecting his original programme, Carnap himself admits this redundancy. At the end of the book, in a section entitled 'Are Laws Needed for Making Predictions?' (pp. 574 ff.), he writes, after giving a summary of his results, and after showing that universal 'laws' or theories are not really needed by the scientist (Mr X):

[1]*Logical Foundations of Probability*, pp. 192 f. (Italics mine.) The next two passages are from pp. 194f., and 196.

'Thus we see that X need not take the roundabout way through the law l at all, as is usually believed; he can instead go from his observational knowledge . . . directly to the singular prediction . . . .' He continues:

'Customary thinking in everyday life likewise often takes this short cut, which is now justified by inductive logic.'[2] And he concludes:

'We see that the use of laws is not indispensable for making predictions. Nevertheless it is expedient, of course, to state universal laws in books on physics, biology, psychology, etc.'

Since according to his philosophy of science there is no other task for pure science than to predict, this means that scientific laws have no function at all; they are completely redundant. In fact, Carnap's argument shows that they are unnecessary detours, and should therefore be abandoned by applying to them either Ockham's razor or Mach's principle of economy (which, it so happens, was used by Mach originally to show the usefulness of universal laws). But all this means the redundancy of the hypothetico-deductive method. There is no longer any need for 'ingenuity and luck' in order 'to hit upon a suitable hypothesis': the kind of suitable hypothesis meant was a universal theory, and universal theories are no longer 'needed'; they are no longer 'indispensable'. In short, the belief that the method of science is the method of hypotheses was all a mistake.

I am inclined to apply to this result the transcendental argument, if only in its negative use. A theory of science which rejects the hypothetico-deductive method by telling us that scientific theories are redundant is 'entirely inadequate'; in precisely the same way as is a theory which tells us that we cannot learn from experience.

18. *No Point in Testing a Theory.*

The result which I rejected in the preceding section was explicitly formulated by Carnap, but it is not at all peculiar to his theory.[1] In one way or another the various probability theories of induction all

---

[2] I need not say how much I disagree with this assertion, but it may be worth referring here to a similar claim by J. S. Mill, *Logic*, book ii, Chapter iii, 3: 'All inference is from particulars to particulars.'

[1] Russell comes near to an explicit formulation in his *Human Knowledge*, 1948, p. 435 (part v, vii, E).

establish too much, and in most cases more than is intended; they do not merely give an estimate of the degree to which a theory has been tested, but they dictate in every case which theory we ought to accept as the best theory, that is to say, the most probable theory. As a result, the question of *testing* a theory no longer arises. It becomes pointless. Induction turns out not to be a method of testing a conjecture or of assessing the result of the test; rather, it becomes a method of calculating the best theory by an inductive inference.[2]

This is due to the application of the simple rule of induction, in any of its various forms (see section 5, note 1, above).

The simple rule of induction tells us what the probability of the next case is. If by '*a*' we mean the statement that the next case or instance has the property *A*, and by '*b*' our past experience which involves $m/n = r$ instances of *A*, then the simple rule tells us that we have, approximately

$$p(a,b) = m/n = r$$

and that the approximation gets better with increasing *n*.

As far as this formula interests a scientist, its main point is that it may allow us to derive a *frequency hypothesis*. This may be done, if we can assume independence, by way of Bernoulli's theorem. Without this theorem, it would have to be postulated that '*the most probable frequency*' of instances bearing the property *A* in any class of instances is equal to the probability we have calculated by the inductive rule, i.e., equal to *r*.

We may thus call *r* the 'best estimate of the frequency of *A* in any future class'.

If we wish by our formalism to distinguish the inductive probability

$$p(a,b) = r$$

from the best estimate of the corresponding frequency (which is

---

[2][*(Added 1980) According to Carnap's account (as developed in his *Logical Foundations of Probability*), probability logic, although a generalization of deductive logic, does not consist 'in making inferences, but rather, in assigning probabilities' (see his remarks in I. Lakatos, ed.: *The Problem of Inductive Logic*, 1968, p. 311). However, the point of assigning probabilities is, very clearly, to choose that hypothesis from some class of possible hypotheses which has the higher probability assigned to it. But this process of choosing can of course be described as one of drawing an inductive inference.]

equal to it), then we may denote the *inductive probability,* calculated by the inductive rule, by 'p_i', and we may write

$$p_i(a,b) = r$$

for 'The inductive probability (given b) that the next instance will have the property A is equal to r'.

In contradistinction we may write 'p_s' for the best estimate of the *statistical probability,* or relative frequency. In order to simplify our symbolism, we can agree to read the letter 'a' in a p_s-expression slightly differently from the letter 'a' in a p_i-expression. That is to say, we shall write

$$p_s(a,b) = r$$

to mean: 'The best estimate (given b) of the statistical probability or the relative frequency, of instances bearing the property A in any future class[3] of instances is equal to r.'

Now if Bernoulli's theorem is applicable, we obtain

(1) $$p_s(a,b) = p_i(a,b);$$

and if we cannot prove this (because we do not know enough about the independence of the conditions described by b, for example, whether b demands mixing of the balls in an urn), then we can still introduce the formula (1) as a *postulate,* or perhaps by a definition. It can be expressed, more briefly, by

(2) $$p_s = p_i.$$

This formula is the basis of all inductive estimates of frequencies, so far as I am aware.[4] (Of course, if $p_s = p_i$ is introduced as a postulate, it amounts to postulating that b demands independence.)

Now all our objective hypotheses *may* be interpreted as fre-

---

[3] I shall not discuss the problem raised by the phrase '*any* future class' (although this phrase leads to paradoxical consequences) because other criticisms which I wish to make seem to me more fundamental.

[4] The formula T 106-1, +c, on p. 551 of Carnap's *Logical Foundations,* means exactly the same as our (1) or (2), if translated into our formalism. In this formula, Carnap does not make use of his expressions 'probability$_1$' (or 'p_1') which corresponds to our 'p_i'; and 'probability$_2$' (or 'p_2') which corresponds to our 'p_s', if we consider only the '*best estimate*'. The formula T 106-1, +c, means in effect something like '$p_1 = p_2$,' (more precisely, 'the best p_2 equals p_1') as my translation (2) shows. Carnap's definition establishing this equation is + D 100-1 on p. 525; cf. (3) on p. 169.

quency estimates. (Not that I now recommend this particular interpretation: I prefer the propensity interpretation to the frequency interpretation. But this point does not matter here; moreover, we would obtain precisely the same result for 'p_p' instead of 'p_s', i.e., for the best estimate of the propensity.) Thus (1) or (2) may be said to *explain* (or 'explicate') *objective probability hypotheses in terms of subjective probability.* To do this was, after all, the main task of the subjective theory; and (1) expresses the (alleged) fact that its task has been carried out. Thus a formula like (1) is essentially implied in all subjective theories, even though it is seldom explicitly stated (as it is, indeed, in Carnap's book[5]).

Inductive logicians believe in an inductive logic, and accordingly interpret inductive probability as *logical* probability. This means that in their view, every *true formula* of the form

$$p_i(a,b) = r$$

must be tautologous or analytic, and every *false* formula of this form self-contradictory.

Now we can derive from (1) immediately

(3) $$p_s(a,b) = r \text{ if and only if } p_i(a,b) = r.$$

This formula must be analytic if (1) is analytic; and so we find: if p_i is not only an inductive but also a logical probability, then every formula of the form

(4) $$p_s(a,b) = r$$

must either be tautologous (analytic) or self-contradictory; and if true, it must be tautologous.

But this implies that *all our 'best' objective probability hypotheses are tautologous.*

This result is clearly absurd.[6] For it destroys the possibility of

---

[5] See the preceding footnote. The derivation of the formula in Carnap's book is highly complex.

[6] It is precisely analogous to the situation in other subjectivist epistemologies, for example to that in Carnap's *Der Logische Aufbau der Welt*, as I have shown in my paper 'The Demarcation Between Science and Metaphysics' in the Carnap volume of the *Library of Living Philosophers*, ed. by P. A. Schilpp; see there section 2, text to note 27; see also *Conjectures and Refutations*, Chapter 11; the language of the *Logische Aufbau* also contains only statements which are either tautologous or contradictory.

genuine hypotheses: estimates such as these cannot be *tested* by experiments.

We may take this as another refutation of the view that inductive probability is logical probability, or in other words, that all true formulae of the inductive calculus are tautologous.

Against this criticism, it may be objected that we can obtain from (4) a genuinely empirical statement if we apply to it the 'rule of absolution' discussed above in section 11. This objection contains the tacit suggestion[7] that the 'rule of absolution' should be extended to *statistical estimates*. This extension, which would be a most dubious procedure, would mean that, *b* being our total *present* knowledge, we would obtain from *b* and $p_s(a,b) = r$, by the rule of absolution, the following:

(5) 'At the present moment, the best estimate of the frequency (within any class of instances) of instances bearing the property *A* is that this frequency equals *r*.'

Let us briefly discuss this suggestion and its results.

(i) The statement (5) is extremely queer from the point of view of the frequency theory, for owing to the application of the rule of absolution, *no conditions for the result A are stated*. (In other words, any reference class is admitted, which either leads to contradictions or makes the frequency theory inapplicable. This difficulty is fundamental and cannot be removed by taking *a* to be a conditional statement, and *A* to be a conditional property.)

(ii) Even if we overlook all these problems, the following situation remains.

Admittedly (5) becomes empirical upon the suggestion discussed: it has been obtained from (4) by applying the rule of absolution together with an empirical report—the report that *b* is our total present knowledge. But this means that (5) is *logically entailed by this empirical report*. It is therefore no hypothesis, no bold invention, no conjecture, no guess which may be submitted to tests. Rather it is *imposed* upon us by our past experience; it can be *inferred* from our past experience.

But if this is so, if (5) is logically entailed, with certainty, by our past knowledge, then it can be no more than part of this knowledge.

---

[7] I find this tacit suggestion contained in R. Carnap's paper 'Probability as a Guide in Life', *Journal of Philosophy* 44, 1947, p. 144 (end of the second paragraph).

Therefore, (5) *does not speak about future cases at all*—in spite of its verbal formulation. It is simply a result of attaching, by a completely arbitrary decision, the name 'best estimate of the future' to a summary report about the past.

All these are consequences of the attempt to escape from the unpalatable consequences of the view that (4) is analytic. If we give up this view then we arrive at a theory which is essentially that of Reichenbach. According to Reichenbach, we must accept the inductive rule, or the formula (1), on the basis of a transcendental argument. The whole procedure of arriving at an estimate $p_s(a,b) = m/n = r$ is not, as he emphasizes, a tautological procedure; on the contrary, he calls it '*a wager*'. By this he wishes to stress that we incur a grave risk in adopting $p_s(a,b) = m/n = r$ as our best estimate. Yet, he argues transcendentally, we *must* adopt it, for it is the only reasonable way open to us, the *best* way. Every other estimate would be worse, and arbitrary. This is a view that makes

$$p_s\,(a,b) \ = \ r$$

a genuine hypothesis, with the uncertainty of a genuine hypothesis; and it thus escapes our criticism that it is analytic and *therefore* untestable.

Yet although Reichenbach's 'wagers' are certainly not analytic, they are no more testable than they would be if they were analytic.

For what would a test consist of? It would have to produce another instance, which would either have the property A or not. If it has the property A, then we obtain a newly determined r such that $r = (m + 1)/(n + 1)$. If it does not have the property A, then r becomes equal to $m/(n + 1)$. In both cases we again have to adopt the value determined by (1). No test can ever get us anywhere else. A new test only changes b, leading automatically to that value of $p_s(a,b)$ which corresponds to the new b.

Thus again, the hypothetical method is abandoned. We cannot freely conjecture r, and then test our conjecture; instead, the value of r is uniquely determined by past experiences. Theories are not freely invented hypotheses—they are inferred by means of an inductive rule. So the concluding transcendental criticism of the foregoing section applies here again. We must reject an analysis of scientific method which makes nonsense of the method of hypotheses.

19. *Summary of this Criticism.*

In my criticism of the subjective theory of probability, I have tried to ignore minor blemishes and to concentrate on its main ideas: its distinction between *complete knowledge* (if we can derive either *a* or *ā* from the evidence *b*) and *partial knowledge or probable knowledge* (if the evidence *b* is insufficient for complete knowledge); its interpretation of probability as a generalization of deductive logic; its further and very different interpretation of *induction* as a generalization of deduction (and thus of logic); its consequent identification of inductive probability and logical probability; its attempt to obtain a principle of induction, in the form of a principle of inductive probability (such as the simple inductive rule), and to interpret it as a principle of logical probability; and ultimately, its perhaps somewhat unintended result that predictive probabilistic hypotheses are fully determined by our past observations, in accordance with the simple inductive rule (provided we want, as undoubtedly we do, to have the best hypotheses or 'estimates').

That is the subjectivist programme. With one exception, every single one of these ideas is mistaken. I am ready to concede that there is a tenable *logical interpretation* of the probability calculus which takes it as a generalization of deductive logic. But this logical interpretation offers no support for induction; and all the other ideas belonging to the subjectivist programme lead to contradictions and to paradoxes.

I have omitted here many points of detailed criticism which I have given elsewhere,[1] and a number of further paradoxes which are of comparatively limited interest.

I have undertaken this task, not because I wished to combat certain philosophical theories, but mainly because of the surprising part played by the subjective theory within physics itself: in physics and in the philosophy of physics, the mode of discussing probability problems consists very largely of a continuous shift from objective problems to subjective arguments, and back again to objective solutions. Without this continuous shift from one interpretation to

---

[1] See my paper 'The Demarcation Between Science and Metaphysics', referred to above, note 3 to section 19; my three papers in *Brit. Journ. Philos. Science* **5**, 1954, pp. 143 *ff*; **6**, 1955, pp. 157 *ff*. and **7**, 1956, pp. 249 *ff*; and Appendices *vii to *ix of *L.Sc.D.* See also 'On Carnap's Version of Laplace's Rule of Succession', *Mind* **71**, 1962, pp. 69–73, and 'The Mysteries of Udolpho: A Reply to Professors Jeffrey and Bar-Hillel', *Mind* **76**, 1967, pp. 103–110.

the other, the weakness of the subjective theory would have been discovered long ago. This oscillation is as frequent among physicists (Einstein included) as it is among philosophers.

The part played by the subjective theory in physics can be largely explained by the fact that determinist views or remnants of determinist views are still influential, even among advocates of indeterminism: determinism leads to a subjective interpretation of probability in physics, simply because it does not leave any room for an objective interpretation of probability.[2]

The weakness of the subjective theory of probability is shared by every subjectivist (sensualist, phenomenalist, solipsist, etc.) epistemology. By a subjectivist epistemology I mean the attempt to answer the question 'How do you know?'[3] in the sense of[4] 'What is the basis of your assertion? What observations led you to [it]. . . ?' Subjectivist and inductivist questions like these beg the usual subjectivist and inductivist answers. My own answer would be, 'I do *not* know: my assertion was merely a guess. Never mind the observations which may have led me to it. Instead, you may help me, by criticizing my assertion, and by using your ingenuity in designing some experimental tests which may refute my assertion if it is mistaken, as it well may be.'[5]

Mere guesses are, however, still unfashionable. Scientific theories are supposed to be more directly 'based' on past experience, and at best it is admitted that induction involves a gamble. I fail to see why a gamble should be better than a guess; especially if we remember that the subjective-inductive theory begins to look a little like a blind gamble.

But the general topic of subjectivism is one which would lead us away from our real task in this Part: the analysis of probability.

---

[2][See Volumes II and III of the *Postscript*, where problems of determinism and physics are discussed in detail. Ed.]

[3]This question is asked by Jeffreys, in the sense stated here, in *Theory of Probability*, 1st edition, p. 34; 2nd edition, p. 33.

[4]The two following leading questions are asked by Carnap in his *Logical Foundations*, p. 189; see also section 4, above, from footnote 1 to the end of the section.

[5]See also section 27 of *L.Sc.D.*, especially notes 2 and *1, and the corresponding text. [See also 'On the Sources of Knowledge and of Ignorance', in *Conjectures and Refutations*, pp. 3–30. Ed.]

Addendum (January 1981). A Brief Summary of the Criticism of Probabilistic Induction.

I

What seduces so many people to accept probabilistic induction are two things:

(A) The *valid* but *misinterpreted* intuitive idea that *it cannot be just due to an improbable accident* if a hypothesis is again and again successful when tested in different circumstances, and especially if it is successful in making previously unexpected predictions. (If not 'due to an improbable accident', we are inclined to reason, invalidly, then it must be due to the *high probability of the hypothesis*. See below.)

(B) The indubitable fact that, according to the probability calculus, the probability of a statement increases (unless it was zero to start with) with the accumulating evidence in its favour, especially with the accumulation of successful predictions. In terms of the probability calculus: Let h be a hypothesis whose initial probability (or prior probability) $p(h)$ may be as small as we like, as long as it is different from zero. Let e be some evidence in its favour. Then we have not only

$$p(h,e) > p(h)$$

but also

$$p(h,e_2) > p(h,e_1)$$

if e_2 contains favourable evidence not yet contained in e_1. Thus the probability of h continues to increase with accumulating favourable evidence.

This undeniable and seductive fact which follows from the calculus of probability has seduced many people into thinking that the probability $p(h,e)$ of any hypothesis h must tend to 1 with accumulating favourable evidence e.

The Arguments (A) and (B) have been powerful with many people, including many philosophers of science.

I will first show that (B), even though it is valid, is totally misleading. (B) relies on the comparison of the change of the probability of *one* hypothesis h with changing evidence. Things look utterly different if we compare *two or more* hypotheses with one constant evidence e, favourable to all of them. We then find, immediately, that the influence exerted by the evidence upon the probability has nothing whatsoever to do with induction.

This will be shown here in section II.

II

Let h_1, h_2, ... , h_n be a sequence of competing and pairwise incompatible hypotheses. Let e be the relevant evidence (the successful predictions) derivable from each of the n hypotheses so that we have $p(h_i, e) = p(h_i)$ and $0 \neq p(e) \neq 1$. We then get, immediately, from the general multiplication theorem of the calculus of probability (see *L.Sc.D.*, p. 324, (1)), as here in section I above,

$$(1) \qquad p(h_i,\ e) > p(h_i);$$

that is to say, the favourable evidence e increases the probability.

But we also get the trivial yet fundamental theorem (2)

$$(2) \qquad p(h_i,\ e) < p(h_j,e) \text{ if, and only if, } p(h_i) < p(h_j).$$

Theorem (2) is shattering. It shows that the favourable evidence e, even though it raises the probability according to (1), nevertheless, against first impressions, leaves everything precisely as it was. It can never favour h_i rather than h_j. On the contrary, the order which we attached to our hypotheses before the evidence remains. It is unshakable by any favourable evidence. The evidence cannot influence it.

Why is this shattering? Because h_1 may be a typical 'inductive generalization' of e, while h_2 may make assertions, in addition to e, which are completely unsupported by e, so that h_2 is far from being a generalization of e; or vice versa.

Take the following examples.

e reports of a million of observed swans that they are white, and that no swans other than white ones have been observed.

h_1 says 'All swans are white.'

h_2 says 'Swans in Greece and Italy and France are white; in

England and Scandinavia they are red; in Central Asia they are green; in Africa they are blue; and in Australia they are black.'
b, the background knowledge, implies that no swan has been observed outside Greece, Italy and France.

Now by making h_1 apply to the whole universe (remember that we conjecture that the electronic charge is constant in the whole universe) may make it *initially* equally or even less probable than h_2. (Or we can weaken h_2 to make it more probable.) So we can assume $p(h_1) \leq p(h_2)$; and we can also assume $p(h_1, b) \leq p(h_2, b)$, from what we have said about b. At any rate, we shall have, from the multiplication theorem, since e follows from both h_1, b and h_2, b,

$$\frac{p(h_1, eb)}{p(h_2, eb)} = \frac{p(h_1, b)}{p(h_2, b)}$$

and thus

$$p(h_1, eb) < p(h_2, eb) \text{ if and only if } p(h_1, b) < p(h_2, b).$$

Thus the evidence e does not support the straightforward generalization h_1 any better than the fantastic and arbitrary hypothesis h_2 (fantastic and arbitrary in the light of evidence e).

This shows that the argument (B) in section I and the relation (1) in this section are wrongly interpreted if they are taken as supporting *induction*, or *inductive generalization*. And it shows that the calculus of probability is of no use whatever as a *theory of induction*.

Of course, the evidence e can change the probability of a hypothesis to which it is *not* favourable. It can, more especially, knock out a hypothesis if it is incompatible with it. But this is not what we are interested in at this point. (And although the evidence will reduce to zero the probability of a hypothesis which it refutes, we obviously do not need the probability calculus for describing this situation.)

III

To make matters more concrete, let h_1 and h_2 be Newton's and Einstein's theories of gravitation, and e the evidence available in 1917 (we can include even the evidence concerning the minute movement of the perihelion of Mercury, which nobody took then as a serious problem for Newtonian theory), so that we have $p(h_1e)$

$= p(h_1)$. We do not know whether to put $p(h_1)$ greater, smaller or equal to $p(h_2)$; but as long as e can be explained by both theories, the empirical evidence e cannot, according to (2) have any bearing on the comparative probabilities.

Now this situation is typical: there always exists for every theory h_1 and for any evidence e a theory h_2 which is related to h_1 as is Einstein's theory to Newton's. The theory h_2 may not be 'compact' or 'beautiful' or 'simple', but this is a different question. It can always be constructed.[1] By the same method we can construct a further h_3, incompatible with h_1 and also with h_2, etc. This consideration makes it quite clear that if we make the prior (or 'absolute') probabilities $p(h_1)$, $p(h_2)$, not too different to start with, we would never obtain a probability as large as 1/2 for any hypothesis given any supporting evidence. (Indeed if we start, as Bayesians should, with something like $p(h_1) = p(h_2) = p(h_3) = \ldots$, an assumption defended by prominent Bayesians by the principle of maximising 'probabilistic entropy', we should get $p(h_1) = 0$; but this is not my point here, although it can be used to support a point made in L.Sc.D., Appendix *vii.) My point here is that as we can never rationally attribute a probability even approaching 1/2 to any hypothesis, given the most excellent evidence; probability in the sense of the probability calculus cannot be the instrument we are after in order to explain our intuition formulated here in section I, (A). It should be clear that we were misled by (B). For a probability less than 1/2 is, of course, an *improbability*. So we cannot explain our inductive intuition by a probability calculus which shows us that our hypotheses will always remain improbable, and that the empirical evidence does not affect the order of the probability of the hypotheses except of those hypotheses which cannot explain the evidence.

Note:
Nothing that has been said here is affected if we introduce throughout, in addition to e, a variable b that covers our background knowledge and the initial conditions, so that we obtain

---

[1] Remember that e is finite and therefore refers to a limited range of velocities, or of forces, or of energies, or what not. Thus we only need to choose an h_2 which within that limited range agrees with h_1 but deviates a little from h_1 outside that range.

(1') $\qquad p(h, be) > p(h,b)$

and

(2') $\qquad p(h_1, be) < p(h_2, be)$ if and only if $p(h_1, b) < p(h_2, b)$.

IV

Our theorem (2) of section II shows, I think, that the results of section 15 are sound. Indeed, the evidence *e* gives no (probabilistic) *inductive* support to the hypothesis. It supports nothing within the hypothesis *except* that part of the content of the hypothesis which covers precisely *e* and no more. It is exactly as Hume saw it. *Probability theory does not have anything ampliative in it: there is no probabilistic ampliative induction.*

Nevertheless, the intuition formulated in section I, (A), is valid. If a theory *h* has been well corroborated, then it is highly probable *that it is truth-like. That is to say, that it agrees well with some of the facts.* This is highly probable in the sense that it is extremely improbable that the predictive success of a powerful and well-tested theory is a mere accident.

But it does not make *h* 'probable': for to say that *h* is probable is to say that *it is more probable than not* that *h* is *true*. This would mean that it is more probable than not that *h* agrees with *all the facts in the world*: that there exists no counter example, no fact that contradicts it. But no finite evidence *e* can ever tell us that.[2]

[2]See my *Objective Knowledge*, p. 102.

REMARKS ON THE OBJECTIVE THEORIES OF PROBABILITY

I N this chapter, I propose to explain the way in which the frequency theory of probability has been superseded by the propensity interpretation and also by the 'measure-theoretical approach', as it is now often called. I shall try to show that, even though this approach supersedes the frequency theory, it also lends it a kind of *post mortem* justification; for the frequency theory becomes *'almost deducible'* from the measure-theoretical approach. That is to say, within the measure-theoretical approach it can be shown that the probability is zero that we may accidentally hit upon a random sequence that violates the demands of the frequency theory of von Mises. In other words it is *'almost certain'* that these demands will be satisfied.[1]

The main significance of this new approach lies in the fact that measure-theoretical probability statements are *singular* probability statements: statements that assert what we may call a *'singular probability'*. Yet *from the point of view of physics, a singular probability which 'almost entails' a frequency can be best interpreted as a physical propensity.* Thus the mathematical transition from frequency to measure theory corresponds, I suggest, to *a transition from the statistical to the propensity interpretation of objective physical probabilities.*

I begin this chapter by a section explaining more fully the propensity interpretation and the reasons for adopting it. But chiefly the

---

[1]These results are due to F. P. Cantelli (1916 and 1917) from whose 'strong law of large numbers' the limit axiom of Venn and von Mises 'almost follows' (practically the same result was found by Harold Jeffreys and Dorothy Wrinch in 1919 and G. Pólya in 1921); and to J. L. Doob who proved in 1936 the same for the axiom of randomness (after a more restricted theorem had been proved by E. Hopf in 1934). Retrospectively one can say that both theorems are foreshadowed by E. Borel's theorems on 'normal' and 'entirely normal' numbers (1909).

chapter consists of historical remarks on some of the developments, since 1934, of the frequency theory of probability, which played such a prominent role in *L.Sc.D.*, more especially in connection with the problems of the consistency of von Mises's theory.[2] It is shown that A. H. Copeland's and my own assumptions (which I had tried to reduce to a minimum) were strengthened in 1935 by A. Wald, who first succeeded in proving that a theory was consistent which was approximately as strong as von Mises's theory. Wald's theory was further strengthened by A. Church, who established in 1940 the consistency or non-emptiness of a class *C* of collectives which are insensitive to selections according to *any effectively calculable gambling system* (or set of gambling systems).[3] This development seems to me to be of some importance since it establishes that a very satisfactory frequency theory can stand on its own feet, and that the situation is not such as to force us to adopt the measure-theoretical approach; though we may adopt it freely for its superior merits. Moreover, it is valuable to know that we may speak of frequencies, even in unlimited sequences, without any fear of inconsistency; for frequency predictions—which 'almost follow' from propensity hypotheses—remain decisively important for us: only the predicted frequencies allow us to *test* these hypotheses.

20. *The Case for Propensities.*

The subjective interpretation of probability may *perhaps* be tenable as an interpretation of certain gambling situations—horse racing, for example—in which the objective conditions of the event are

[2]My theory of *n*-free or *n*-insensitive sequences (*'suites indifférentes'*) was shown, by Jean Ville, in his *Étude Critique de la Notion de Collectif* (Paris, 1939) pp. 70–83, to be equivalent to the theory of admissible numbers, first developed by A. H. Copeland, 'Admissible Numbers in the Theory of Probability', *Am. J. of Math.* **50**, 1928, and in many papers since. Copeland operates with what I have called (in section 21) *'ordinal selection'*, in contradistinction to 'neighborhood selection' which is the basis of my '*n*-insensitiveness' or '*n*-freedom'.

[3]Denoting the set of Copeland's 'admissible numbers' by *A* (these are also my own 'absolutely free' sequences); Wald's collectives by *W*; von Mises's collectives—a somewhat vague concept—by *M*; and Church's random sequences by *C*, we find that *A*, *W*, *M*, and *C* form a decreasing sequence of sets. (I could not find any reference to Church in von Mises's later writings on the subject, such as the third German edition of his *Probability Statistics and Truth*, 1951, or his discussion with Doob, *Ann. Math. Statist.* **12**, 1941, pp. 191–217.)

ill-defined and irreproducible. (I do not really believe, however, that it is applicable even to situations like these: a strong case could be made—if it were worth making—for the view that what a gambler, or a 'rational better', tries to find out, in order to bet, are the *objective* conditions, the *objective* propensities, the *objective* odds of the event. Thus the man who bets on horses is anxious to get more information about horses—rather than information about his own state of belief, or about the logical force of the information in his possession.) Yet in the typical game of chance—roulette, say, or dice, or tossing pennies—and in all physical experiments, the subjective interpretation fails completely, as we have seen. For in all these cases probabilities depend upon the *objective conditions of the experiment.*

In *L.Sc.D.*, I considered only *one* objective interpretation of probability—the purely statistical or frequency interpretation. (I take these two designations to be synonymous.) And I attempted to reconstruct this interpretation by taking account of all the criticisms usually advanced against it. Here I wish to suggest that there is a second objective interpretation, and a better one—the propensity interpretation. My suggestion is not motivated by the belief that any of the usual criticisms of the frequency theory are justified. On the contrary, I do not doubt the consistency of the frequency theory, and I believe that it is possible to work with it. But I also believe that the propensity interpretation is decidedly preferable. Some of my reasons for believing this will be discussed in the present section, while others will be left for later.

In this section, the discussion will be confined solely to the problem of interpreting the probability of 'singular events' (or occurrences), and it is the frequency theory of the probability of *singular events* which I have in mind whenever I speak here of the frequency interpretation of probability, in contradistinction to the propensity interpretation.

It will be remembered that from the point of view of the frequency interpretation the probability of an *event of a certain kind*—such as obtaining a six with a particular die—is *nothing but* the relative frequency of this kind of event in an extremely long (perhaps infinite) sequence of events. And if we speak of the probability of a *singular* event (that is, an 'occurrence', in the sense of section 23 of *L.Sc.D.*; see also section 71), such as the probability of obtaining a six in the third throw made after 9 o'clock this morning with this die, then, according to the purely statistical interpretation, we mean to say *only* that this third throw may be regarded as a member of a sequence of throws, and that, in its capacity as a member of this sequence, it shares the probabilities of that sequence; that is to say, the probabilities which are *nothing but the relative frequencies* within that sequence.

In the present section I propose to argue against this interpretation, and in favour of the propensity interpretation, by making use of the results reached in the preceding chapter concerning *the significance of the objective experimental conditions for the interpretation of probabilities.* I propose to proceed as follows. (1) I will first show that, from the point of view of the frequency interpretation, objections must be raised against the propensity interpretation which seem to make the latter unacceptable. (2) I will next give a preliminary reply to these objections; and I will then present, as point (3), a certain difficulty which the frequency interpretation has to face, though it does not, when first raised, look like a serious difficulty. (4) Ultimately I will show that in order to get over this difficulty, the frequency interpretation is forced to adopt a modification which appears to be slight at first sight; yet the adoption of this apparently slight modification turns out to be equivalent to the adoption of the propensity interpretation.

(1) From the point of view of a purely statistical interpretation of probability it is clear that the propensity interpretation is unacceptable. For propensities may be explained as possibilities (or as measures or 'weights' of possibilities) which are endowed with tendencies or dispositions to realize themselves, and which are taken to be responsible for the statistical frequencies with which they will in fact realize themselves in long sequences of repetitions of an experiment. Propensities are thus introduced in order to help us to explain, and to predict, the statistical properties of certain sequences; and *this is their sole function.* Thus (the frequency theorist will assert) they do not allow us to predict, or to say, *anything whatever* about a singular event, except that its repetition, under the same conditions, will generate a sequence with certain statistical properties. All this shows that the propensity interpretation can add nothing to the frequency interpretation except a new word—

'propensity'—and a new image or metaphor which is associated with it—that of a tendency or disposition or urge. But these anthropomorphic or psychological metaphors are even less useful than the old psychological metaphors of 'force' and 'energy' which became useful physical concepts only to the extent to which they lost their original metaphysical and anthropomorphic meaning.

This, roughly, would be the view of the frequency theorist. In defending the propensity interpretation I am going to make use of two different arguments: a preliminary reply (2), and an argument that amounts to an attempt to turn the tables upon the frequency theorist; this will be discussed under (3) and (4).

(2) As a preliminary reply, I am inclined to accept the suggestion that there is an analogy between the idea of propensities and that of forces—especially fields of forces. But although the labels 'force' or 'propensity' may both be psychological or anthropomorphic metaphors, the important analogy between the two ideas does not lie here; it lies, rather, in the fact that both ideas draw attention to *unobservable dispositional properties of the physical world,* and thus help in the interpretation of physical theory. Herein lies their usefulness. The concept of force—or better still, the concept of a field of forces—introduces a dispositional physical entity, described by certain equations (rather than by metaphors), in order to explain observable accelerations. Similarly, the concept of propensity, or of a field of propensities, introduces a dispositional property of singular physical experimental arrangements—that is to say, of singular physical events—in order to explain observable frequencies in sequences of repetitions of these events. In both cases the introduction of the new idea can be justified by an appeal to its usefulness for physical theory. Both concepts are 'occult' in Berkeley's sense, or 'mere words'. But part of the usefulness of these concepts lies precisely in the fact that they suggest that the theory is concerned with the properties of an *unobservable* physical reality and that it is only some of the more superficial effects of this reality which we can observe, and which thus make it possible for us to test the theory. (See above, sections 11 to 15.) The main argument in favour of the propensity interpretation is to be found in its power to eliminate from quantum theory certain disturbing elements of an irrational and subjectivist character—elements which are more 'metaphysical'

than propensities and, moreover, 'metaphysical' in the bad sense of the word. It is by its success or failure in this field of application that the propensity interpretation will have to be judged.[1]

Having made this preliminary reply, I proceed to my main argument in favour of the propensity interpretation. It consists in pointing out certain difficulties which the frequency interpretation must face. We thus come to point (3), announced above.

(3) Many objections have been raised against the frequency interpretation of probability, especially in connection with the idea of infinite sequences of events, and of limits of relative frequencies. I shall not refer to these objections here because they will be discussed in the next section. (They will be found to be invalid, in the main.[2]) Yet there is a simple and important objection which has not, to my knowledge, been raised in this form before.

Let us assume that we have a loaded die, and that we have satisfied ourselves, after long sequences of experiments, that the probability of getting a six with this loaded die very nearly equals $1/4$. Now

[1] [See Volume III of the *Postscript*. Ed.]

[2] Thus it is not the usual criticisms of the frequency interpretation which have induced me to change my mind. The opposite has been suggested by W. C. Kneale (*Observation and Interpretation*, edited by S. Körner, 1957, p. 80). In his comments on an extract from the present section (*ibid.*, pp. 66–8), he made the following supposition concerning my reasons for advocating a propensity interpretation: 'More recently the difficulties of the frequency interpretation, *i.e.* the muddles, if not the plain contradictions, which can be found in von Mises, have become well known, and I suppose that these are the considerations which led Professor Popper to abandon that interpretation of probability.' I am not aware of any well-known 'muddles' or 'contradictions' in the frequency theory other than those discussed in *L.Sc.D.* in 1934. And these have all been cleared up long ago—by Wald, Copeland, Church, myself and others. (See also the next chapter.) I do not think that Kneale's criticism of the frequency theory in his *Probability and Induction*, 1949, presents a correct picture of the logical situation prevailing at any time since 1934. Yet there is certainly one point criticized by Kneale (see especially p. 156) which I did not discuss. It is the following: in the frequency theory, a probability equal to 1 does not mean that the event in question will occur without exception (under the conditions given). But this point does not constitute a weakness of the frequency theory, as Kneale asserts. Rather, it *necessarily holds good in every adequate probability theory which allows for application to infinite classes*. (This has been shown in *L.Sc.D.*, note 14, Appendix *vii.) It holds good, of course, in the propensity interpretation. All the other criticisms mentioned in Kneale's book had been dealt with in mine, as far as I can see; and I feel confident that none of his points constitutes a valid criticism of either Wald's or Church's or my version of the frequency theory. My reasons for adopting the propensity interpretation are therefore quite different from those supposed by Kneale.

consider a sequence *b*, say, consisting of throws with this loaded die, but including a few throws (two, or perhaps three) with a homogeneous and symmetrical die. Clearly, we shall have to say, with respect to each of these few throws with this fair die, that the probability of a six is 1/6 rather than 1/4, in spite of the fact that these throws are, according to our assumptions, *members of a sequence* of throws with the statistical frequency 1/4, and in spite of the fact that two or three throws cannot possibly influence the frequency 1/4 of the long sequence.

I believe that this simple objection is decisive, even though there are various possible rejoinders.

One rejoinder need be mentioned only in passing, since it amounts to an attempt to fall back upon the subjectivist interpretation of probability. It amounts to the assertion that it is our special *knowledge*, the special *information* we have concerning these throws with the fair die, which changes the probability. I need hardly say that I consider this reply unjustified, in view of my general discussion of the subjective theory. Moreover, the case before us suggests a further argument (although not a very important one) against this subjective theory. For we may not know which of the throws are made with the correct die, although we may know that there are only two or three such throws. In this case it will be quite reasonable to bet (provided we are determined to bet on a considerable number of throws) on the basis of a probability very close to 1/4, even though we do know that there will be two or three throws on which we should not accept bets on these terms, if only we could identify them. We know that in the case of these throws, the probability of a six is less than 1/4—that it is, in fact, 1/6; but we also know that we cannot identify these throws, and that their influence must be very small if the number of bets is large. Now it is clear that as we nevertheless attribute to these unknown throws a probability of 1/6, we do not mean by the word 'probability', and cannot possibly mean by it, a 'reasonable betting quotient in the light of our total actual knowledge' (as the subjective theory has it).

But let us now leave the subjective theory aside. What can the frequency theorist say in reply to our objection?

Having been a frequency theorist myself for many years, I know fairly well that my own reply would have been along the following lines.

The description given to us of the sequence b shows that b is composed of throws with a loaded die and say three with a fair die. We estimate or, rather, we conjecture (on the basis of previous experience, or of intuition—it never matters what is the 'basis' of a conjecture) that the side six will turn up in a sequence of throws with the loaded die with the frequency $1/4$, and in a sequence of throws with the fair die with the frequency $1/6$. Let us denote this latter sequence, that of throws with the fair die, by 'c'. Then our information as to the composition of b tells us (i) that $p(a,b) = 1/4$, or very nearly so, because almost all throws are with the loaded die, and (ii) that bc—that is, the class of three throws belonging to both b and c—is not empty; and since bc consists of throws belonging to c, we are entitled to assert that the singular probability of a six, among those throws which belong to bc, will be $1/6$—by virtue of the fact that these singular throws are members of a sequence c for which we have $p(a,c) = 1/6$.

I think that this would have been my reply, by and large; and I now wonder how I could ever have been satisfied with a reply of this kind, for it now seems plain to me that it is utterly unsatisfactory.

Of course there is no doubt as to the compatibility of the two equations (pertaining to infinite sequences)

(i) $$p(a,b) = 1/4$$

(ii) $$p(a,bc) = 1/6;$$

nor is there any question that these two cases can be realized within the frequency theory: we *might* construct some sequence b such that equation (i) is satisfied, while in a selection sequence bc—a very long and virtually infinite sequence whose elements belong both to b and to c—equation (ii) is satisfied. *But our case is not of this kind.* For bc is not, in our case, a virtually infinite sequence. It contains, according to our assumption, exactly three elements. In bc the six may come up not at all, or once, or twice, or three times. But it *certainly* will not occur with the frequency $1/6$ in the sequence bc because we know that this sequence contains at most three elements.

Thus there are only two infinite, or very long, sequences in our case: the (actual) sequence *b* and the (virtual) sequence *c*. The throws in question belong to both of them. And our problem is this. Although they belong to both of these sequences, and although we only know that these particular throws *bc* occur somewhere in *b* (we are not told where, and we are therefore not able to identify them), we have no doubt whatever that in their case the proper, the true singular probability, is ¹/₆ rather than ¹/₄. Or in other words, although they belong to both sequences, we have no doubt that their singular probability is to be estimated as being equal to the frequency of the sequence *c* rather than *b*—simply because they are throws with a different (a fair) die, and because we estimate or conjecture that, in a sequence of throws with a fair die, the six will come up in ¹/₆ of the cases.

(4) All this means that the frequency theorist is forced to introduce a modification of his theory—apparently a very slight one. He will now say that an admissible sequence of events (a reference sequence, a 'collective') must always be a sequence of repeated conditions. Or more generally, he will say that admissible sequences must be either virtual or actual sequences which are *characterized by a set of generating conditions*—by a set of conditions whose repeated realization produces the elements of an independent sequence.

If this modification is introduced, then our problem is at once solved. For the sequence *b* will no longer be an admissible reference sequence. That part of it (i.e., $b\bar{c}$) which consists of throws with the loaded die will make an admissible sequence, and no question arises with respect to it. The other part, *bc*, consists of throws with a regular die, and belongs to a virtual sequence *c*—also an admissible one—of such throws. There is again no problem here. It is clear that, once the modification has been adopted, the frequency interpretation is no longer in any difficulty.

Moreover, it seems that what I have here described as a 'modification' only states explicitly an assumption which most frequency theorists (myself included) have always taken for granted.

Yet if we look more closely at this apparently slight modification, we find that it amounts to a transition from the frequency interpretation to the propensity interpretation.

The frequency interpretation always takes probability as relative to a sequence which is assumed as 'given'; and it works on the assumption that a probability is *a property of some given sequence*. But with our modification, the sequence in its turn is defined by its set of *generating conditions*; and in such a way that probability may now be said to be *a property of the generating conditions*.

But this makes a very great difference, especially to the probability of a singular event (or an 'occurrence'). For now we can say that the singular event a possesses a probability $p(a,b)$ owing to the fact that it is an event produced, or selected, in accordance with the generating conditions $b\bar{c}$, rather than owing to the fact that it is a member of a sequence b. In this way, a singular event may have a probability even though it may occur only once; for its probability is a property of its generating conditions: it is generated by them.

Admittedly, the frequency theorist can still say that the probability, even though it is a property of the generating conditions, is equal to the relative frequency within a virtual or actual sequence generated by these conditions. But if we think this out more fully it becomes quite clear that our frequency theorist has, inadvertently, turned into a propensity theorist. For if the probability is a property of the generating conditions (say, of the experimental arrangement) and if it is therefore considered as depending upon these conditions, then the answer given by the frequency theorist implies that the virtual frequency must also depend upon these conditions. But this means that we have to visualize the conditions as endowed with a tendency, or disposition, or propensity, to produce sequences with frequencies equal to the probabilities; which is precisely what the propensity interpretation asserts.

It might be thought that we can avoid the last step—the attribution of propensities to the generating conditions—by speaking of mere possibilities rather than of propensities. In this way one may hope to avoid what seems to be the most objectionable aspect of the propensity interpretation: its intuitive similarity to 'vital forces' and similar anthropomorphisms which have so often been said to be barren pseudo-explanations.

The interpretation of probabilities in terms of possibilities is of course very old. We may, for the sake of the argument, suppress the well-known objections (exemplified by the case of the loaded die) against the classical definition of probability in terms of *equal* possibilities, as the number of the favourable possibilities divided by the number of all the possibilities; and we may confine ourselves to cases such as symmetrical dies or pennies, in order to see how this definition compares with the propensity interpretation.

The two interpretations have a great deal in common. Both refer primarily to singular events, and to the possibilities inherent in the conditions under which each event takes place. And both regard these conditions as reproducible in principle, so that they may give rise to a sequence of events. The difference, it seems, lies merely in this: the one interpretation introduces those objectionable metaphysical propensities, while the other simply refers to the physical symmetries of the conditions—to the equal possibilities which are left open by the conditions.

Yet this agreement is only apparent. It is not difficult to see that mere possibilities are inadequate for our purpose—or that of the physicist, or the gambler—and that even the classical definition assumes, implicitly, that equal *dispositions, or tendencies, or propensities to realize the possibilities in question*, must be attached to the equal possibilities.

This can be easily shown if we first consider equi-possibilities very close to zero. An example of an equi-possibility very close to zero would be the probability of any definite sequence of 0's and 1's of the length n: there are 2^n such sequences, so that in the case of equi-possibility, each possibility has the value $1/2^n$ which for a large n is very close to zero. The complementary possibility is, of course, just as close to one. Now these possibilities close to zero are generally interpreted as 'almost impossible', or as 'almost never realizing themselves', while, of course, the complementary possibilities, which are close to one, are interpreted as 'almost necessary', or as 'almost always realizing themselves'.

But if it is admitted that possibilities close to zero and close to one are to be interpreted as predictions—'almost never happening' and 'almost always happening'—then it can be easily shown that the two possibilities of getting heads or tails, assumed to be exhaustive, exclusive, and equal, are also to be interpreted as predictions. They correspond to the prediction 'almost certain to realize themselves, in the long run, in about half of the cases'. For we can show, with the help of Bernoulli's theorem (and the above example of sequences of the length n) that this interpretation of possibilities $1/2$ is *logically equivalent* to the interpretation, just given, of possibilities close to zero or to one.

To put the same point somewhat differently, mere possibilities could never give rise to any prediction. It is possible, for example, that an earthquake will destroy tomorrow *all* the houses between the 13th parallels north and south (and *no* other houses). Nobody can calculate this possibility, but most people would estimate it as exceedingly small; and while the sheer possibility as such does not give rise to any prediction, the estimate that it is exceedingly small may be made the basis of the prediction that the event described will not take place ('in all probability').

Thus the estimate of the *measure* of a possibility—that is, the estimate of the probability attached to it—always has a predictive aspect, while we should hardly predict an event upon being told no more than that this event is possible. In other words, we do not assume that a possibility as such has any tendency to realize itself; but we do interpret probability measures, or 'weights' attributed to the possibility, as measuring its disposition, or tendency, or propensity to realize itself; and in physics (or in betting) we are interested in such measures, or 'weights' of possibilities, as license to make predictions. We cannot therefore get round the fact that we treat measures of possibilities as dispositions or tendencies or propensities. My reason for choosing the label '*propensity interpretation*' is that I wish to emphasize this point which, as the history of probability theory shows, may easily be missed.

This is why I am not intimidated by the allegation that propensity is an anthropomorphic conception, or that it is similar to the conception of a vital force. (This conception has indeed been barren so far, and it seems to be objectionable. But the disposition, or tendency, or propensity, of most organisms to struggle for survival is not a barren conception, but a very useful one; and the barrenness of the idea of a vital force seems to be due to the fact that it promises to add, but fails to add, something important to the assertion that most organisms show a propensity to struggle for survival and, in doing so, develop other propensities, like that of investigating their surroundings, and occupying new ecological niches.)

To sum up, the propensity interpretation may be presented as retaining the view that probabilities are conjectured or estimated statistical frequencies in long (actual or virtual) sequences. Yet by drawing attention to the fact that these sequences are defined by the manner in which their elements are generated—that is, by the generating conditions—we can show that we are bound to attribute our conjectured probabilities to these generating conditions: we are bound to admit that they depend on these conditions, and that they may change with them. This modification of the frequency interpretation leads almost inevitably to the conjecture that probabilities are dispositional properties of these conditions—that is to say, propensities. This allows us to interpret the probability of a *singular* event as a property of the singular event itself, to be measured by a conjectured *potential or virtual* statistical frequency rather than by an *actual* or by an observed frequency.

Like all dispositional properties, propensities exhibit a certain similarity to Aristotelian potentialities. But there is an important difference: they cannot, as Aristotelians might be inclined to think, be inherent in the individual *things*. They are not properties inherent in the die, or in the penny, but in something a little more abstract, even though physically real: they are relational properties of the total objective situation; hidden properties of a situation whose precise dependence on the situation we can only conjecture. And if we wish to test our conjecture, we have to try to keep *the relevant situation* constant, by keeping some conditions constant in every repetition of the event. In this respect propensities again resemble forces, or fields of forces: a Newtonian force is not a property of a thing but a relational property of at least two things; and the actual resulting forces in a physical system are always a property of the whole physical system. Force, like propensity, is a relational concept.

This relational aspect of a propensity (say, of an experimental arrangement) is one we may easily miss: we may think that the propensity to turn up heads or tails in one half of the tosses is an inherent property of a penny. But quite apart from the fact that it is not a property of the penny but of the *tossing* of a penny, we shall find that we get lower probabilities if we let the penny drop upon the surface of soft sand or mud (where it can come to rest upright) rather than, say, a tennis court; which shows that there may be several experimental conditions to be considered even in the simplest cases.

These results support, and are supported by, the results of our analysis of the role of b—the second argument—in '$p(a,b)$'; and they show that, although we may interpret '*b*' as the name of a (potential or virtual) sequence of events, we must not admit every possible sequence: the only sequences to be admitted are those which may be described as repetitions of a situation generating certain possible outcomes and which may be characterized by the method of their generation, that is to say, by a generating set of experimental conditions.

There is a possibility of misinterpreting my arguments, and especially those of the present section. For they might perhaps be taken as illustrating the method of *meaning analysis*: what I have done, or tried to do, it could be said, is to show that the word 'probability' is used, in certain contexts, to denote propensities. I have perhaps even encouraged this misinterpretation, especially in the present section, by suggesting that the frequency theory is, partly, the result of a mistaken meaning analysis, or of an incomplete meaning analysis. Yet I do not suggest putting another meaning analysis in its place. This will be seen clearly as soon as it is understood that what I propose is *a new physical hypothesis* (or perhaps a metaphysical hypothesis), analogous to the hypothesis of Newtonian forces. It is the hypothesis that every experimental arrangement (and therefore *every state of a system*) generates propensities which can sometimes be tested by frequencies. This hypothesis is testable, and it is corroborated by certain quantum experiments. The two-slit experiment, for example (*cf.* Volume III of the *Postscript*, section 18), may be said to be something like a crucial experiment between the purely statistical and the propensity interpretation of probability, and to decide the issue against the purely statistical interpretation.

The propensity interpretation will be discussed more fully in remaining volumes of this *Postscript*. It will have to be judged in the light of these discussions.

In the present chapter it will be shown that the transition from the frequency interpretation to the propensity interpretation corresponds to the transition from the *mathematical frequency theory*, developed by von Mises, Copeland, Wald, Church (and myself) to the *neo-classical or measure-theoretical treatment of probability* which, I am satisfied, is superior to the frequency theory, not only from a philosophical but also from a purely mathematical point of view. At the end of Volume II of the *Postscript*, which is devoted to the problem of determinism, I intend to show that what has stood so long in the way of a conscious acceptance of the propensity interpretation has been the belief in metaphysical determinism. In Volume III of the *Postscript*, on quantum theory, the usefulness of the propensity interpretation will be put to a test. In my epilogue (also in Volume III) I intend to show that, with the help of the propensity interpretation, a new metaphysics of physics can be constructed—a new research programme for physics which unifies most of its older programmes and which, in addition, seems to offer possibilities for a unification of the physical and the biological sciences.

21. *Where the Frequency Theory Succeeds.*

Whatever interpretation of scientific probability statements we may adopt, there is no doubt that the frequency interpretation remains of fundamental importance, since it is always frequency statements which we submit to empirical tests. For this reason, I shall begin here again with the problems which I discussed in the longest chapter (chapter 7) of *L.Sc.D.*

When I wrote that chapter, the discussion of von Mises's so-called axiom, or postulate, of *randomness* (or of the futility of gambling systems) was approaching its climax.

This climax is described by Karl Menger as follows:[1] 'At that time, there occurred a second event which proved to be of crucial importance in Wald's further life and work. The Viennese philosopher Karl Popper . . . tried to make precise the idea of a random sequence, and thus to remedy the obvious shortcomings of von Mises's definition of collectives. After I had heard (in Schlick's Philosophical Circle[2]) a semi-technical exposition of Popper's ideas, I asked him to present the important subject in all details to the Mathematical Colloquium. Wald became greatly interested[3] and the

---

[1] See K. Menger, 'The Formative Years of Abraham Wald . . .', a contribution to the 23rd volume of the *Annals of Mathem. Statistics*, 1952, which was dedicated to the memory of Abraham Wald.

[2] I was not present when this exposition of my ideas was given, since I was not a member of Schlick's 'Vienna Circle'. (All these footnotes to the passage quoted from Menger are added by me. K.R.P.)

[3] Actually, Wald read my book, at Menger's instigation, about two months before the 84th Colloquium at which both he and I read our papers; and he had all his main results ready by then. *Cf.* the next footnote but one.

result was his masterly paper on the self-consistency of the notion of collectives . . . in the *Ergebnisse* . . . It was through this work on collectives and a study of time series . . . undertaken at Morgenstern's[4] suggestion that Wald became interested in the foundation of statistics.'[5]

I hope I may be excused for referring to this incident. It was important not only for Wald but also for myself. For Wald's work was a generalization of mine, of a range and a depth which went far beyond what I could ever aspire to in this field. (In consequence I never published certain minor papers which I had prepared on the subject, and to which I had alluded in *L.Sc.D.*) Wald's method can be described, roughly, as employing a kind of 'diagonal argument' which allows us to construct sequences insensitive to any denumerable set of gambling systems.

Wald's results were, briefly, these. He showed that, given any denumerable set of gambling systems, there exist collectives—in fact, a whole continuum of collectives—which are insensitive to all these gambling systems (in other words, there exist sequences with converging frequencies which the gambling systems fail to affect). He showed, moreover, that if the set of gambling systems is 'defined in a constructive manner', then one can *effectively construct examples of collectives* which are insensitive to all these gambling systems.[6]

Wald's result[7] received its finishing touches, as it were, in a paper

---

[4] O. Morgenstern was then the director of the Institute for Business Cycle Research in Vienna.

[5] Abraham Wald's long paper in the *Ergebnisse eines Mathematischen Kolloquiums*, edited by Karl Menger, was published in no. 8 (1937); but he submitted his results in February 1935, in Kolloquiums 84 and 85; see *Ergebnisse* no. 7, 1936, p. 12. A final result (corresponding to Part 3, pp. 70–3 of the long paper, and described here in the next footnote) was added in Kolloquium 86, on 1st March, 1935.

[6] A minor result of Wald's, although a very surprising one, was the following. (See preceding footnote.) If we give 'neighbourhood-selection' a meaning somewhat different from that given in *L.Sc.D.* (where it is to depend only upon the element whose selection is to be decided), by cutting up the sequence into nonoverlapping segments of a definite length and taking each such segment to constitute a 'neighbourhood' of its elements, then collectives or Bernoullian sequences (including mine) are *not* insensitive to 'neighbourhood selections' in this special sense. See also notes *4 and *5 to section 58 and *3 to section 60 of *L.Sc.D.*

[7] Wald's result was, essentially, a generalization of results previously reached by A. H. Copeland. A related generalization was achieved at almost the same time by

by Alonzo Church. Wald had used the words 'defined in a constructive manner' and 'can be effectively constructed' in an unsophisticated sense: he had asserted, simply, that whenever a (denumerable) set of gambling systems was presented to us with the help of some indication or method for constructing all the systems belonging to the set, then this indication or method could be used for constructing collectives which are insensitive to selection according to all the gambling systems of the set.

Now Church pointed out[8] that this was a case where the concept of *effective calculability,* for which he had proposed a formal definition in 1936, might be applied. He recalled that von Mises had severely criticized[9] all those who, like Copeland, had worked with collectives[10] that *might* be constructed with the help of a rule; and even more severely those who, like myself, had given methods for such constructions.[11] For such sequences, successful gambling sys-

---

J. L. Doob, 'Note on Probability', *Annals of Mathematics* (Second Series) 37, pp. 363–367 (published April 1936; received September 16th, 1935). Wald's first brief publication of his results was a little earlier (*Comptes rendus de l'Académie des sciences*, Paris, *tome* 202, pp. 180–183, January 20th, 1936; and Wald's paper was first read in February 1935, as mentioned in the footnote before last. See further A. H. Copeland, 'Consistency of the Conditions Determining Collectives', *Trans. Am. Math. Soc.* 42, 1937, pp. 333 *ff.*, and W. Feller, '*Ueber die Existenz von sogenannten Kollektiven*', in *Fundam. mathem.* 32, 1939, pp. 87 *ff.*

[8]Alonzo Church, 'On the Concept of a Random Sequence', *Bull. Am. Math. Soc.* 46, 1940, pp. 130–135.

[9]See especially *Probability, Statistics and Truth* (1939), p. 136 (2nd German edition, p. 117; 3rd German edition, pp. 105 *ff.*). A. H. Copeland had discussed as early as 1928)*Am. J. of Math.* 50, 1928; 53, 1931; etc.) 'admissible numbers' which were identical with what Reichenbach later called 'normal sequences'. They are sequences insensitive to 'normal ordinal selection'; and Copeland had proved the existence of such numbers without, however, constructing an example. I, on the other hand, had started from a requirement which appeared to me intuitively more important—insensitiveness to *selections according to predecessors.* I proved that this requirement was sufficient to establish insensitiveness to both, 'normal ordinal' and 'pure neighbourhood' selections; and Ville proved shortly afterwards the equivalence of 'normal ordinal' selection to my requirement. I also gave a method of construction for sequences insensitive to selections, according to n predecessors.

[10]Copeland's term is 'admissible numbers'.

[11]Wald believed (*cf. op. cit. (Ergebnisse)*, p. 44) that mine was the first method of actually constructing sequences which were insensitive to predecessor selections. I too believed that it was the first (see Appendix iv); but later I heard from von Mises that he himself had given a method for constructing Bernoullian sequences in 1933. (*Math. Annalen* 108, p. 769.) Neither Wald nor I had been aware of this fact. However, my sequences (*cf.* the new note 2 to section 55, and Appendix iv) happen to be in several respects different from those of von Mises; in particular, they

tems must always exist, since the sequences are subject to a mathematical construction; and von Mises pointed out that there may always be many more successful gambling systems for such sequences.

For my part, I never regarded this as a serious objection. All I wanted was to derive the formalism of probability theory from *the assumption that the probability of the nth element of the sequence was independent of the properties of all its predecessors.* Church, however, found von Mises's objections important. He answered them by pointing out that *any* practicable system of selection (gambling system) must be one which allows us *to calculate effectively* the elements to be selected (for gambling upon them). He therefore proposed to define a *random sequence* by leaving von Mises's first condition ('axiom of convergence') unaltered, and by changing von Mises's second condition ('axiom of randomness', or 'axiom of excluded gambling systems') in such a way as to demand of random sequences *insensitivity to all effectively calculable selection functions.*

By thus excluding all effectively calculable, that is, all practicable gambling systems (all those which are susceptible of a precise mathematical formulation) Church has, in my opinion, succeeded in characterizing precisely the kind of collective which von Mises had in mind. Church showed that Wald's proof was applicable to this case.[12] Thus the existence of collectives—or 'random sequences', as Church called them—was demonstrated.

These results seem to me to establish the most complete justification of von Mises's frequency theory which could reasonably be demanded. They should silence all its critics. Even those who (like myself) objected to the 'axiom of convergence' (or 'limit axiom') received an effective answer.

For Church showed that a most important result of Borel's[13] could be extended to Church's 'random sequences'—or at least to

become sooner *n*-insensitive for a given *n*, which appears to make them, from the beginning, better copies of empirical random sequences. (*Cf.* Appendix *vi.)

[12]This follows from Wald's result in view of the fact that the set of effectively calculable functions is (non-effectively) denumerable.

[13]E. Borel, *Leçons s. l. Théorie des Fonctions* (ed. 1914, 1928) note v. (Copeland had earlier used Borel's result in connection with his own 'admissible numbers'.) Borel published his results first in his paper *Les probabilités dénombrables et leur applications arithmétiques, Rend. Palermo* **27**, 1909.

random sequences with two properties, '0' and '1', provided they have equal distribution, i.e., provided $p(0) = p(1) = 1/2$.[14]

The result of Church's I have in mind is this. If we consider all possible infinite 'alternatives'—that is, all possible infinite sequences of 0's and 1's—then *almost all* of them are random sequences in Church's sense.

This clearly implies, first, that almost all alternative sequences, if continued for ever, are convergent; and secondly, that almost all of them have 'chance-like' or 'random' character. And it further implies that such random sequences *exist*.

By 'almost all', the following is meant here. Interpret the sequences of 0's and 1's as binary fractional expansions of the real numbers between zero and one. It is then found that those which are not random, i.e., which do not satisfy Church's two conditions of randomness, form a set of measure zero.

In order to understand this theorem intuitively, for alternatives with the probability $1/2$, we may consider all the possible sequences of length 2, 4, 6, . . . ordered according to magnitude. (For reasons of space, I give only the first two sets.)

Length 2: Number of sequences: 4

00	10
01	11

Length 4: Number of sequences: 16

0000	1000
0001	1001
0010	1010
0011	1011
0100	1100
0101	1101
0110	1110
0111	1111

[14]By $p(0)$ and $p(1)$ I denote the (absolute) probability of the occurrence of 0 or 1; similarly, by $p(0,a)$ and $p(1,a)$ I denote the corresponding relative probabilities, given a. Church asserts the theorem only for alternatives with probability $1/2$. However, Borel's results indicate a more general theorem, as Copeland has pointed out. (See, for example, his paper in *Erkenntnis* **6**, 1936, pp. 189–203. See also the last paragraph of J. L. Doob's paper 'Note on Probability', *Ann. of Maths.*, Second Series **37**, 1936, pp. 363–367.)

In general, the number of different sequences of length n is 2^n.

As Bernoulli noticed a long time ago, the following holds:

(1) The relative frequency of the sequences which have *exactly* equal distribution decreases with their length. (It is $1/2$ among the sequences of length 2, $3/8$ among those of the length 4, $5/16$ among those of length 6.)

(2) Nevertheless, the relative frequency of those with almost equal distribution increases. (This cannot very well be illustrated if we do not at least proceed to sequences of the length 6.)

(3) The relative frequency of those sequences which have intuitively a random-like character, and which are, with good approximation, insensitive to 'normal ordinal selection', increases with the length of the sequences. (This can only be properly illustrated if we consider sequences of greater length—at least up to 6 and 8.)

Now if the sequences become longer and longer, a greater proportion will deviate very little from $p(0) = p(1) = 1/2$, and a greater proportion will become more nearly insensitive to more and more selection methods. In this way, we get our theorem.

Now let us consider the usual criticism of the axiom of convergence in the light of this result. It has usually been said:

(a) that it is pointless to postulate of a sequence which cannot be computed in accordance with a mathematical rule that it converges;

(b) that according to probability theory all sequences are possible and compatible with any probability assumption, and therefore also a sequence like the following:[15]

01 0011 000000111111 00000000000000000001111111111111111111

which oscillates between the frequencies $1/2$ and $1/3$ and is therefore not convergent. But it is inadmissible to exclude this divergent sequence, since it is clearly a possible sequence.

The answer to these objections can now be given:

(a) Almost all sequences which are *not* effectively computable in accordance with a mathematical rule have convergent frequencies.

(b) Although there are divergent sequences, they may therefore be neglected. This will give us a theory which very slightly idealizes, and simplifies, the situation.

Seen in this light, the 'axiom of convergence' or ('limit axiom' as

---

[15]In this sequence, $a_1 = 0$, $a_2 = 1$, and each block of zeros is followed by an equally long block of ones, followed in turn by a block of zeros whose length equals twice the number of all previous occurrences of zeros.

von Mises called it) completely loses its apparently objectionable character. For instead of being laid down as an arbitrary postulate, it now assumes the character of an idealized or simplified version of a theorem. And instead of appearing to apply to empirically non-existent infinite sequences, it may now be seen merely to idealize a property shared by *almost all finite* sequences of great length, and by ever more finite sequences as their length is increaed.

All this is very straightforward and satisfactory; and I can only repeat that, in my opinion, it completely justifies the frequency approach.

Yet this very justification of the frequency approach supersedes it: the frequency theory becomes obsolete at the very moment at which it can be mathematically fully justified. For the theory which justifies it is not, in its turn, a frequency theory in von Mises's sense: it is, essentially, a theory that measures possibilities, or sets of possibilities, like the classical theory (originally of Bernoulli's making). It may perhaps be called the 'neo-classical' theory. By its ability to justify the frequency theory, it proves to be the stronger theory; indeed, it makes the frequency theory superfluous. In other words, once Bernoulli's aim has been realized, and a bridge built across the gap separating the classical, or rather the neo-classical, theory from the frequency theory, the latter loses, through its justification, its independent existence, and becomes part of the former.

22. *Where the Frequency Theory Fails.*

The frequency theory does not *need* the neo-classical theory: it is quite self-contained. And yet it fails because it is not sufficiently general. There are problems, and solutions, of the greatest interest which cannot be brought within the scope of the frequency theory.

In order to show this with the help of an example I shall discuss in more detail a simple form of the theorem mentioned in the preceding section according to which in *almost all* alternatives the relative frequency of the 1's has a limit.

Let a be an alternative (that is, a sequence of 0's and 1's). Let n' be the number of 1's occurring up to the nth place of a, so that n'/n is the relative frequency of the 1's up to the nth place of a.

A mathematician will say that n'/n has a limit—we may call it

$p(1,a)$—if and only if, for every small fraction ε, chosen as small as we like but with $\varepsilon > 0$, there exists a number m such that, from the mth place on, n'/n deviates from $p(1,a)$ by less than ε; or in symbols:

(*) There is an m such that, for every n so chosen that $n > m$,
$$|n'/n - p(1,a)| < \varepsilon .$$

Now for an empirical *random* alternative a (for a collective), we can never effectively compute the number m corresponding to any chosen ε (simply because the collective is not determined by a mathematical rule). What we can do, however, is this: we can calculate effectively an m for every chosen ε, however small, such that the *probability that* (*) *will fail* is smaller than ε; or what amounts to the same, that the *probability that* (*) *will hold* is greater than $1 - \varepsilon$. In fact, it more than suffices to make m equal to $1/\varepsilon^3$, which will be a very large number if ε is a small fraction. (If $\varepsilon = 1/1,000$, then $m = 1,000$ million; which means that there is a probability smaller than $1/1,000$ that after the 1,000 millionth place of an alternative, there will ever again occur a deviation of the relative frequency from $p(1,a)$ exceeding $1/1,000$.)[1]

---

[1] The simplified corollary of Cantelli's theorem here used in the text

(1) $$m \geqslant 1/\varepsilon^3$$

is valid under the condition that $\varepsilon \leqslant 0.037$ (if we take (3), below, as our best estimate of m). This entails $m = 19,742$. Formula (1) can be obtained from (2), and also from that version of Cantelli's theorem which von Mises discusses (see the next footnote) and which may be written,

$$m \geqslant 1/\varepsilon^2 \, \eta.$$

We can prove (1) by putting $\eta = \varepsilon$ in either this version, or in the following obvious yet interesting corollary of Cantelli's theorem (whose conditions entail $m \geqslant 16,166$):

(2) $m = (\varepsilon + \eta)/2\varepsilon^3\eta$ \qquad (provided $\varepsilon \leqslant 0.037$ and $\eta \leqslant 0.058$).

This corollary in its turn can be easily obtained from a result due to J. V. Uspensky, *Introduction to Mathematical Probability*, 1937, who gives on pp. 101–3 a beautifully simple proof of a strengthened version of Cantelli's theorem. This version may be written:

(3) $m \geqslant (2\varepsilon^2 - 4 \log \varepsilon + 4 \log 2 - 2 \log \eta)/\varepsilon^2$.

From (3), we obtain (2) under the condition that the two inequalities, $1/\varepsilon \geqslant 4\varepsilon^2 - 8 \log \varepsilon$ and $1/\eta \geqslant 8 \log 2 - 4 \log \eta$, are satisfied.

Cantelli's theorem (sometimes—for example, by von Mises—called the 'strong

In this form, the theorem can be interpreted within the frequency theory, as von Mises has shown,[2] as follows.

We take the alternative a and form out of it a new collective b, in the following way: we cut up a into very long (non-overlapping) segments of some chosen length n, where $n > m$. The elements of the new collective b are these long segments of a. What the theorem asserts is that, within the segments which are the elements of b, a greater deviation from $p(1,a)$ than $1/1{,}000$ will, in the average, be found not more than once in every 1,000th segment; so that the relative frequency of these deviations tends, within the collective b, to a limit which is less than $1/1{,}000$—no matter how large we have chosen n.

No objection can be offered to this frequency interpretation of the theorem.

However, the theorem discussed can be made the basis of a corollary asserting that *almost all alternatives have convergent frequencies*. And the reasoning leading from the theorem to this corollary cannot be reproduced within the frequency theory.

Within the framework of the classical theory, the theorem itself can be given an interpretation which is not so very different from the frequency interpretation. It may be put like this. (I confine myself, for the sake of simplicity, to the case of equal distribution, $p(1) = \frac{1}{2}$.)

Put in a bag one specimen each of the various segments of the length n (where n is chosen so that $n > m$) so that there are 2^n segments in the bag, all different. Then among these segments there will be a fraction of at most $1/1000$ in which deviations from $\frac{1}{2}$ exceed $\varepsilon = 1/1000$, after the nth, i.e., the 1,000 millionth place of each segment.

Given this formulation of the theorem, we may then proceed to reason as follows:

We can choose ε as small as we like, and if we make ε smaller and smaller, then, since $m = 1/\varepsilon^3$, m will go to infinity (and therefore n

law of large numbers', although stronger laws have been found since by Khinchine and Kolmogorov) allows us, in the form (3), to calculate effectively for any positive ε and η, however small, a number m such that the probability that (*) fails for some n (greater than m) will be less than η.

[2] *Probability, Statistics and Truth*, pp. 184–5. (2nd German edition, pp. 154–7; 3rd German edition, pp. 151–3.)

also). In the end, i.e., with $\varepsilon \to 0$, our bag will no longer contain the set of all possible *segments of some finite length n*, but *the set of all possible infinite sequences* (incidentally, a non-denumerable set, as Cantor's 'diagonal argument' shows). At the same time, the probability of finding a non-convergent sequence will have become zero.

Now this argument cannot possibly be reproduced within the frequency interpretation in which von Mises expressed our theorem. First of all, he started from *one* sequence, the alternative *a* which he dissected into an infinite number of long segments *n*. But one cannot dissect *a* into infinitely many segments of infinite length—not even into two such segments: any dissection of *a* into segments of which one at least is infinite can produce at most *one* infinite 'segment' of *a*, that is *a* itself (less some commencing segment). Thus there is no possibility of constructing *b* in a manner like the one used before; and it was only within *b* that the probability of finding (or not finding) the deviation was defined by von Mises.

Secondly, the idea of a collective *b* whose elements are infinite sequences cannot be entertained within the frequency theory. In the frequency theory, the elements of a collective are, essentially, observable events, or the results of experiments. They also may be *finite* sequences of events, because a *finite* sequence of events may be interpreted, in its turn, as a complex event. But an element of a collective clearly cannot be an infinite sequence of events.

Thus the actual transition to the limit is completely blocked to von Mises's interpretation. In most cases, this would not matter at all: the content of a limit theorem can, as a rule, be fully expressed without actually making the transition, simply by speaking about longer and longer finite sequences. Even in the case before us, this may be said to be so *if we consider the classical interpretation of the theorem*—precisely because it does not allow us to proceed to the limit. But within a framework which does not allow us to proceed to the limit, the theorem cannot have the full force which it has within a framework which makes the transition possible.

To this criticism of mine a frequency theorist may offer the following reply. I admit, he may say, that von Mises's interpretation, although correct in itself, does not give the full force of the theorem. But this may be remedied by a more direct translation of the classical theorem into the language of collectives. Corresponding to your bag of 2^n segments (with $n > m = 1/\varepsilon^3$) there is a collective *b*—one of segments picked at random from the bag. The probability, i.e., the limit of the relative frequency, of picking a segment from this bag in which deviations from ½ occur which exceed ε, will be smaller than ε, exactly as in the classical model. Moreover, this probability will be zero if *m*, and therefore *n*, becomes infinite.

This reply must be rejected, but it deserves a careful analysis. What it shows, correctly, is that probability may always be linked with a sufficiently flexible idea of frequency: an idea comprising frequency in finite classes; its generalization, i.e., frequency limits in infinite sequences; and certain further generalizations, such as measures defined for continuous sets. But in the frequency theory of von Mises, *only* frequency limits in infinite sequences of observable events are admitted. In view of this, the following must be said.

(a) If we admit the method, suggested by my hypothetical opponent, of re-interpreting a relative frequency in a finite class by constructing the collective of random draws from that finite class, then this kind of re-interpretation becomes trivial *and* redundant. For it only re-states an ordinary finite ratio in terms of a limit of an infinite sequence of ratios; and it has, *in addition*, to make the assumption of 'random' draws.

(b) Our original bag, and its frequencies, had nothing chance-like about them. This was a purely mathematical model, with every frequency exactly calculable for every *n*. It does not contribute to clarity to interpret *these same frequencies* as the results of chance-like or random (that is, independent) draws.

(c) Within a frequency theory of von Mises's type, the statement that the random draws will yield these frequencies must retain the character of a *hypothetical estimate* about frequencies of chance events. But the corresponding statement about the contents of the bag were demonstrable mathematical theorems.

(d) Thus the statement that *there is a zero probability* of picking, from the bag of all possible sequences of 0's and 1's, a sequence with a non-converging frequency (a statement which could hardly be interpreted within von Mises's theory) is certainly not equivalent to the theorem under discussion, according to which *almost all* such sequences converge (the measure of the set of the non-converging ones being zero).

From what has been said it will be clear that the theorem we have

371

been discussing—a form of the 'strong law of great numbers'—does not amount to a derivation of von Mises's 'axiom of convergence'. It does not (and cannot) establish that *all* sequences converge, only that *almost all* converge. The 'axiom of convergence' characterizes the 'collectives' as belonging to those which converge. In other words, von Mises's theory singles out, quite properly, a particularly interesting class of sequences.

The situation with von Mises's second axiom, the axiom of randomness, is closely analogous. Again, it cannot be derived as it stands. But it can be replaced by Doob's theorem[3] that under the assumption that the sequences in question consist of *independent* events, every gambling system fails in *almost all* sequences (in all sequences of independent events, with the exception of a set of measure zero).

The assumption of independence used here is, in effect, the same as my assumption that the nth element is independent of all its predecessors, or n-1-insensitive to predecessor selection; and to this extent, it turns out that my attempt to base the theory on the assumption of independence or n-insensitiveness alone (discarding the axiom of convergence) was, in the main, on the right lines. This is important in so far as independence or n-insensitiveness has a clear intuitive sense: we do try to assure, by shaking the die or by mixing or by similar means, that the result of the nth throw or draw is not influenced in any way by the results of the previous throws or draws.

23. *The Significance of the Failure.*

Here we touch a crucial point. It is here that the failure of the frequency theory turns out to be really significant.

The problem which I have called in *L.Sc.D.*, section 49, the *'fundamental problem of the theory of chance'* is completely soluble in terms of the neo-classical theory. This theory, if joined with the propensity interpretation, can *explain why* sequences of independent events (events which are n-insensitive to predecessor selection) behave in the strange way that they do: *why* they behave as do von Mises's collectives: *why* their frequencies show a tendency towards

---

[3]*Cf.* J. L. Doob, 'Note on Probability', *Annals of Maths.* **37**, 1936.

convergence: *why* they are at the same time random-like, so that (almost) all gambling systems fail. The fundamental problem of the theory of chance is solved when we can understand this strange behaviour, this regularity *cum* irregularity which, in a way, we all expect of chance events, but which must create considerable difficulties for anybody who seriously reflects upon it.

The neo-classical theory, combined with the propensity interpretation, shows that we can expect such behaviour with a probability equal to 1 in an infinite sequence of independent or chance-like systems.

The problem is solved, fundamentally, along the lines of Bernoulli, Poisson, Borel, and other classical theorists—in fact the lines along which I tried to solve it in *L.Sc.D.* For in this respect I never agreed with von Mises. While I stressed the problem, perhaps more strongly than anybody had done before, von Mises believed that there was no problem here, and that we have to accept the existence of random sequences as an ultimate empirical fact. Probability theory, in his view, merely took note of this fact, and described it in an idealized form. That was how von Mises arrived at his two axioms which he considered as irreducible.

He was confirmed in his view when he found that the classical attempts to solve the problem had all been circular.[1] Partly because he despaired of breaking out of this circularity, partly because, as a positivist, he did not believe in 'explanation', he gave up the idea of a derivation: here were irreducible facts of nature which could be described but not explained.

I tried to reconstruct the theory because I felt, on the one hand, that von Mises's criticism of the classical theory was justified, but that, on the other hand, he *had assumed more than the mathematical theory needed*, and that he had thereby made the solution of the 'fundamental problem of chance' impossible. I thus attempted to reduce his assumptions to the bare minimum, by trying to show that a theory of chance-like sequences can be derived, essentially, from the idea of *n*-insensitivity, an idea equivalent to the classical idea of independence.[2] But I now believe that von Mises's objections against the classical theory no longer hold if we consider the form

[1] See sections 48 (note 6) and 62 of *L.Sc.D.*

[2] This is the reason for my choice, on intuitive grounds, of *n*-insensitivity to predecessor selection as distinct from insensitivity to normal ordinal selection: the first is the frequency form of the idea of *independence*.

which the theory has assumed, partly thanks to von Mises's own ideas, in the work of Cantelli, Kolmogorov, Wald, Church, and J. L. Doob. And I further believe that with the help of Doob's theorem of the excluded gambling systems, the problem has reached, or very nearly reached, a final solution (provided we assume propensities).

24. *The Neo-Classical and the Frequency Theory Contrasted.*
Like its classical predecessor, the neo-classical theory may be said to treat probability as a measure of possibilities (or of properties, or of classes or sets). But it differs from the classical theory on the following points.

(a) It does not start from a definition of probability. Instead it takes 'probability' as anything that satisfies the rules of a certain calculus.

(b) It does not need to treat equal distributions or equi-probabilities as more fundamental than non-equal distributions.

(c) It replaces, in a number of decisively important cases, certain classical *limit-theorems* (asserting that certain possibilities *tend* to the limit zero or one) by theorems asserting that certain sets of possible sequences have *the measures* zero or one.

(d) It allows other interpretations, but it strongly suggests, especially by its theory of independence, an interpretation that *attributes probabilities to single occurrences or events,* to be *tested* by frequencies within sequences of repetitions of the event in question; that is to say, it suggests the propensity interpretation of probabilities.

I shall briefly discuss these points in turn.

(a) It will be remembered (*cf. L.Sc.D.*, section 48) that the classical theory[1] *defined* probability as the ratio of the number of the favourable to the equally possible cases. Seeing that 'equally possible' here also means 'equally probable', this definition amounts to an attempt to define non-equal probabilities in terms of equal probabilities; or in other words, it amounts to the proposal that the calculus of probabilities should take equi-probability as its funda-

---

[1]The classical definition is often attributed to Laplace, but it was anticipated in essence by De Moivre (1718).

mental concept and should construct the general calculus upon these foundations. This suggestion will be more fully criticized under (b).

By contrast, the neo-classical theory does not attempt to give a definition of 'probability', either on the lines of Laplace or of von Mises. It clearly separates the formal task of constructing a mathematical calculus of probability from the task of interpreting this calculus with a view to its usual applications in games of chance.

Both Laplace and von Mises, when defining probability, had applications in mind; Laplace had in mind the six possibilities of a die, and von Mises the strange fact that, in long sequences of throws, the sides of a die fall irregularly, but with equal frequencies. For the neo-classical theory, 'probability' means whatever satisfies the rules of the formal mathematical system. Thus the system should be developed first (both with an eye to mathematical generality and to possible applications); and the question of its various interpretations should be raised only afterwards.

(b) The neo-classical theory does not assume preferential status for equi-probability; more especially, it does not attempt to construct all probabilities as sums of ultimate equi-probable 'units' or 'bits', or in other words, as the results of actually *counting possibilities*.

Since the opposite view is still very popular[2]—one often hears that no other method exists of assessing the actual numerical value of a probability—it may be worth while to add here some further critical comments. But before doing so I wish to recall my admission (*cf. L.Sc.D.*, section 57) that hypothetical estimates of equal probabilities, which may suggest themselves by symmetry considerations, are of the greatest importance in physics. (These, of course, are *neither derived nor derivable* from a principle of indifference; but like all hypotheses they may recommend themselves to our intuition by anything whatever (including the interesting fact that equidistributions maximize the uncertainty of predictions).

My first point is that (as von Mises often stressed) even the simple case of the loaded die—clearly a problem of relevance to physics—transcends the problem of equi-probability.

Secondly, it may be remembered that many of the classical writers on probability (including Laplace himself) who started from De

[2]See for example L. Vietoris, in *Dialectica* 8, 1954, p. 37 *ff.*, esp. p. 43, note 1.

Moivre's definition, later (sometimes only a few pages later) began to develop a more general theory quite independent of the assumption of equal probabilities; deriving, for example, the binomial formula immediately in its general form.

As a third point, Janina Hosiasson's criticism[3] may be mentioned. It can be formulated by considering the difference between two very similar games of chance.

Description of the first game:

We have one bag and two urns, I and II. In the bag, there are *three counters,* two of them marked 'I' and one marked 'II'. In urns I and II, there are three balls each; in urn I, two balls are white and one is black; in urn II, one ball is white and two are black.

We draw a counter at random from the bag. If it is marked 'I', we next draw a ball at random from urn I; while if it is marked 'II', we next draw a ball at random from urn II. The game ends when we have drawn a ball from either urn I or II. The question is to determine the probability of drawing a white ball.

The answer is, of course, $(\frac{2}{3} \cdot \frac{2}{3}) + (\frac{1}{3} \cdot \frac{1}{3}) = \frac{5}{9}$.

Description of the second game:

The second game is exactly like the first except that urn II contains two balls only, one white and one black.

In the second game, the answer to the question of the probability of drawing a white ball is, of course, $(\frac{2}{3} \cdot \frac{2}{3}) + (\frac{1}{3} \cdot \frac{1}{2}) = \frac{11}{18}$.

[3]Janina Hosiasson (born 1899, imprisoned by the Gestapo in September 1941 and executed in April 1942; her husband, Adolf Lindenbaum, a most distinguished mathematician, and forty other distinguished Polish philosophers suffered a similar fate) communicated this criticism to Sir Harold Jeffreys who reports it in his *Theory of Probability,* 1939, p. 301. There is a reply to Jeffreys's criticism of the theory of counting the favourable possibilities by J. Neyman, in his admirable *First Course in Probability and Statistics* (1950), pp. 21–24. Although Neyman's discussion and solution are, of course, perfectly correct, he has, in my opinion, missed Dr. Hosiasson's and Professor Jeffreys's point. I am therefore going to present this point anew, in a slightly modified and more elaborate form.

Comparing the two games, we find that the first may be represented by the diagram:

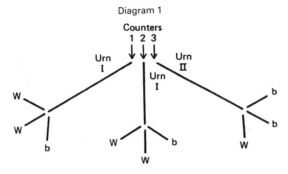

Diagram 1

The second may be represented by the diagram:

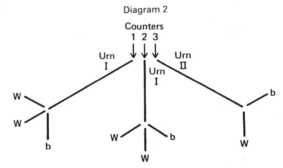

Diagram 2

Our calculation of the result of the second game may be represented by a third diagram,

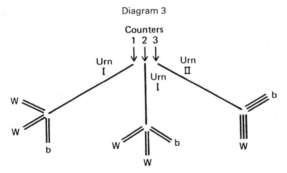

Diagram 3

or, rather, by the assertion that this diagram is *'equivalent'* to the second diagram.

Now the point made by Janina Hosiasson may be put as follows:

In the first game we have, essentially, nine equal possibilities. (As the diagram clearly indicates, we might just as well have no bags and, instead of the one urn numbered I, two different urns, both numbered I, and thus three urns with nine balls altogether, five of them white.) And we can represent our result by saying that we first *count all* the equal ways or possibilities—nine altogether—of bringing the game to an end, and then *count* the number of those which end with the drawing of a *white ball* (i.e., five). The ratio, $5/9$, is the solution.

But this method is inapplicable to the second case. Its result is $11/18$. Yet it is simply untrue that we have eighteen possibilities to consider of which eleven are favourable: no *counting* of possibilities will yield this result. Although the second game is very simple (there is not the slightest difficulty about its calculation) and although it operates at every stage with equal probabilities or equal possibilities, its result cannot be represented within a theory that *counts* (equal) possibilities. It consists, rather, in a kind of fictitious *construction* of eighteen equal probabilities which we can *calculate*, no doubt, but which do not really occur in the second game at all. For even though our third diagram describes a game which, as everybody who knows anything about the probability calculus must see at a glance, is '*equivalent*' to the second game, it really describes a new *third game*, totally different from the second; and the '*equivalence*' of the second game (with eight non-equal possibilities) and the third game (with eighteen equal possibilities) is not a thing that can be 'seen' prior to the construction of the calculus, or that may be 'assumed' as basic to the calculus.

The argument may be summed up like this. In the second game, there are eight different ways of drawing balls, five of which yield a white ball. But the possibilities are assessed as unequal (no matter for what reason) and it is the *calculus* which determines the result $11/18$, rather than the counting of alleged equal possibilities. Or to put it in another way, even though the problem is formulated in terms of existing equal possibilities, its solution, in spite of its simplicity, cannot be formulated in this way; which shows the inappropriateness of any definition based on the *counting of equal possibilities*.

(c) The replacement, in the neo-classical theory, of certain important limit-theorems by theorems showing that certain sets have the measure zero (or the measure one) is of considerable philosophical importance. It solves, on the one hand, certain of the alleged paradoxes of probability theory, and it helps, on the other hand, to build a 'bridge' leading from probability hypotheses to tests in terms of relative frequencies.

The alleged paradoxes have recently been much discussed.[4] They are connected with what has sometimes been called 'Cournot's principle' (or 'Cournot's lemma'); it may be formulated as 'Events whose probability is very small are practically impossible'.[5] The difficulty connected with this (as with any similar principle) arises from the fact that events whose probability is very small do happen, if only very rarely; or in the words of Paul Bernays, 'One can of course produce counter-examples to Cournot's Lemma (such as my win in a lottery)'.[6] (Moreover, every sufficiently complex actual occurrence has a very low probability.)

Within probability theory this problem arises in two clearly distinct forms. First it arises as the *problem of the bridge*—the transition from probability to frequency; secondly it arises *within* the frequency theory (as the problem of the testability of probability statements, discussed at length in *L.Sc.D.*).

Now the first of these two problems can be completely solved by the new derivation of the frequency theory from the neo-classical theory; that is to say, by the derivation of the strong law of large numbers, and Doob's theorem of the futility of gambling systems as 'almost certain'.

For these are derived with a probability *exactly* equal to 1; so that exceptions have a probability *exactly* equal to 0, rather than very close to 0. And there can be no empirical counter-examples for a quasi-Cournot principle which refers to probabilities *equal* to 0: there are no winning tickets with a probability *equal* to 0.

Admittedly, this lack of counter-examples is connected with the fact that the new derivations and zero probabilities refer to *infinite*

---

[4]*Cf. Dialectica* **8** (no. 30), 1954, pp. 125–144.

[5]*Cf.* Padrot Nolfi, *loc. cit.*, p. 143.

[6]*Cf.* P. Bernays, *loc. cit.*, p. 140. In fact one occurrence of an arbitrarily improbable event (of non-zero probability) within a sequence of events can be made as probable as we like by making the sequence sufficiently long; *cf.* for example, *L.Sc.D.*, section 67.

sequences, and that we cannot produce an infinite sequence of empirical events; nor can we run a lottery with an infinity of tickets. Nevertheless, the laws of convergence and of randomness for infinite sequences constitute the basis of the frequency theory; and to their derivation, the avoidance of probabilities which are merely *close to* one, or *close to* zero, makes all the difference.

This may be seen very clearly if we remember von Mises's criticism of the 'bridge'.[7] He points out that, in order to use Bernoulli's or Poisson's theorems as a justification for the deduction of frequency statements from classical probability statements, it would be necessary to assume *ad hoc* some auxiliary principle such as the following: 'Whenever we have obtained by our calculation a probability which is only a little less than 1, then the event will occur in *almost all* repetitions of our experiment.' But, von Mises points out, if we are to interpret a probability of 0.999 by 'almost always', why not 'concede at once that a probability of 0.50 means that the event occurs on the average in 50 cases out of 100?' Or in other words, why not adopt at once the frequency definition?

Yet the new derivation makes a difference here.

First, it removes the following difficulty: a probability of 0.999, for example, cannot be satisfactorily interpreted to mean 'almost always'. However close to 1, something like a frequency *limit* equal to 0.999 would be needed, and also a statement that this probability 0.999 is insensitive to gambling systems. In other words, we need the whole frequency theory if we wish to interpret 0.999 in the frequency sense. The case is different if we obtain a probability that is *exactly* equal to 1 (or to 0, as in the case of a measure zero). Admittedly, even in this case, 'probability' has to mean something connected with frequency if we are to obtain the required result. But no precise connection need be assured—no limit axiom and no randomness axiom; for these have been shown to be valid except for cases which have a probability (a measure) zero, and which therefore may be neglected. Thus all we need to assume is that zero probability (or zero measure) means, in the case of random events, *a probability which may be neglected as if it were an impossibility.*

---

[7]*Cf. Probability, Statistics and Truth,* Fourth Lecture, the section entitled 'Supplementary Adoption of the Definition of Probability, etc.' English translation (1939), and German edition, p. 135*f.*; 3rd German edition, p. 129. See also sections 48 and 62 of *L.Sc.D.,* (especially notes 6 and 3, respectively).

Secondly, it is quite true that if we interpret 'probability' in this way we are bound to concede that a probability of 0.50 implies, with probability 1, that the event occurs in 50 cases out of 100. Indeed, the fact that this conclusion is justified is now demonstrable. But it is now demonstrable without assuming a frequency definition. Thus there is no question of the frequency definition being *inadequate*. It has merely become *unnecessary*: we can now derive consequences concerning frequency limits even if we do not assume that probability means a frequency limit; and we thus make it possible to attach to 'probability' a wider and vaguer meaning, without threatening the bridge on which we can move from probability statements on the one side to frequency statements which can be subjected to statistical tests on the other.

However, it is clear that any such 'bridge' leading from the classical theory to statistics can only be built if the classical theory is interpreted *objectively*—say, in the sense of the propensity interpretation. To a subjective interpretation, my old criticism applies. (See section 48, note 6, and section 62, note 3, of *L.Sc.D.*)

(d) We have seen that the neo-classical theory *does not define* 'probability', and does not, therefore, attempt to derive the calculus from a definition of 'probability' as did both the classical theory and von Mises. Instead, it first constructs the calculus (either in an axiomatic fashion, or as part of the theory of measure). Afterwards, various interpretations of the calculus may be considered, either subjective or objective. But the failure of the subjective interpretation in the theory of independence almost forces us to adopt the propensity interpretation.

However, a frequency interpretation, in the precise sense suggested by von Mises (although as I said before, a consistent and highly satisfactory theory) does *not* furnish one of the possible interpretations of the neo-classical theory: the latter is genuinely more general, and contains the frequency theory as a kind of 'first approximation'.

Since there does not seem to be any objective interpretation other than the frequency and propensity interpretations, and since the frequency interpretation cannot be the 'bridge' to itself, I do not see any possibility other than to interpret the (neo)classical theory in the sense of the propensity interpretation.

The main point in this connection is that the new theory genuinely attributes probabilities to *single events*.[8] And although it considers sequences of events, and frequencies within these sequences, *the probability of an event may radically differ from its frequency in some of the observed segments of the sequences.* (It only agrees with the frequency *limits* of *almost all* sequences.)

I have tried to show that a theory of the von Mises type cannot be regarded as one of the possible interpretations of the formalism of the neo-classical (set-theoretical) probability theory.[9] Yet it may seem, at first sight, that a reconciliation between the two approaches—the neo-classical and the frequency approach—might be possible. For it may seem that the following *frequency interpretation* of the *neo-classical theory* might realize the main intention of a frequency theorist of the von Mises type:

(*) We interpret the phrase 'the probability (or the measure) of the event *x*', as used within the neo-classical theory, to mean 'the limit of the frequency of events of the type *x* within almost all sequences of an infinite set of (random, or random-like) sequences'.

It is clear that, given a probability distribution (or field, or label-space), the interpretation (*) can always be carried through. This follows, of course, from the strong law of great numbers and from Doob's law. But I shall try to show that (*) is utterly unsatisfactory: that it really means putting the cart before the horse; and that it badly obscures the situation.

[8] I do not think that the point was adequately brought out in the highly interesting discussions of 1940 between von Mises and Doob; *cf. Ann. of Math. Statistics* **12**, 1941, pp. 191–217.

[9] Although there are, as we have just seen, results of the neo-classical (set-theoretical) theory which cannot be interpreted in a frequency theory of the von Mises type, it is possible to interpret a *purely formal system* (as opposed to a set-theoretical interpretation of a formal system), such as the one expounded in Appendices *ii to *v of *L. Sc. D.* in terms of a frequency theory, for example that of Appendix *vi. For let S be a set of shortest random sequences (collectives) such as $a = a_1, a_2, \ldots, b = b_1, b_2, \ldots$, where each element of the sequences, a_i or b_i, equals either 1 or 0, and let S include two alternatives consisting of 1's and of 0's only. Let

$$
\begin{aligned}
p(a,b) &= \lim((\Sigma((a_n b_n)/\Sigma\, b_n); \\
p(ab,c) &= \lim((\Sigma\, a_n b_n c_n)/\Sigma\, c_n) ; \\
p(\bar{a},b) &= \lim((\,\Sigma\,(1 - a_n)\, b_n)/\Sigma b_n); \\
p(a) &= \lim((\Sigma\, a_n)/n) ;
\end{aligned}
$$

then all the postulates and axioms of *L.Sc.D.*, Appendix *iv (p. 332f.) are satisfied (and beyond these, a postulate or definition of independence).

The interpretation (*) is not, of course, identical with the frequency theory, but it is so close to it in spirit that a frequency theorist might well accept it as a kind of generalization of his ideas; and he might argue in favour of (*) as follows.

'Probability' means, in the first instance, relative frequency within a finite class of events, and in the second instance, the limit of relative frequencies within an infinite sequence of events. With respect to an infinite sequence, we may speak of a 'distribution', i.e., of the various fundamental probabilities whose sum is 1 which are nothing but the frequency limits of the fundamental properties, or kinds of events. Now there is no reason why we should not consider, instead of a distribution which is relative to *one* sequence (as we frequency theorists did so far, following von Mises), a distribution which is relative to a *class* of sequences—those, namely, which have the same given distribution. This does not essentially change the frequency doctrine according to which probabilities and distributions have meaning only relative to some given reference sequence. In this new form, they have meaning relative to *all* those reference sequences which belong to some set of reference sequences. This set might be called the 'reference set'. As a last step we may even extend this 'reference set' of reference sequences so as to include 'exceptional sequences' with frequencies different from the probabilities, provided the measure of the reference set is one, and that of the set of exceptional sequences is zero. In this way we find that (*) is a perfectly natural interpretation of the probability calculus from the point of view of the frequency theory.

But this argument confuses the issue completely.

For we can speak of probabilities or measures in the sense of the neo-classical theory only relative to a fundamental distribution, also called a field, or a 'space'; and without having set up this distribution, we cannot say anything at all about any measure of a set of sequences. But the distribution is a distribution of probabilities. We thus start with certain probabilities when we set up our system of measurement, which is a system of other probabilities. (As von Mises himself always stressed, we are given probabilities and derive others from them.) The 'reference set' thus has the measure one, and the 'exceptional set' the measure zero, relative *only* to our initial distribution, that is to say, to the probabilities which are given to start with; and it is putting the cart before the horse to explain these probabilities with respect to a set of sequences of measure one which, for a different initial distribution (i.e., arranged in a different space) would become a set of measure zero. Or in other words, if we are given the continuous set of all possible alternatives, then, with respect to one initial distribution, the subset A of alternatives with the corresponding frequencies has measure one, while with respect to another distribution, the subset A may have measure zero.[10] It is therefore impossible to use (*) as an *explanation* (or as a definition) of 'probability' in terms of the frequencies of *almost all* sequences; for the 'measure zero', translated by 'almost all', turns out to be relative to the distribution, that is to say to the initially assumed *probabilities*. Thus (*) offers a possible interpretation, no doubt; but it is unsatisfactory.

25. *The Structure of the Neo-Classical Theory.*

This point is of cardinal importance. It indicates that a satisfactory interpretation will not be one which explains $p(x)$ in terms of frequencies; or in other words, it indicates that *probabilities of single events or occurrences, although somehow linked to frequencies, have consequences that are not exhausted by the frequency interpretation.*

This is borne out by the way in which sequences and probabilities of events are connected in the neo-classical theory. The probability that a certain event—say, the mth event in a sequence of events—has the property P, is introduced in the following way.

As a first step, we consider a fundamental set of mutually exclusive and exhaustive properties, $P, P', P'' \ldots$ which the event may exhibit, and co-ordinate with each of them (freely, or if you like,

---

[10]If we start with $p(0) = p(1) = 1/2$ then we obtain the measure one for the set A of alternatives with the frequency limits $p(0) = p(1) = 1/2$. But if we start with $p(0) = p(1) = p(2) = 1/3$, then the measure of A becomes zero while the set B of sequences with the three fundamental properties 0, 1 and 2, and the frequency limits $p(0) = p(1) = p(2) = 1/3$ has measure 1. And if we decide not to distinguish between the properties 1 and 2, but to denote both by 1, so that we start with the distribution $p(0) = 1/3, p(1) = 2/3$, then the set A again has measure zero while another set, A', of alternatives with the frequency limits $p(0) = 1/3, p(1) = 2/3$, has measure 1; but in our first case A' has measure zero. Thus the transition to a new distribution amounts to a transformation which is not measure-preserving.

hypothetically) a positive number smaller than 1, in such a way that the sum of all these numbers equals 1.

These freely assigned (or hypothetically estimated) numbers will turn out to be the probabilities of the properties belonging to the fundamental set. Since they can be freely chosen (or hypothetically estimated), subject to the condition that their sum equals 1, we have here a generalization of the older classical approach (according to which the numbers were all equal).

The next two steps are designed to introduce the theory of (probabilistic or stochastic) *independence* or of the probabilities of joint occurrences. They are based, essentially, upon the consideration of *all possible sequences*, and the ascribing of measures to them.

Since we do not know the actual sequence, we construct all possible sequences, by writing down:

(1) all possible properties which may be exhibited by the first event of the actual sequence;

(2) all possible combinations of properties which may be exhibited by the first two events of the actual sequence;

(3) all possible combinations of properties which may be exhibited by the first three events of the actual sequence; etc.

We may call these different possible combinations of properties the *possible sequences*.

It is clear that, if there is more than one fundamental property,

(a) the number of possible sequences must increase more quickly than their length—in fact, at least as quickly as 2^n;

(b) on the mth place, each property must occur in at least 2^{m-1} different possible sequences.

But this means that if we consider an infinite sequence of events, and accordingly a set of possible sequences which are each of infinite length, then

(a) the set U of all possible sequences will be an infinite set (of the cardinality of the continuum);

(b) the set $S(m,P)$ of all the possible infinite sequences whose mth element has the property P will also be an infinite set (of the same cardinality).

But in the sense of the old classical theory, we should interpret the number of the set $S(m,P)$ divided by the number of the set U as the *probability* of the mth event having the property P.

Making use of the fact that these numbers are infinite, we associate with the set U the measure one and with the set $S(m,P)$ a measure equal to the number which, in the first step, we co-ordinated with the property P. Thus the classical ratio of possibilities becomes equal to the latter number which we may now recognize as the probability of the mth event having the property P.

This is the second step. It associates the same probability with the mth and nth event, for every m and n, and thus ensures insensitivity to place-selection.

The third step lays it down that the measure of the sequences whose lth, mth, nth . . . events have the properties P, P', P'' . . . respectively is the product of the measures of the sets of (1) the sequences whose lth events have the property P, (2) the sequences whose mth event has the property P'; (3) the sequences whose nth element has the property P'' . . . etc. That is to say it establishes the product rule of independence. .

From this it follows mathematically, among other more important theorems, that:

(i) the set of the sequences in which frequencies do not converge to probabilities has the measure zero. (Consequently, the set of sequences with non-convergent frequencies has the measure zero.)

(ii) The set of sequences which are sensitive to selection according to any given gambling system has the measure zero. (This is a corollary of Doob's theorem.)

Since we interpret the measure zero as a probability zero, but not as an impossibility, we have not excluded the possibility of sequences which contradict von Mises's theory; but we have shown that the probability of hitting upon such sequences is zero.

In this theory, probabilities are generalized measures of possibilities; but it is shown, with the help of what is essentially Bernoulli's method, that sequences with frequencies that deviate from the probability distributions are so rare that their occurrence can be neglected.

Thus the interpretation of probabilities as measures of possibilities is rooted in the very structure of the neo-classical theory.

In the neo-classical theory both probability and independence are ideas which are logically prior to the calculation of frequencies; and they cannot be reduced to frequencies. Yet the neo-classical statements about probability, or independence, allow us to assert, with probability 1, the crucial statements about frequencies which we need in all applications to physics.

26. *Singular Probability Statements.*

The most important point of difference between the frequency theory and the neo-classical theory is in the interpretation of singular probability statements. The frequency interpretation of singular probability statements has been fairly fully discussed in section 73 of *L.Sc.D.* It amounts to the assertion that the statement 'the probability that the next toss will be heads is one-half' *means* the same as the hypothesis 'the relative frequency of the heads in a sequence (whether finite or infinite) of tosses with this coin is one-half'; that is to say, the sentence only seems to be singular, but should be properly interpreted as one about a sequence.

As opposed to this, the neo-classical view interprets singular probability statements as statements that attribute probabilities to single events, or, more precisely, to a single event *and* a set of circumstances under which the event in question is supposed to happen or not to happen.

I shall call a theory of probability a *'single event theory'*[1] if it attributes probabilities to single events or singular statements directly, rather than by way of a detour through sets or sequences.

One often hears it said that if we want a 'single event theory' it must be a theory which interprets probability as degree of rational belief, that is to say, either a subjective or a logical theory; especially if a classical (Laplacean) definition, based upon probability or equipossibility, is to be avoided as being too narrow.[2] It appears that not a few of those who believe in degrees of reasonable belief do so because they wish to adopt a 'single event theory', and one not based upon equi-probability. They believe (for some not very cogent reason) that this wish can be gratified only by adopting a subjective theory or a logical one.

But there is no reason whatever why the classical or neo-classical view according to which probability is an assessment (or measure)

[1]*Cf.* E. C. Kemble, *Am. Journ. of Physics* 10, pp. 6 *ff.* Kemble speaks of 'the probability of single events'.

[2]See, for example, E. C. Kemble, *loc. cit.,* and I. J. Good, *Probability and the Weighing of Evidence,* 1950.

of possibilities should be inescapably linked to equi-probability. On the contrary, it seems clear that equi-probability was used only as a means of establishing an assessment or measure, and that it always played only a minor role in the actual mathematical development of the theory.

I shall make a few further critical remarks on the subjective and logical theories of probability in a later section. Here I only wish to make clear that there is no reason whatever to believe that, if we discard the equi-probability approach, our choice is confined to the frequency theory on the one hand, or the subjective and logical theories on the other. I shall try to explain my point by first discussing an equi-probability case—a perfect die—and then varying it by loading the die.

In the case of the perfect die, we attribute equal probabilities to each of the six possible results of the next throw. This is to say something about the next throw—a single event. The problem is to find out exactly what it says about *this single event* (apart from asserting that it belongs, potentially, to a sequence with a certain frequency distribution).

I suggest that what we assert about the single event—the next throw—may be analysed along the following lines.

(1) We decide in advance to be interested in the result of the throw only as far as it is characterized by one of the six sides turning up. (Thus we shall neglect, for example, the question of which side turns west.) In other words, we delimit in advance the 'possible results' under consideration.

(2) The objective conditions under which the event is to occur (or the experiment is to be performed) are such that we cannot predict the result. (This point is of minor importance since we may interpret full predictability as an assertion made with probability one.)

(3) We imagine that to each of the possible results, and their logical combinations (I have especially in mind their disjunction or logical sum) a number can be attached which satisfies the axioms of the calculus of probability, especially the addition theorem, so that the number in question may be interpreted as an additive measure; thus the number one will be attached to the disjunction (join) of all possible results considered, and the number zero to the conjunction (meet, intersection) of two exclusive results.

(4) These numbers are intended as measures of the various possi-

bilities which are left open by the conditions of the event or experiment: *if these conditions are, objectively, symmetrical* with respect to these results, the numbers can be found by assuming that they are equal.

(5) It is important to stress that in this analysis we attribute equal probabilities on the basis of an assumed objective symmetry of conditions. If we are mistaken in the case of a particular die (and every actual physical die will be at least very slightly asymmetrical), then to that extent, our attribution of equal numbers, i.e., of equal probabilities, will also be mistaken. But we assumed that:

(6) to very slight deviations from symmetry or homogeneity, there correspond very slight deviations from equi-probability.

This is an analysis of the case of a die supposed to be (approximately) homogeneous. It attributes the numbers (measures of possibility or probability) to the *objective circumstances* of the event; and neither to our *subjective knowledge*, nor to the (objectively rational) degree of belief which our knowledge may warrant.

To show the difference, assume we are shown a die and asked to determine the probability distribution of the results of throws. According to my analysis, the proper answer is this: 'I don't know. All I can say is that, *if* the die is approximately homogenous, the probabilities should be about equal.'

Now let us turn to a die with a load which can be adjusted, by a mechanism, to move from the centre towards the side opposite to the side marked '1'.

(1) We shall say that for a central position of the load, equi-probability will approximately hold; and that to a very small shift of the load, a small (and perhaps negligible) deviation from equi-probability will correspond; and our knowledge of mechanics will suggest to us that the deviation will mean an increase in the probability (possibility) of the side '1' turning up.

(2) If we are asked 'What do you call a small difference?' the proper answer is, in my opinion, 'I do not know'. I do not think that we can calculate the deviation from equi-probability from the known eccentricity (as Weyl suggests[3]); for the probability distribution is a characteristic, or property, not of the loaded die alone, but of all the relevant conditions; and it may partly depend, for exam-

[3]*Cf.* H. Weyl, *Philosophy of Mathematics and Natural Science*, 1949, p. 197.

ple, upon the surface on which the die falls—whether it is, say, steel, or rubber, or a flat cushion covered in velvet, or sand, or mud. (*Cf.* section 21 above.)

(3) But although we cannot apply symmetry considerations to the calculation of the probability distribution in the case of a loaded die, we do know something about it. For example, we know that the shift of the load, whether 'small' or 'large', has increased the probability of '1'; and this entails that there *is* a probability distribution, even though we may not know it, attached to the particular conditions of the single event.

(4) There is no difficulty about the assertion that a physical event is characterized by numbers which we do not know, and which we often cannot calculate. In mathematics and mathematical physics, we quite often obtain important results by discussing the behaviour of a mathematical function whose numerical values we cannot (or cannot precisely) specify.

(5) The very uncertainty which attaches to the more exact determination of probability numbers appears to me a clear indication of the *objectivity* of probability. Take again the case of a loaded die. We may be very exactly informed about the eccentricity and other relevant conditions. But this *knowledge* of ours may not suffice to determine the probability with the degree of precision we wish to attain, although it may enable us to assert, for example, that the distribution is unequal. There will be an objective probability, but we do not know it, or do not know it yet. But we may know how to measure it, although we may not, for example, have time to carry out the long sequence of experiments which would satisfy our desire for a precise determination.

On the basis of the subjective theory, our state of knowledge determines the probability *exactly* at every moment. There may be grave difficulties in actually calculating the probability from our state of knowledge, but it is an exact number for every state of knowledge, since it is a measure of our state of knowledge or lack of knowledge. On this view, there is no sense in speaking of measuring *the* probability by way of repeated experiment, i.e., by means of acquiring further knowledge; for further knowledge will, in general, *alter* the probability. (See the previous chapter.)

In my view, repetition of the experiment will leave the probability unchanged; thus we may utilize it to improve the precision of our estimate of the objective probability.

27. *Further Criticism of the Subjective and Logical Theories.*

Before commencing my analysis of the objective probability of single events, I have a little more to say by way of criticism of the subjective and logical theories, from a point of view slightly different from that in the foregoing chapter.

I am prepared to admit that without running into trouble one can interpret the probability calculus as one of *degrees of rational expectation,* or something of the kind, and that one can do so by adopting either a more subjective or psychological view (with Ramsey or Good or Kemeny) or a more logical view (with Jeffreys or Keynes or Carnap).[1]

What I am going to criticize here is the view that these interpretations describe the use of probability in the physical sciences.

From my point of view, the statements of physics are objective and do not in any way refer to our state of information: nor do they 'express' our information, or our nescience. They are assertions— conjectural assertions, of course—about the world.

This holds good for the probability statements of the physical sciences. *They do not result from our lack of knowledge.* Lack of knowledge does not miraculously produce a knowledge of frequencies—not even with the help of any laws of great numbers.

That frequency hypotheses—say, about the intensity of spectral lines—are as objective as any other physical hypotheses will, perhaps, be admitted by some who nevertheless reject the objectivity of single-event probabilities. But my point is that the difference is comparatively slight. Objective single-event probabilities give rise (by the Borel-Cantelli law and by Doob's law) to objective frequency statements. On the other hand, probabilities which express a state of reasonable belief could only give rise to statements about reasonably expected frequencies; 'reasonably expected' in a subjective sense if the interpretation is subjective, and 'reasonably expected' in a logical sense if the original interpretation was logical or tautological.

In the case of objective probabilities of single events, we assess

---

[1] I am a little more doubtful whether the interpretation in terms of degrees of rational *confidence* is not a different matter altogether; it may be quite correct to say (1) that the measure of my expectation in a toss of heads is $1/2$; (2) that the measure of my confidence in the event is zero; and (3) that the measure of my confidence in the assessment (of the propensity $1/2$) is one.

first the objective conditions under which the single event will occur; and then, by a mathematical derivation, the conditions under which certain frequencies will occur. In the case of subjective and logical probabilities, we assess first the logical relation between the 'data', the statement of the information 'given', and the statement of the event in question; and then, by a mathematical derivation,[2] we assess the logical relation between these same data and certain frequency statements. Accordingly, what we obtain remains linked to our information; we do not derive, as we do from the objective probability hypothesis, a *frequency hypothesis,* to be tested and, if necessary, rejected; but we obtain (provided the calculation was correct) a *true statement* (in fact a tautological one) about the degree to which our given knowledge warrants the expectation of a certain frequency: instead of a physical conjecture, we obtain a truism about the state of our own knowledge, and what it implies. (See the previous chapter.)

Consider again the case of a die. The subjective and logical theories do not really deal with the problem which interests the physicist: how will this die behave? They do not really raise the (admittedly unanswerable) problem: what will turn up at the next throw? The question which they ask is a different one; it is this: 'To what degree do our data permit us to make the one or the other statement about the behaviour of this die?' And concerning the next single event they can only ask: 'To what degree does our knowledge (or nescience) permit us to predict that the next throw will turn up a one?'[3]

I am interested in the behaviour of the die, and so, I believe, are most physicists. I want to know something about it, and I am prepared to propose conjectures—for example, that the sides will turn up about equally often in a long sequence of throws *made under certain conditions*; or that the conditions establish a symmetry with respect to all sides.

By contrast, the subjective and the logical theorists are interested

[2]The derivation is very questionable in this case, because the condition of objective independence, if interpreted as subjective irrelevance, cannot 'reasonably' be believed to be satisfied, as was first emphasized by Keynes (see above).

[3]Very often, I think more often than not, the difference between these two ways of asking questions is not seen; and when it is seen (as it is by Jeffreys, and Ramsey, and perhaps Good) it often leads to the view that science is nothing but an instrument for the transformation of our 'data'.

in how much they know already about the die, and not in conjectures about its behaviour. They find, let us say, that their nescience concerning the result of the next throw is symmetrical with respect to the six sides, and that *therefore* the six probabilities are equal.

We now begin a series of tosses, and obtain, say, 3, 1, 5, 1, 2, 2, 3, 5. The subjective and logical theorists are bound to assert that these new data must affect the probability of the next throw. Whether much or little, the probabilities will change: those for 4 and 6 will decrease (much or little), those for 1 and 2 will increase, because 4 and 6 have not occurred at all, while 1 and 2 occurred twice each.

Now assume that the die is homogeneous, and that the sequence turns out to be a beautiful example of a normal sequence. Then the probabilities—in the sense of the subjective and logical interpretations—will always change a little, but will come in time to settle nearer and nearer to an equidistribution. The total impact of the long sequence of throws and new 'data' was nil. Thus the data turned out to be 'irrelevant on the whole'.[4]

The situation looks very different from the point of view of the objective theory—whether of the single-event or the frequency type. Seen from this point of view, the sequence of throws constituted a statistical test of a hypothesis. Had we originally suspected that the die was biased, the test would have refuted our suspicion. Thus we can claim that it corroborates our original conjecture of equal probability.

Seen from this point of view, the subjective and logical interpretations confuse two things: physical statements, about objective physical systems, and epistemological estimates of the degree to which these statements are 'founded in our experience'.

It is not only the erroneous belief that every single-event theory must be either subjective or logical which accounts for these mis-

[4]This problem was always clearly seen by Keynes who described it as the problem of the *weight of evidence*. It was attacked by a few adherents of the subjective and logical theories, but without any success. It clearly is quite unreasonable to say that the probability statement attributing the probability of $\frac{1}{6}$ to the outcome h of the throw of a die after very many observations, was not better founded in our experience than the original statement about the outcome h of the throw, based on symmetry of nescience. But in Keynes's (and also in Carnap's) theory, this kind of foundation in experience is not expressed. Good's theory, for example (*Probability and the Weighing of Evidence*, 1950, pp. 62 f.), attributes to the example considered here a '*net gain of the weight of evidence*' (p. 64) or of the 'amount of information' (p. 63) (or a gain in 'plausibility') that is zero.

takes. It seems that the undoubted fact that new 'information' *may* change probabilities in various ways has led some thinkers to believe that probabilities cannot describe the objective properties of the experimental conditions: if we are 'informed' that the throw results in an even number, then the probability of throwing a 2 rises from $\frac{1}{6}$ to $\frac{1}{3}$; in symbols:

$$p(2) = \frac{1}{6}$$
$$p(2, \text{ even}) = \frac{1}{3}.$$

This is undoubtedly so. But by interpreting the word '*even*' in our formula as 'information', we have already adopted a subjective or logical interpretation. (This does not mean that we cannot use the term 'information' even in the context of the objective theory, as long as we are not misled by our terminology into mixing two interpretations.)

From the objective point of view, '$p(2)$' and '$p(2, \text{ even})$' refer to two different experimental arrangements: the first refers to one in which every throw is considered, the second to one in which we decide to ignore a throw if its outcome happens to be odd. We could also put it thus: the 'information' expressed by the symbol '*even*' tells us that we are asking a *different question*—that we are no longer asking: 'What is the probability of throwing 2?', but instead: 'What is the probability of throwing 2, taking into account only those throws which result in even numbers?'[5]

28. *The Propensity Interpretation of the Probability of Single Events.*

In the two previous sections, I have tried to show that the usual arguments against the possibility of an objective single-event theory are unfounded, even if the single-event theory does not build upon equi-probability.

But the idea of probability as a measure of possibilities remains, I

---

[5]This seems to me a complete answer to the arguments and examples given by Erwin Schrödinger in the first of his papers 'The Foundation of the Theory of Probability', *Proc. Royal Academy of Ireland* **51**, 1945, section A, pp. 51–66 and 141–6; see esp. pp. 63–6. The fact that in Schrödinger's examples new information changes some probabilities but not others and thus establishes certain new 'dependencies' or 'relevancies' in no way affects our argument.

admit, somewhat thin; especially since it is obvious that this formula cannot be considered as a definition: 'possibility', we all remember, is hardly anything else here but another word for 'probability'; at any rate, it is not (like 'frequency') a word whose meaning is clearer than that of 'probability'.

It is not my intention to *define* 'probability'—still less 'the probability of single events': we do not need a definition (since we have an axiom system) but merely an interpretation; and I shall attempt to give an interpretation, and to make it intuitively more acceptable than the phrase 'measure of possibilities'. I propose to interpret the objective probability of a single event as a measure of an objective *propensity*—of the strength of the tendency, inherent in the specified physical situation, to realize the event—to make it happen.

I have not the slightest doubt that this announcement of my intentions will be received with horror by many of my positivist friends, who will see in it a proof of my own metaphysical— 'tendencies or propensities', I nearly wrote, but I must not; so let us say, perhaps, 'disposition'.

I am not a believer in the magic of words, and I do not mind using the term 'disposition' instead of 'propensity' or 'tendency'. But I wish to stress that, like most explanatory hypotheses, a hypothesis about objective probability is *transcendent* in the sense that it goes far beyond what can be known on the basis of observations. (*Cf.* the end of section 25 of *L.Sc.D.*) This holds for objective frequency hypotheses; and it holds to an even higher degree (because their degree of universality is higher), for hypotheses about objective probabilities of single events since they can *explain* frequencies.

In order to elucidate the interpretation of objective probabilities of single events, it may be helpful first to point out that a propensity 1 will mean that the event is certain, or at least almost certain, to happen, and a propensity 0 will mean that the event is certain, or almost certain, not to happen: in these limits, both objective interpretations agree fairly well. But a propensity p with $0 \neq p \neq 1$ means, first, that an event of the kind contemplated may or may not happen, under the circumstances considered; and together with $p > 1/2$ it means that the circumstances specified are such as to make it more likely to happen than not.

But this at once raises the question whether it is not *our lack of knowledge of the precise circumstances*, rather than the circum-

395

stances themselves, which gives rise to probabilities other than one or zero. If the answer to this question is 'yes', then we must give up the objective theory. If the answer is 'no', then, it seems, we can apply the theory only to events which are indeterministic in the sense that not even the most complete knowledge of the 'circumstances' would make the outcome predictable. But in fact we wish to apply the theory to penny tosses and such like macro-physical events which nobody believes to be indeterministic in this sense.

This is an important objection, and one which forces the defenders of the objective theory to go to the root of the matter.

Take a machine for tossing pennies: one puts a penny in a slot, presses a button, and the penny falls out of the machine, flat, on to a soft cushion. If I have seen this being done by the machine once or twice, and I am then asked whether I should be prepared to gamble on heads, the reasonable reply would be, I think: 'I do not know whether this machine is one that randomizes its result (as a well-constructed roulette table would). For all I know, it may be so constructed that by pressing the button in a special way (or by some similar means) one can produce at will the result one desires.' It is a very different thing if the penny is, before our eyes, allowed, say, to roll over an uneven surface, until it gathers speed, and then drops into a beaker. In the first case we shall say that we are doubtful whether the objective conditions of the experiment are such as to ensure a certain 'randomness' of the mechanical initial conditions, while in the second case we shall be confident that they are.

In both cases, there is lack of knowledge. In the first case, the unknown objective conditions *may* easily be such that those who know them can make precise predictions. The first experiment may thus be predicted, and perhaps even controlled, by those who know more about it than I do.

In the second case, the situation is different. It is part of the conditions of the experiment that those initial conditions which might be used to predict the outcome are 'randomized'. By this I mean that they are arranged in such a way (rough surface, etc.) that most of us would conjecture that in a long sequence of experiments carried out under the same specified conditions (including the 'randomization'), the mechanical initial conditions are likely to change in a random manner.

It is this positive conjecture that the specified conditions of the experiment ensure randomness of the initial conditions which is the basis of our objective probability hypothesis: we conjecture that our experimental specifications are such as to produce random initial conditions provided they are repeated.

But what if there is a man who rapidly measures and calculates the actual initial conditions of the rolling coin and the rough surface, and every time makes the correct prediction just before the coin falls into the beaker? My reply is that his predictions do not conflict with our estimates of the objective probabilities of single events any more than they would with a frequency interpretation: the frequencies retain stability and randomness whether we know the result of the experiments a little ahead or not. *And so do the propensities of the single experiments*—or, more precisely, the propensities of the single experimental set-up. For by 'propensity' I mean precisely the disposition (or whatever you may wish to call it) of the set-up *to produce these frequencies,* if only the experiment is repeated sufficiently often. Propensities are dispositions to produce frequencies: this is the interpretation suggested by the neo-classical theory. But 'propensity' does not *mean* 'frequency', for there are events too rarely repeated to produce anything like a good segment of a random sequence (or a 'frequency'); yet these rare events may well have a propensity.

The number 1/6, in the case of the die, is thus interpreted as characterizing the experimental arrangement, and remains a valid characteristic of this arrangement even though we might succeed, by precognition or rapid calculation, in predicting every result in a long sequence of throws. The number is attributed to the single experimental set-up on the basis of what we know, and not on the basis of our lack of knowledge. And additional knowledge does not interfere with the probability or propensity which characterizes the experimental set-up.

An interesting question remains—why do we believe that such processes as the rolling of the coin over a rough surface will randomize the initial conditions? I intend to come back to this question in Volume II of the *Postscript*, sections 29 and 30. (Landé's Blade.)

Here I wish to make only two more remarks. The first is that by speaking of a propensity I wish to suggest an intuitive idea akin to that of a Newtonian force, yet distinct from a force in that it produces frequencies rather than accelerations. Frequencies change when propensities change. Propensities are as 'transcendent' or 'metaphysical' as Newtonian forces (which Berkeley denounced as 'occult'). Mathematically they are perfectly definite—they are interpretations of a simple calculus. And as to their testability, we have to appeal to the frequency statements derivable from them (with probability 1). But even the component which transcends frequency statements is, in a sense, testable—by frequency statements of a different kind. (This can be shown by an analysis of quantum theory.)

My second and last point is this. The propensity interpretation is, I believe, that of classical statistical mechanics. Boltzmann spoke of a *tendency*. But I think that it describes best what classical authors really had in mind when speaking of the quotient of the equally likely to the possible cases: they had in mind that this quotient was a measure (a particularly important and convenient measure, though not the most general one) of the propensity, characteristic of certain specified conditions, to produce a given event.

Although the idea of logical probability played a considerable part in *L.Sc.D.*, I was, admittedly, a frequency theorist when I wrote it. I stressed that there was a plurality of interpretations of the calculus of probabilities; yet I continued to believe that the frequency interpretation was of central importance for all applications in physics.

I could still uphold this position, I think, in view of the crucial significance of the frequency theorems derivable from the neo-classical theory. Yet I feel that it is perhaps more correct to stress the discontinuity rather than the continuity of my views; for I have changed my mind. Historically, the change occurred first as a consequence of the continued attempt to understand the situation in quantum theory (about which more in Volumes II and III of the *Postscript*): it was here that I first realized the need for a propensity interpretation. From here I went back to probability theory, finding, to my great satisfaction, that the neo-classical theory had indeed provided the mathematical basis for a propensity interpretation by building that 'bridge' between the classical and the frequency theories which I had previously considered, under von Mises's influence, to be impossible to construct.

Indeed, the introduction of the propensity interpretation proves to be very far from *ad hoc*. It solves the 'fundamental problem of the theory of chance' (*cf. L.Sc.D.*, sections 49 and 64). That is to say, it explains *why* almost all finite sequences of penny tosses behave in the surprising way they do: why long sequences look as if their relative sequences were tending to a limit; why they combine this strange regularity with a characteristic irregularity; and why their segments appear to obey Bernoulli's law so well.

Concluding Summary, 1982.

(1) The main message of the last chapter of this book is that *the fundamental problem of the theory of chance* (as I called it in *L.Sc.D.*) has *now been solved,* in a far more satisfactory manner than in *L.Sc.D.* This disturbing problem can be put:

How can we explain that every recorded sequence of tosses with a coin, or of throws with a die, exhibits, on the one hand, a typically random character, and on the other hand, a stable relative frequency seemingly tending to a limit?

This riddle can, I suggest, now be solved completely if we assume the propensity interpretation of the probability calculus. The calculus itself turns out to be no more than a metric generalization of the propositional logic (see *L.Sc.D.*, new Appendices *iv and *v) to which a well-known and highly satisfactory definition of the independence of events can be added.

The calculus, interpreted by the propensity interpretation, allows us to deduce:

(A) Almost all infinite sequences of independent events occuring under constant conditions will tend to exhibit relative frequencies that tend to limits which are equal to the probabilities (propensities) of the repeated single events.

(B) Almost all such sequences will have a random character describable in various ways, for example by the failure, in the long run, of all gambling systems.

These two characteristics of sequences of independent events (A) and (B) were *postulated* in the 1920s by the mathematician Richard von Mises,[1] who called them axioms and founded upon them a mathematical theory of probability. But they can now be deduced,

[1]Late Professor of Applied Mathematics at Harvard.

as indicated, from a simple, almost logical, system, provided we interpret probabilities as propensities of single events to realize themselves in long runs. But this turns out to be an almost necessary demand on other grounds.

This important work of pure mathematics was begun by Jacob Bernoulli (*Ars Conjectandi*, 1713), who worked on it for twenty years, a marvellous achievement. It was carried on by many mathematicians. I regard it as a tremendous success of pure mathematics, and of the greatest possible philosophical interest. These ideas helped to solve one of the great riddles of the world, a riddle that Richard von Mises still regarded as insoluble.

(2) It is also shown in this chapter that this development is incompatible with the subjective interpretation of the probability calculus.

(3) The view that we need to appeal to probability only because of our lack of knowledge has been widely held by many of the greatest physicists since Laplace; it had been defended by Einstein in two letters to me (one reprinted in *L.Sc.D.*). This was, no doubt, the main reason why I collected in this book so many arguments (often, I fear, repeating myself) in an attempt to show, as exhaustively as I could, that the objective interpretation solves the problem, while the subjective theory is incapable of doing so.

(4) Einstein's famous objection to 'the dice-playing god'[2] is undoubtedly based on his view that probability theory is a stopgap due to our lack of knowledge, our human fallibility; in other words, to his belief in the subjectivist interpretation of probability theory, a view that is clearly linked with determinism. I tried, at our meeting, to draw his attention to the fact that this view should be dropped; and according to Pauli's letter to Born of 31 March 1954, Einstein (who certainly held it at the beginning of our meetings) did drop it.[3]

---

[2]Letter to Max Born of 7 November 1944, quoted in Max Born: *Natural Philosophy of Cause and Chance* (Oxford, 1949), p. 122. See also letter from Max Born of 9 October 1944, in *Albert Einstein, Max Born Briefwechsel 1916–1955* (Munich, 1969), pp. 207–208.

[3]See the discussion in *The Open Universe*, Vol. II of the *Postscript to the Logic of Scientific Discovery*, pp. 2n–3n.

(5) It may be added that the solution of the fundamental problem of the theory of chance opens at least one possible way to the solution of the problem of causation. That it opens the way in those cases where (apparent) causal laws can be derived as statistical mass effects (as, for example, Boyle's law or the causal laws based on the smallness of Planck's quantum) is fairly obvious. But there does not seem to be any reason why we should not treat all effects of causal character (all *prima facie* deterministic effects, such as the Compton effect) in the same manner; that is, treating forces as propensities equal to 1.

(6) Of course, I have found some new arguments since this volume was conceived (approximately in 1953). Some of them will be found in the new Appendices recently published in the seventh edition, 1982, of the German edition of *Logik der Forschung*. I hope to include them in the next English edition of *The Logic of Scientific Discovery*.

INDICES

compiled by Michael J. O'Donnell

References in italics indicate passages of special importance. A page number followed by '*t*' indicates that the term is discussed in the place referred to; '*n*' means 'footnote', and '*q*' (in the Index of Names) means 'quoted'. The reader is also advised to consult the Indices to the other volumes of the *Postscript*.

Index of Names

Index of Subjects

and classical empiricist view, 46; un-influenced by philosophy and theology, 86; statements of, and reality of physical bodies and existence of laws of nature, 128; clash with, 131; science as, 260

Concepts, author's instrumentalist interpretation of, 262; essentialist interpretation of, 263; occult, 107, *see also* Occult; operational analysis of, 109*n*; metaphysical character of, 187, 214; scientific, 108, 214; universal, 109; theoretical, 107–111; versus statements, 109*n*, 215; constitution of, 215

Conceptual scheme, xxxiii&*n*

Conditions, 318*t*, 392; generating, 355, 356, 359, 360; initial, 57, 74, 76, 134, 184, 196, 236, 252, 345, 396, 397; known, 318; objective, 312, 313, 316, 317, 318, 322, 349, 388–389, 396, 400; relevance, 290, 291*t*, 318–319, 389–390; reproducibility, 312, 357. *See also* Experimental conditions arrangements; Independence

Confirmation, 243, 259, 332; and Carnap's views, 127*n*, 252*n*; and corroboration, 227–230, 245*n*; degree of, 228–229, 253*n*, 282, 332&*n*, 333; and Hempel's views, 234*n*, 257–258*n*; and verificationism, 230, 234. *See also* Verification

Conjecture, *see* Hypothesis

Conservation, laws of, xxvii*n*, xxx

Content, xxxvi*ff*, 133–134, 139, 216, 292, 326, 346; empirical, 239, 245–246, 249; logical, 203–205&*n*; and probability, 224, 227, 231, 232, 233, 239, 256, 284. *See also* Monotony, law of

Context, theoretical, 178–179, 187

Continuum, and cardinal numbers, 273; theory of, 271

Contradiction, 245–246, 266, 270–271, 285

Conventionalism, 24, 212&*n*, 152, 180, 276; of Duhem and Poincaré, 112; strategems of, 116*n*, 168, 276

Convergence, axiom of (von Mises), 196&*n*, 372, 380; Church's answer to criticisms of, 364; author's answer to criticisms of, 366–367; and possibility of sequences that contradict, 386

Copernican theory, and Bacon, 107; and instrumentalism, 116

Correspondence, principle of, 145; and mathematical transition from fre-quency to measure theory, and from statistical to propensity interpretation, 347, 360; of statement with the facts, 274; *see also* Truth, correspondence theory of, concept of, Tarski on

Corroborability, 245*t*, 246, 249; equals testability and empirical content, 245, 250; inversely proportional to logical probability, 245; versus probability, 250

Corroboration, Part I, Chapter 4; 11–12, 217–223, 257*n*, 332, 346; degree of, 11, 58*t*, 64–65, 72, 220*t*, 223–228&*n*, 230*t*–233, 243*t*–244, 255, 261–262, 282–283, 332*n*; author's desiderata for, 242, 244–255; complete logical analysis impossible, 236; and confirmation, 227–230, 245*n*; contrasted with probability, 223–227, 231, 248–251, 255; as contribution to methodology, 254–255; definitions of, 221–232&*n*, 233, 240–255, 275; due to emergence of stronger theory, 231–232&*n*; of inconsistent hypothesis, 251*n*; not goal attained, 261; does not satisfy probability calculus, 227, 232, 243; as report of severity of tests, 250; support provided by, 71, 240; and topological equivalence, 242–243

Cosmology, 59, 74, 78

Counting, of favourable possibilities, criticism by Hosiasson and Jeffrey, 376&*n*, 278

Cournot's principle, 379

Critical approach, xxxii, xxxv, Part I, section 2; 29, 31, 35, 60, 70, 133, 156, 164, 233, 240, 261

Critical argument or discussion, 7, 20–28, 55, 59–62, 71, 77, 153–157, 175

Criticism, Part I, section 2; 24; Freud's way of rejecting, 168; immanent, 29*t*, 30; method of, 163*t*; never conclusive, 28; non-justificationist, 27, Bartley on, 27; openness to, 27; philosophical, 17; problem of, 20, author's solution of, 20, and problem of justification, 20; proceeding from competing theory, 30; rational, in ancient and modern science, xxxii, 81, 180, author's emphasis on, xxxii, 27, function of, 24, and growth of knowledge, 27, no knowledge exempt from, 28; scientific, 13, 27; standards of, 25, 52*t*, 57, 65, 272*n*; transcendent, 29*t*, 30, 339; versus positive argument, 28–29; with-

415

ABOUT THE AUTHOR

Sir KARL (RAIMUND) POPPER was born in Vienna in 1902. He studied mathematics, physics, psychology, education, the history of music, and philosophy at the University of Vienna from 1918 to 1928. At the same time he worked as a cabinet maker's apprentice and as a teacher. In 1978, fifty years after he took his Ph.D. degree, the University of Vienna 'renewed' this degree in a solemn ceremony and gave him an honorary doctorate in the natural sciences.

In 1934, when still a schoolteacher in Vienna, he published *Logik der Forschung*, a book that became a classic after its translation into English. (*The Logic of Scientific Discovery* was published in 1959.) It has now been translated into many languages and, after 47 years, is still frequently reprinted. Popper, who lives in England, has lectured in Europe, in New Zealand, Australia, India, Japan and, since 1950, when he delivered the William James Lectures in Philosophy at Harvard University, often in America. Among his books are *The Open Society and Its Enemies* (for which he received the Lippincott Award of the American Political Science Association); *Conjectures and Refutations; The Poverty of Historicism; Objective Knowledge; Unended Quest.* Together with Sir John Eccles he published *The Self and Its Brain.*

He holds honorary degrees from the Universities of Chicago, Denver, Warwick, Canterbury (New Zealand), Salford, The City University (London), Vienna, Mannheim, Guelph (Canada), Frankfurt, Salzburg, Cambridge, Oxford, and Brasilia; and from Gustavus Adolphus College.

He is a Fellow of the Royal Society (London), and of the British Academy; a Foreign Honorary Member of the American Academy of Arts and Sciences; Membre de l'Institut de France; Socio Straniero dell'Accademia Nazionale dei Lincei; Membre de l'Académie Internationale de Philosophie des Sciences; Associate, Académie

Royale de Belgique; Membre de l'Académie Européenne des Sciences, des Arts, et des Lettres; Hon. Member, the Royal Society of New Zealand; Membre d'Honneur, Académie Internationale d'Histoire des Sciences; Ehrenmitglied, Deutsche Akademie für Sprache und Dichtung; Ehrenmitglied, Österreichische Akademie der Wissenschaften.

He is also an honorary member, Harvard Chapter of Phi Beta Kappa; Ehrenmitglied, Allgemeine Gesellschaft für Philosophie in Deutschland; Hon. Fellow, London School of Economics and Political Science; and Hon. Fellow, Darwin College, Cambridge.

He was awarded the Prize of the City of Vienna for Moral and Mental Sciences; the Sonning Prize of the University of Copenhagen; the Dr Karl Renner Prize of the City of Vienna; and the Dr Leopold Lucas Prize of the University of Tübingen.

He received the Grand Decoration of Honour in Gold (Austria); the Gold Medal for Distinguished Service to Science of the American Museum of Natural History, New York; the Ehrenzeichen für Wissenschaft und Kunst (Austria); and the Order Pour le Mérite (German Federal Republic).

In 1965 he was knighted by Queen Elizabeth II, who invested him in 1982 with the insignia of a Companion of Honour (H.C.).

ABOUT THE EDITOR

WILLIAM WARREN BARTLEY, III, a graduate of Harvard and the University of London, is a former student and colleague, and a long-time associate of Sir Karl Popper. He has been Lecturer in Logic at the London School of Economics, Lecturer in the History of the Philosophy of Science at the Warburg Institute, and S. A. Cook Bye-Fellow at Gonville and Caius College, Cambridge. Formerly Professor of Philosophy and of the History and Philosophy of Science at the University of Pittsburgh, he has since 1970 been Professor of Philosophy at California State University, Hayward, where, in 1979, he was named Outstanding Professor in the 19-campus California State University system.